Lattice Boltzmann and Gas Kinetic Flux Solvers

Theory and Applications

Advances in Computational Fluid Dynamics

Published

Vol. 4 *Direct Modeling for Computational Fluid Dynamics: Construction and Application of Unified Gas-Kinetic Schemes*
by Kun Xu (Hong Kong University of Science and Technology, Hong Kong)

Vol. 3 *Lattice Boltzmann Method and Its Applications in Engineering*
by Zhaoli Guo (Huazhong University of Science and Technology, China) and Chang Shu (National University of Singapore, Singapore)

Vol. 2 *Adaptive High-Order Methods in Computational Fluid Dynamics*
edited by Z. J. Wang (Iowa State University, USA)

Vol. 1 *Lattice Boltzmann and Gas Kinetic Flux Solvers: Theory and Applications*
by Liming Yang (National University of Singapore, Singapore), Yan Wang (Nanjing University of Aeronautics and Astronautics, China), Zhen Chen (National University of Singapore, Singapore) and Chang Shu (National University of Singapore, Singapore)

Forthcoming

Computational Methods for Two-Phase Flows
by Peter D. M. Spelt (Imperial College London, UK), Stephen J. Shaw (X'ian Jiaotong – University of Liverpool, Suzhou, China) and Hang Ding (University of California, Santa Barbara, USA)

Advances in
Computational Fluid Dynamics Vol. 1

Lattice Boltzmann and Gas Kinetic Flux Solvers
Theory and Applications

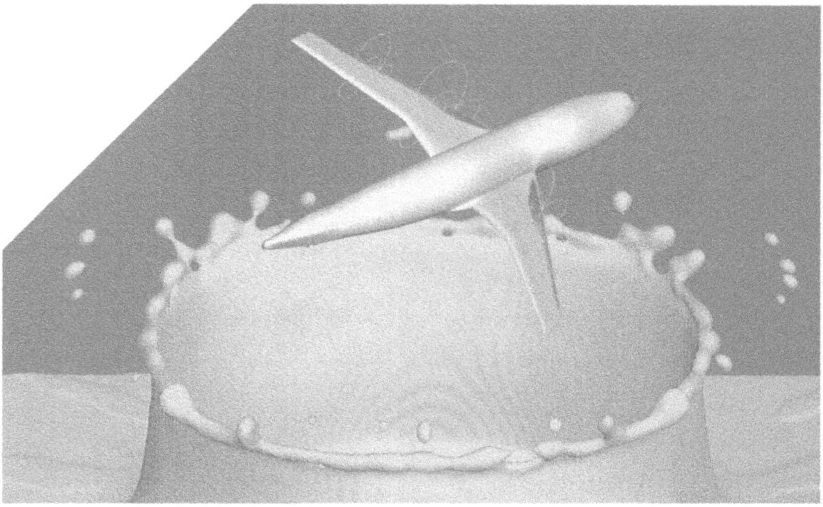

Liming Yang
National University of Singapore, Singapore

Yan Wang
Nanjing University of Aeronautics and Astronautics, China

Zhen Chen
National University of Singapore, Singapore

Chang Shu
National University of Singapore, Singapore

World Scientific

NEW JERSEY • LONDON • SINGAPORE • BEIJING • SHANGHAI • HONG KONG • TAIPEI • CHENNAI • TOKYO

Published by

World Scientific Publishing Co. Pte. Ltd.
5 Toh Tuck Link, Singapore 596224
USA office: 27 Warren Street, Suite 401-402, Hackensack, NJ 07601
UK office: 57 Shelton Street, Covent Garden, London WC2H 9HE

British Library Cataloguing-in-Publication Data
A catalogue record for this book is available from the British Library.

Advances in Computational Fluid Dynamics — Vol. 1
LATTICE BOLTZMANN AND GAS KINETIC FLUX SOLVERS
Theory and Applications

ISBN 978-981-122-468-3 (hardcover)
ISBN 978-981-122-469-0 (ebook for institutions)
ISBN 978-981-122-470-6 (ebook for individuals)

For any available supplementary material, please visit
https://www.worldscienti ic.com/worldscibooks/10.1142/11949#t=suppl

Dedication

The authors would like to devote this book to their families for their infinite patience and encouragement.

Preface

Over the years, we have two concerns in the numerical simulation of fluid flows. The first concern is the possible combination of different flow solvers in order to take their advantages, and in the meantime, to remove their disadvantages. For a given fluid flow problem, there are a variety of flow solvers, and each of them can be independently applied to provide numerical solutions. It is naturally expected that their combination should also be able to solve the same problem, and possibly, the combination can keep their advantages and remove their disadvantages. Another concern is the solution reconstruction at certain locations during the computational process. For the cell-centered finite volume method, it is well known that the numerical solution is defined at the cell center, at which discrete governing equations are solved. However, the discrete form of governing equations would involve numerical fluxes on the cell interface, which are not defined. To obtain the numerical solution, one needs to reconstruct the solution/numerical flux on the cell interface from the solutions at cell centers. We usually term this process as flux solver. It is noted that most of existing flux solvers are either based on mathematical reconstruction (smooth function approximation) or partially physical reconstruction (solution to one-dimensional Euler equation). One may be interested to develop a flux solver, which is based on fully physical reconstruction. The answer towards the above two concerns motivates the authors to develop the lattice Boltzmann flux solver (LBFS) and the gas kinetic flux solver (GKFS) shown in this book. As demonstrated in the book, the kinetic solver and Navier-Stokes solver can be combined through the solution reconstruction process on the cell interface.

This book provides a deep insight into the intrinsic connections between the Boltzmann/lattice Boltzmann equations and the macroscopic equations in the continuum flow regime. The new perspective allows connecting these two systems of interpreting fluid behaviors and removing each other's limitations. The highlight of each chapter is shown below. Chapter 1 starts with the physical conservation laws and the finite volume discretization of macroscopic governing equations. The mathematical and partially physical reconstruction of numerical fluxes on the cell interface is discussed, and then the need to develop LBFS and GKFS is introduced. Chapter 2 describes the kinetic theory and the Boltzmann equation. The connection between the Boltzmann equation and the Navier-Stokes equations is presented, which allows reconstructing fluxes on the cell interface with the local solution to the Boltzmann equation. Chapter 3 introduces the LBFS for isothermal and thermal flows, while Chapter 4 extends the LBFS for incompressible multiphase flows. In Chapter 5, the Maxwellian function-based gas kinetic scheme (M-GKS) and the Maxwellian function-based GKFS (M-GKFS) are presented. By replacing the Maxwellian distribution function with the circular/sphere function or the D1Q4 model, the circular/sphere function-based GKFS (C-GKFS/S-GKFS) and the compressible LBFS can be established, which is detailed in Chapter 6. Chapter 7 proposes discrete models for the circular and the sphere functions, based on which the circular/sphere function-based GKFS (C-DGKFS/S-DGKFS) are constructed. In Chapter 8, the Maxwellian function-based incompressible gas kinetic scheme (M-IGKS) and the circular/sphere function-based incompressible GKFS (C-IGKFS/S-IGKFS) are described.

This book is written for different levels of readers. It provides step by step implementation of algorithms, sample applications and computer codes. The comprehensive references enable the reader to find more sources if further information is needed.

Finally, the authors would like to thank their families for their support during the writing of this book. The contents of Chapter 5 are mainly from the PhD dissertation of Dr Sun Yu.

L. M. Yang, Y. Wang, Z. Chen and C. Shu

Contents

List of Tables

List of Figures

Chapter 1

Introduction

Computational Fluid Dynamics (CFD) has emerged as one of the major solvers to fluid problems in recent decades [Peyret and Taylor, 2002; Wendt and Anderson, 2008; Cebeci *et al.*, 2009; Lomax *et al.*, 2013]. Compared with the analytical and experimental approaches, CFD possesses more flexibility and economic advantages. It is applicable to fluid problems with various initial/boundary conditions and is not constrained by the available substance or experimental facility. In nowadays, CFD has grown up as a popular topic in both academia and industry. As an important solution candidate for almost all subjects that are related to fluid mechanics, it is now routinely employed in the areas of aerodynamics, hydrodynamics, biological mechanics, chemical engineering, energy industry, manufacturing, etc.

The prosperity of the CFD community attributes to the developments of both the computer hardware and computational algorithms. On one hand, the continuous upgrade of the processer greatly enhances the computational speed. And the construction of high-performance computing or super-computing facilities makes the computation of large-scale problems be possible. On the other hand, the advancement of computational algorithms aims to recover true physical behaviors and improve numerical accuracy, stability and efficiency.

The computational theories for CFD compose of robust governing equations for the fluid system and the associated numerical resolutions. The establishment of governing equations relies on the understanding of physical principles, while the numerical solutions can be achieved by various numerical schemes and discretization strategies. In this chapter, these two aspects will be elaborated.

1.1 Conservation Laws

The conservation law is the fundamental principle of mathematical interpretations of the physical world. For an isolated physical system, the conservation law states that a measurable property of the system does not change with time. In more general conditions, the conservation law reads that the time evolution of the conservative variable of a control volume depends on the fluxes across its surfaces and the sources inside itself. This principle can be mathematically expressed as

$$\int_{\Omega} \frac{\partial \phi}{\partial t} \mathrm{d}\Omega + \oint_{\partial\Omega} \mathbf{F} \cdot \mathbf{n} dS = \int_{\Omega} Q \mathrm{d}\Omega, \tag{1.1}$$

where Ω and $\partial\Omega$ denote a control volume and its surface, respectively; ϕ is a conservative variable; $\mathbf{F}$ represents the fluxes on the surface and $\mathbf{n}$ is the unit normal vector pointing outwards the surface; Q is the source term inside the control volume.

1.2 Governing Equations of Fluid Dynamics

The governing equations are essentially mathematical interpretations of the fluid system. And the construction of governing equations must fulfill basic physical principles and rational assumptions. The CFD investigates macroscopic fluid behaviors that state the collective appearance of large amounts of fluid particles; thus, the continuum assumption is often applied. In the meantime, the governing equations certainly fulfill basic conservation laws. More specifically speaking, the total mass, momentum and energy of a closed or isolated fluid system are conserved. Thus, the variation of these quantities within a given volume can be evaluated by the combined effects of the quantities transported across the boundary and the forces or sources exerted on the volume. Adopting these conservation laws in the control volume gives the following continuity, momentum and energy equations in differential forms [Versteeg and Malalasekera, 2007; Toro, 2009; Blazek, 2015]:

$$\frac{\partial \rho}{\partial t} + \nabla \cdot (\rho \mathbf{u}) = 0, \tag{1.2}$$

$$\frac{\partial \rho \mathbf{u}}{\partial t} + \nabla \cdot (\rho \mathbf{u}\mathbf{u}) = -\nabla p + \nabla \cdot \boldsymbol{\tau} + \mathbf{f}, \tag{1.3}$$

$$\frac{\partial \rho E}{\partial t} + \nabla \cdot (\rho E \mathbf{u}) = \dot{q} + \nabla \cdot (k \nabla T) + \nabla \cdot (-p\mathbf{u} + \mathbf{u} \cdot \boldsymbol{\tau}) + \mathbf{f} \cdot \mathbf{u}, \tag{1.4}$$

where ρ, $\mathbf{u}$, p, E and T are the density, velocity vector, pressure, total energy and temperature, respectively; t represents the time; $\mathbf{f}$ is the forcing term; $\dot{q}$ and k are the rate of the volumetric heat addition per unit mass and the heat conductivity, respectively; $\boldsymbol{\tau}$ denotes the stress tensor. For the perfect gas, the relationship between the pressure and the temperature can be constructed through the following equation of state:

$$p = \rho R T, \tag{1.5}$$

where R is the specific gas constant. Conceptually, the establishment of Eqs. (1.2)-(1.4) is to explain general fluid behaviors at the macroscopic scale. Bearing in mind that fluids may exhibit different characteristics at even smaller scales (microscopic or mesoscopic), it should be admitted that the application of these equations has limitations.

In the above equations, the convective terms on the left-hand side originate from the inconsistency between the moving fluid particles and

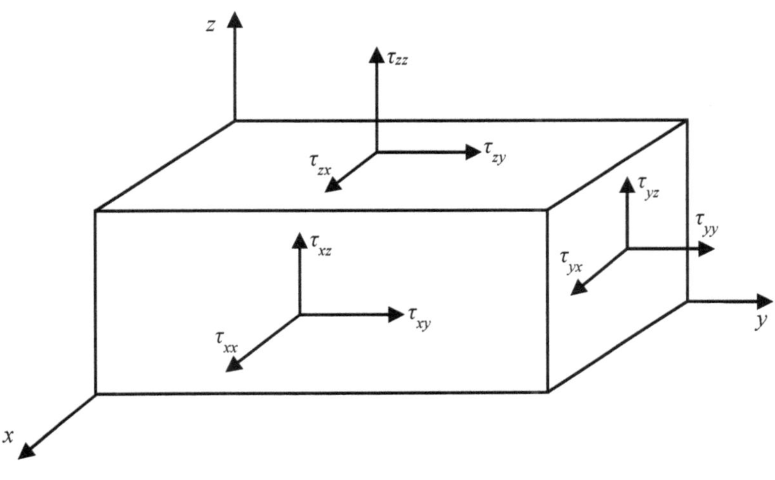

Figure 1.1. Illustrations of stresses acting on the fluid volume.

the fixed control volume in the Eulerian framework. These terms would vanish if the whole system is converted into the Lagrangian framework, and all terms on the left-hand side are merged into the substantial derivative denoted as "D/Dt".

The stress tensor $\boldsymbol{\tau}$ quantifies the viscous effects of the fluid. In the Cartesian coordinate system, it has the following form (see Figure 1.1):

$$\boldsymbol{\tau} = \begin{bmatrix} \tau_{xx} & \tau_{xy} & \tau_{xz} \\ \tau_{yx} & \tau_{yy} & \tau_{yz} \\ \tau_{zx} & \tau_{zy} & \tau_{zz} \end{bmatrix}. \tag{1.6}$$

The tensors compose of the normal and the shear parts, which respectively tend to displace and distort the fluid volume. For a general Newtonian fluid, the shear stress is proportional to the velocity gradient. And the components of the stress tensor can be expressed as

$$\begin{aligned} \tau_{xx} &= \lambda\left(\frac{\partial u}{\partial x}+\frac{\partial v}{\partial y}+\frac{\partial w}{\partial z}\right)+2\mu\frac{\partial u}{\partial x}, \\ \tau_{yy} &= \lambda\left(\frac{\partial u}{\partial x}+\frac{\partial v}{\partial y}+\frac{\partial w}{\partial z}\right)+2\mu\frac{\partial v}{\partial y}, \\ \tau_{zz} &= \lambda\left(\frac{\partial u}{\partial x}+\frac{\partial v}{\partial y}+\frac{\partial w}{\partial z}\right)+2\mu\frac{\partial w}{\partial z}, \end{aligned} \tag{1.7}$$

$$\begin{aligned} \tau_{xy} &= \tau_{yx} = \mu\left(\frac{\partial u}{\partial y}+\frac{\partial v}{\partial x}\right), \\ \tau_{xz} &= \tau_{zx} = \mu\left(\frac{\partial u}{\partial z}+\frac{\partial w}{\partial x}\right), \\ \tau_{yz} &= \tau_{zy} = \mu\left(\frac{\partial v}{\partial z}+\frac{\partial w}{\partial y}\right), \end{aligned} \tag{1.8}$$

where u, v and w are components of the velocity vector $\mathbf{u}$ along the x-, y- and z- coordinates, respectively; λ and μ respectively denote the second viscosity coefficient and the dynamic viscosity coefficient. The dynamic viscosity, which is defined as the tangential force per unit area required to move one horizontal plane with respect to another plane in a unit distance within the same fluid, is a measure of the internal resistance to the shear force. It is linked to the kinematic viscosity v through

$$\mu = \rho \nu \, . \tag{1.9}$$

For incompressible fluid, the continuity equation (1.2) is reduced to an expression that the divergence of the velocity is zero, which naturally closes the evaluation of the normal stresses in Eq. (1.7). For gas, the relationship between the second viscosity coefficient and the dynamic viscosity can be constructed through the Stokes hypothesis which defines the bulk viscosity as

$$\mu_B = \lambda + \frac{2}{3}\mu \, . \tag{1.10}$$

The bulk viscosity identifies the energy dissipation in a fluid of uniform temperature during a change in volume at a finite rate. The practical value of the bulk viscosity varies with both the fluid molecules and their vibrating frequencies. For monatomic gases at low density, the bulk viscosity is zero, and thus the normal stresses are derived as

$$\begin{aligned} \tau_{xx} &= 2\mu\left(\frac{\partial u}{\partial x} - \frac{1}{3}\nabla\cdot\mathbf{u}\right), \\ \tau_{yy} &= 2\mu\left(\frac{\partial v}{\partial y} - \frac{1}{3}\nabla\cdot\mathbf{u}\right), \\ \tau_{zz} &= 2\mu\left(\frac{\partial w}{\partial z} - \frac{1}{3}\nabla\cdot\mathbf{u}\right). \end{aligned} \tag{1.11}$$

The energy equation (1.4) can be physically interpreted as that the rate of energy variation inside the fluid element attributes to the net flux of heat into the element and the rate of work done on the surface and the body of the element. The four terms on the right-hand side of Eq. (1.4) correspond to the volume heating effect, the thermal conductivity through surfaces, the work done by the pressure and the stress tensor on the surfaces, and the work done by the body forces, respectively.

Over hundreds of years, efforts have been made to analytically resolve the governing equations (1.2)-(1.4). Yet these equations show up to be complex partial differential equations, which make their analytical solutions hard to be achieved. Until now, the progress has been limited to cases with certain assumptions (e.g. unidirectional flow, potential flow, symmetricity, etc.) or specific initial and boundary conditions. And the analytical solution to the Navier-Stokes equations (Eq. (1.3)) remains one of the seven Millennium Prize Problems. Thus, finding approximate solutions to these equations becomes an attractive detour in resolving practical fluid problems, which is exactly the aim and scope of CFD.

1.3 Finite Volume Discretization of Governing Equations

In past decades, many numerical methodologies have been proposed, aiming to resolve governing equations of fluid mechanics. Basically, these methods can be categorized into two categories: the mesh-based methods and the meshfree methods. Representative ones in the first category include the finite difference method (FDM), finite element method (FEM), finite volume method (FVM), etc. Interests have also been growing in developing meshfree methods like smoothed particle hydrodynamics (SPH), moving particle semi-implicit (MPS) method, material point method (MPM), diffuse element method (DEM), etc. Although the meshfree methods show advantages in some challenging problems with free surfaces or complex geometries, their accuracy and stability in general problems remain questionable at the present stage. Therefore, the mesh-based methods have been the mainstream solutions in the CFD community until now.

The mesh-based methods, as the name indicates, essentially partition the computational domain with meshes. The quality of the generated mesh may pose a great influence on the final numerical results, due to which the mesh generation is an important subject in the CFD community. Two common categories of meshes are the structured mesh and the unstructured mesh. The structured mesh refers to the type of grid that each cell or mesh point can be identified by index (i, j) in two dimensions or (i, j, k) in three dimensions. And the unstructured grid exhibits irregular connectivity between cells and can only be identified by a specific cell index and its neighboring cells. Although the unstructured mesh demands higher storage space, its flexibility in problems with complex geometries clearly outperforms its structured counterpart. Therefore, from the application point of view, the easy implementation of a numerical scheme on the unstructured mesh would prove advantageous.

The physical properties are defined at the mesh points or the center of the cell enclosed by mesh lines. The governing equations are then numerically resolved at the mesh point or the cell center to update the local properties. Different schemes have been developed to fulfill that requirement. Representative mesh-based CFD methods include the FDM, FEM and FVM. With the same goal of approximating solutions of the partial differential equation (PDE), these methods adopt different strategies:

(a) The FDM is based on the Taylor series expansion and directly discretizes the derivatives in the PDE with a finite difference scheme. By doing so, it converts the nonlinear PDE into a set of algebraic equations that can be resolved by algebra techniques.
(b) The FEM adopts the concept of trial functions into the approximation procedure and presents a resolving process of minimizing the error by adjusting the weight functions that project the residual caused by the trial functions.
(c) The FVM is essentially an integral method that performs volume integrations on the PDE. By using the Green-Gauss theorem (also known as the divergence theorem), the volume integral of the divergence term in the PDE can be converted to surface integrals of numerical fluxes. This conversion reduces the order of space derivatives and thus facilitates numerical approximations.

Among these methods, FVM has always been an attractive one due to its intrinsic advantages. Two well-recognized merits of the FVM are its conservative nature and flexibility on unstructured mesh. Since one cell interface is shared by two cells, the numerical flux entering a cell through one cell interface is identical to the flux leaving its adjacent cell that shares the same cell interface. Therefore, if the boundary condition is properly specified, the FVM is always conservative both locally and globally. Meanwhile, in general cases, the numerical fluxes on a cell interface can be evaluated from the physical properties defined at centers of two adjacent cells that share this cell interface. This characteristic allows flexible implementation of FVM on unstructured mesh which is a preferred space meshing strategy when encountering complex geometries. These appealing characteristics of FVM make it a powerful numerical tool for engineering problems. It has been adopted as the flow solver in many commercial CFD software or open-source packages, among which two most representative ones might be Ansys Fluent and OpenFOAM.

For the continuity, momentum and energy equations considered in this book (Eqs. (1.2)-(1.4)), the finite volume discretization starts from the integration of these equations within the control volume, which gives

$$\frac{\partial}{\partial t}\int_{\Omega}\mathbf{W}\mathrm{d}\Omega+\oint_{\partial\Omega}\left(\mathbf{F}_c-\mathbf{F}_v\right)\mathrm{d}S=\int_{\Omega}\mathbf{Q}\mathrm{d}\Omega. \tag{1.12}$$

Note that the Green-Gauss theorem is used to convert the second term on the left-hand side in the above equation from the volume integral to the surface integral. Here, $\mathbf{W}=(\rho, \rho u, \rho v, \rho w, \rho E)$ is the vector of conservative variables; $\mathbf{F}_c$ and $\mathbf{F}_v$ are the inviscid and the viscous flux vectors, respectively; $\mathbf{Q}$ represents the vector of source terms. Specifically, the inviscid fluxes attribute to the convective and the pressure-related terms, which give

$$\mathbf{F}_c=\begin{bmatrix}\rho u_n\\ \rho u u_n+n_x p\\ \rho v u_n+n_y p\\ \rho w u_n+n_z p\\ \rho H u_n\end{bmatrix}, \tag{1.13}$$

where $\mathbf{n}=(n_x, n_y, n_z)$ is the unit normal vector pointing outwards the surface. u_n and H are respectively the contravariant velocity and the total enthalpy in the following forms:

$$u_n = \mathbf{u} \cdot \mathbf{n}\,, \tag{1.14}$$

$$H = E + \frac{p}{\rho}. \tag{1.15}$$

The viscous fluxes account for the viscous stresses and the heat conduction effects, which give

$$\mathbf{F}_v = \begin{bmatrix} 0 \\ n_x\tau_{xx} + n_y\tau_{xy} + n_z\tau_{xz} \\ n_x\tau_{yx} + n_y\tau_{yy} + n_z\tau_{yz} \\ n_x\tau_{zx} + n_y\tau_{zy} + n_z\tau_{zz} \\ n_x\Theta_x + n_y\Theta_y + n_z\Theta_z \end{bmatrix}, \tag{1.16}$$

where

$$\begin{aligned} \Theta_x &= u\tau_{xx} + v\tau_{xy} + w\tau_{xz} + k\frac{\partial T}{\partial x}, \\ \Theta_y &= u\tau_{yx} + v\tau_{yy} + w\tau_{yz} + k\frac{\partial T}{\partial y}, \\ \Theta_z &= u\tau_{zx} + v\tau_{zy} + w\tau_{zz} + k\frac{\partial T}{\partial z}. \end{aligned} \tag{1.17}$$

And the vector of source terms yields

$$\mathbf{Q} = \begin{bmatrix} 0 \\ f_x \\ f_y \\ f_y \\ \mathbf{f} \cdot \mathbf{u} + \dot{q} \end{bmatrix}. \tag{1.18}$$

The implementation of FVM varies with the definition of the control volume. Based on the locations of defined physical properties, the FVM can be classified into the cell-vertex scheme, the cell-centered scheme, or the hybrid scheme. It is generally believed that the cell-centered scheme bears the advantage of easy implementation. Therefore, it is commonly used in fluid dynamics problems and will also be adopted in this book.

In the cell-centered finite volume scheme, the physical properties are defined at the centroids of cells and represent the averaged quantities in the specific cell. And the cell averaged values are used as the second-order approximation to the local values at the cell center. Eq. (1.12) can thus be derived into the following discretized form:

$$\frac{\partial \mathbf{W}_i}{\partial t} = -\frac{1}{\Omega_i}\sum_{j=1}^{N_f}\left(\mathbf{F}_c - \mathbf{F}_v\right)_j S_j + \mathbf{Q}_i, \tag{1.19}$$

where the subscripts i and j are the indexes of the cell and its interfaces, respectively; Ω is the volume of the cell; N_f represents the number of cell interfaces; and S_j denotes the area of the jth control surface enclosed the cell i. And the numerical fluxes in the above equation are usually calculated at the center of the cell interface and represents the averaged value on that interface in second-order schemes. The above equation can be interpreted as that the evolution of conservative variables at the cell center relies on the numerical fluxes on the cell interface and the forcing terms at the cell center. Since the forcing terms can be evaluated from the physical properties defined at the same location, the essential issue here is the reconstruction of the fluxes.

There are generally two types of strategies in reconstructing the numerical fluxes on the cell interface: the mathematical reconstruction and the physical reconstruction. The mathematical reconstruction essentially utilized the pure mathematical manipulations (interpolation,

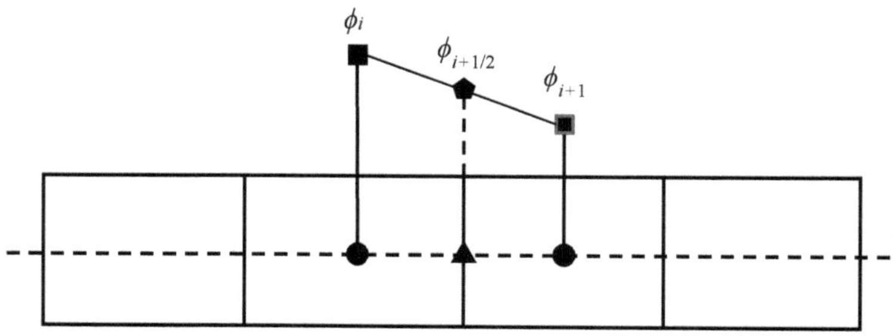

Figure 1.2. Illustration of central difference scheme.

fitting or differentiation) to determine the numerical fluxes at desired locations. The physical reconstruction, on the other hand, obtains the physical properties on the cell interface by locally resolving physical equations. Specific types of physical reconstruction schemes are then categorized by the physical equations constructed on the cell interface. In the following sections, we will look into these two approaches and elaborate their advantages and limitations.

1.4 Mathematical Reconstructions of Numerical Fluxes on Cell Interface in Finite Volume Discretization

As shown in Eq. (1.19), the numerical fluxes on the cell interface can be divided into the inviscid and the viscous parts, for which specific strategies of the mathematical reconstruction may differ. In this section, we will separately discuss the mathematical reconstructions of the inviscid and the viscous numerical fluxes.

1.4.1 *Mathematical reconstruction of inviscid fluxes*

The inviscid fluxes shown in Eq. (1.13) are evaluated from the physical quantities and involve no derivative terms. Therefore, the reconstruction of inviscid fluxes is essentially a process of approximating physical quantities on the cell interface. The approximation is based on the known physical properties defined at cell centers and the relative location between the cell interface and adjacent cell centers.

1.4.1.1 *Central scheme*

An instinct choice of evaluating the quantities on the interface from those defined at adjacent cell centers is to perform the linear interpolation. This approach is generally recognized as the central scheme which is a generalization of the central difference scheme on uniform meshes. As illustrated in Figure 1.2, the physical quantity ϕ on the cell interface $i+1/2$ can be calculated by

$$\phi_{i+1/2} = \phi_i + \frac{\left(x_{i+1/2} - x_i\right)}{\left(x_{i+1} - x_i\right)}\left(\phi_{i+1} - \phi_i\right). \tag{1.20}$$

In the case of the uniform structured meshes, the interfacial quantity simply takes the average value of the properties defined at two adjacent cells. The validity of this approach is established through the Taylor series analysis. The properties ϕ_i and ϕ_{i+1} can be expanded with respect to the interface quantity $\phi_{i+1/2}$, and the approximation in Eq. (1.20) reveals second-order accuracy. The second-order accuracy is sufficient for general flow simulations except for some specific categories of problems requiring higher resolutions.

1.4.1.2 *Jameson-Schmidt-Turkel (JST) scheme*

The above central scheme, although preserving second-order accuracy, is only valid for problems whose solution is sufficiently smooth. In cases with strong nonlinearities (e.g. discontinuities like the shock wave), the basic principle in numerical simulation is to add numerical dissipations (or so-called artificial viscosity) into the discretization process to diffuse the sharp profile into a smoother one. The amount of numerical diffusion is decided by the balance between numerical accuracy and stability. The most famous and perhaps the most successful model for this purpose is the Jameson-Schmidt-Turkel scheme, or JST scheme in short [Jameson, 2017]. Taking the following one-dimensional Euler equation as an example, its semi-discrete finite volume form on the uniform mesh is

$$\frac{\partial \mathbf{W}_i}{\partial t} + \frac{\mathbf{F}_{i+1/2} - \mathbf{F}_{i-1/2}}{\Delta x} = \mathbf{0}, \tag{1.21}$$

where $\mathbf{W}_i$ represents the vector of average conservative variables of the cell; $\mathbf{F}_{i+1/2}$ or $\mathbf{F}_{i-1/2}$ are the flux vectors on the cell interface; i is the index of the cell; and the subscripts $i\pm1/2$ denote the indices of the cell interfaces. Different from the pure central scheme, the numerical flux in the JST scheme incorporates both the second-order and the fourth-order derivatives as dissipative terms. And the flux vector is evaluated by

$$\mathbf{F}_{i+1/2} = \frac{1}{2}\left(\mathbf{F}_i + \mathbf{F}_{i+1}\right) - \mathbf{D}_{i+1/2}, \tag{1.22}$$

where **F** denotes the inviscid flux vector in the one-dimensional scenario; and the dissipative flux vector reads

$$\mathbf{D}_{i+1/2} = \varepsilon^{(2)}_{i+1/2}\Delta\mathbf{W}_{i+1/2} - \varepsilon^{(4)}_{i+1/2}\left(\Delta\mathbf{W}_{i+3/2} - 2\Delta\mathbf{W}_{i+1/2} + \Delta\mathbf{W}_{i-1/2}\right), \tag{1.23}$$

where

$$\Delta\mathbf{W}_{i+1/2} = \begin{bmatrix} \rho_{i+1} - \rho_i \\ (\rho u)_{i+1} - (\rho u)_i \\ (\rho u H)_{i+1} - (\rho u H)_i \end{bmatrix}. \tag{1.24}$$

And the dissipative coefficients $\varepsilon^{(2)}$ and $\varepsilon^{(4)}$ are essentially switch functions for the second and the fourth-order dissipative terms. The principle of setting these coefficients is to implement the fourth order dissipative term in smooth regions of the flow and to adopt the second-order dissipative term at locations where the shock wave is identified. With the notions that the second-order derivative term and the fourth-order derivative term are respectively in the scale of $O(\Delta x^1)$ and $O(\Delta x^3)$, this principle aims to maintain the second-order of accuracy in the most computational area while introducing sufficient numerical diffusions into areas with strong nonlinearities.

1.4.2 *Mathematical reconstruction of viscous fluxes*

The viscous fluxes in Eq. (1.16), in their mathematical forms, consist of space gradients of the velocity components and the temperature. Unlike the diversified strategies for the approximation of inviscid fluxes, the reconstruction of viscous fluxes is usually accomplished through either the finite difference scheme or the Green-Gauss theorem.

The finite difference scheme requires the local transformation between the Cartesian coordinate system and the curvilinear coordinate system, which gives

$$\frac{\partial \phi}{\partial x} = \frac{\partial \phi}{\partial \xi}\frac{\partial \xi}{\partial x} + \frac{\partial \phi}{\partial \eta}\frac{\partial \eta}{\partial x} + \frac{\partial \phi}{\partial \zeta}\frac{\partial \zeta}{\partial x}. \tag{1.25}$$

The derivatives ϕ_ξ, ϕ_η and ϕ_ζ can be evaluated by the finite difference scheme, while ξ_x, η_x, ζ_x are components of the Jacobian matrix that identifies the correlation between the two coordinate systems.

Although being very straightforward, the finite difference scheme might face difficulty in implementation on non-orthogonal grids. Therefore, the other approach, which utilizes the Green-Gauss theorem, is a more general solution to the approximation of space gradient terms in FVM. Performing the Green-Gauss theorem on the control volume i yields

$$\int_{\Omega_i} \nabla \phi \mathrm{d}\Omega = \oint_{\partial\Omega_i} \phi d\mathbf{S}, \tag{1.26}$$

where $\mathbf{S}$ denotes the outward pointing surface vector. Since the properties at the cell centroid represent the mean values over the cell, the above formulation can be written into the discrete form of

$$\nabla \phi_i = \frac{1}{\Omega_i}\sum_{j=1}^{N_f} \phi_j \mathbf{S}_j . \tag{1.27}$$

Note that the physical properties ϕ_j on cell interfaces can be calculated by the simple average of those at the adjacent cells. After obtaining the gradient terms at cell centroids, the gradient on the cell interface can be approximated in a weighted interpolation manner, which gives

$$\overline{\nabla \phi_j} = g_i \nabla \phi_i + g_{i+1} \nabla \phi_{i+1}, \tag{1.28}$$

where g_i and g_{i+1} are geometric interpolation parameters that are determined by the relative locations between the cell interface j and the cell centroids i and i+1. A more precise approximation of the gradient can be given by [Haselbacher and Blazek, 2000]

$$\nabla \phi_j = \overline{\nabla \phi_j} + \left[\frac{\phi_{i+1} - \phi_i}{d_{i,i+1}} - \left(\overline{\nabla \phi_j} \cdot \mathbf{e}_{i,i+1}\right)\right]\mathbf{e}_{i,i+1}, \tag{1.29}$$

where

$$\overline{\nabla \phi_j} = \frac{1}{2}\left(\nabla \phi_i + \nabla \phi_{i+1}\right), \tag{1.30a}$$

$$\mathbf{e}_{i,i+1} = \frac{\mathbf{d}_{i,i+1}}{d_{i,i+1}}, \ \mathbf{d}_{i,i+1} = \mathbf{r}_{i+1} - \mathbf{r}_i, \ d_{i,i+1} = \left|\mathbf{d}_{i,i+1}\right|. \tag{1.30b}$$

Here, $\mathbf{r}$ represents the location vector of the cell centroid. Note that the correction term in Eq. (1.29) aims to offset the difference between the true midpoint of the interface and the intersection point of the interface and the line connecting cell centroids i and i+1 (see Figure 1.3) [Moukalled *et al.*, 2015]. This correction would be valuable when applying the method to unstructured meshes.

1.5 Physical Reconstruction of Numerical Fluxes on Cell Interface in Finite Volume Discretization

1.5.1 *Upwind schemes*

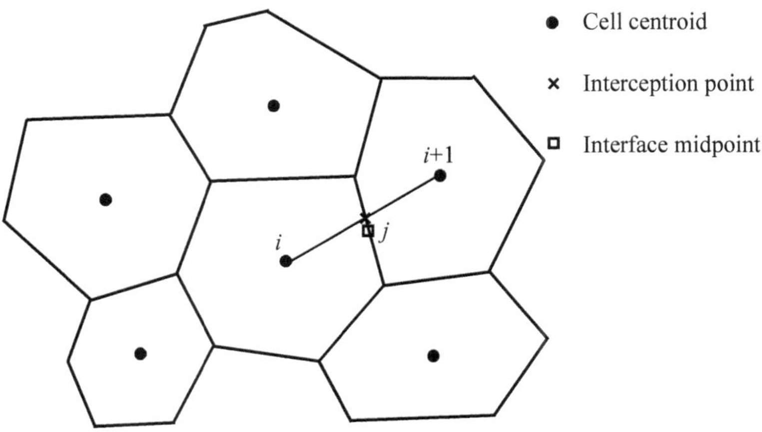

Figure 1.3. Illustration of inconsistency between the interception point and the midpoint on the cell interface.

In either the central scheme or the JST scheme, the interpolation of physical properties on the cell interface is based on the recognition of smooth solution. Therefore, the weighting parameters are evaluated by the relative locations in space and have no directional preference. Such mathematical interpretation works for the non-directional physics such as the viscous effects, but somehow violates the physical nature of the convection phenomenon.

The upwind scheme is thus proposed to reflect the convection nature [Courant *et al.*, 1952; Patankar, 2018]. It proposes that the physical properties on the cell interface should be determined by the values at upwind cell centers. Figure 1.4 presents the simplest upwind scheme which directly uses the properties defined at the nearest upstream centroids as the values on the cell interface. The mathematical formulation gives

$$\phi_{i-1/2} = \begin{cases} \phi_i & \text{if } \dot{m}_{i-1/2} > 0, \\ \phi_{i-1} & \text{if } \dot{m}_{i-1/2} < 0. \end{cases} \tag{1.31a}$$

$$\phi_{i+1/2} = \begin{cases} \phi_i & \text{if } \dot{m}_{i+1/2} > 0, \\ \phi_{i+1} & \text{if } \dot{m}_{i+1/2} < 0. \end{cases} \tag{1.31b}$$

where $\dot{m}$ is the mass flow rate which is used to identify the flow direction

$$\dot{m} = \rho \mathbf{u} \cdot \mathbf{n} S . \tag{1.32}$$

It is noted that this simplest upwind model would reveal a spatial accuracy in the first order, which might be over dissipative in smooth

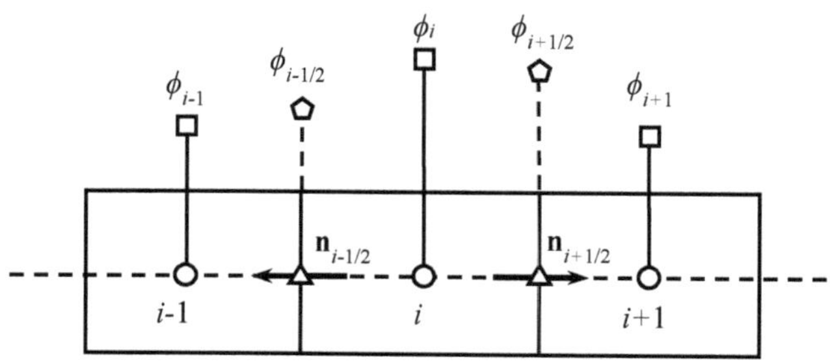

Figure 1.4. Illustration of the first-order upwind scheme.

regions. To improve the accuracy, the higher order of approximations can be constructed through various approaches such as the total variation diminishing (TVD) scheme, the weighted essentially non-oscillatory (WENO) scheme, etc..

The upwind scheme provides an effective reconstruction strategy for the numerical fluxes in the uncoupled convection-diffusion equations. In more complicated situations, such as the Euler/Navier-Stokes equations, where the mass, momentum and energy equations are fully coupled, the upwind scheme is usually utilized to firstly approximate the flow variables on both sides of the cell interface; and the flux-vector splitting schemes or the Riemann solvers are then adopted to reconstruct the inviscid fluxes on the cell interface, while the viscous fluxes are still evaluated by the mathematical strategies presented in **Section 1.4.2**.

1.5.2 *Flux-vector splitting schemes*

The flux-vector splitting scheme considers the direction of the wave propagation when reconstructing convective fluxes of Euler/Navier-Stokes equations. Two typical examples of the flux-vector splitting schemes are the van Leer's scheme [Van Leer, 1997; Anderson *et al.* 1986] and the advection upstream splitting method (AUSM) [Liou, 1993; Liou and Steffen, 1993; Liou, 2001]. In van Leer's scheme, the convective fluxes are split directly into the forward and the backward parts according to the sign of the local Mach number. The AUSM further introduces a decomposition step before performing the splitting. It essentially decomposes the inviscid fluxes into the convective part and the pressure part.

1.5.2.1 *Van Leer's scheme*

For the three-dimensional cases on an arbitrary mesh, the van Leer's scheme approximates the convective fluxes on the cell interface by [Van Leer, 1997; Anderson *et al.*, 1986]

$$\mathbf{F}_{c,i+1/2} = \mathbf{F}_c^+ + \mathbf{F}_c^- , \tag{1.33}$$

where $\mathbf{F}_c^+$ and $\mathbf{F}_c^-$ are the forward and the backward inviscid fluxes, respectively. These fluxes are determined by the local Mach number, which quantifies the Mach number normal to the face of the control volume and can be approximated by

$$Ma = Ma_L^+ + Ma_R^-, \tag{1.34}$$

where the split Mach numbers are defined as

$$Ma_L^+ = \begin{cases} \frac{1}{4}(Ma_L + 1)^2 & \text{if } |Ma_L| < 1, \\ \frac{1}{2}(Ma_L + |Ma_L|) & \text{if } |Ma_L| \geq 1. \end{cases} \tag{1.35}$$

$$Ma_R^- = \begin{cases} -\frac{1}{4}(Ma_R - 1)^2 & \text{if } |Ma_R| < 1, \\ \frac{1}{2}(Ma_R - |Ma_R|) & \text{if } |Ma_R| \geq 1. \end{cases} \tag{1.36}$$

Here, the Mach numbers Ma_L and Ma_R are respectively calculated by the flow variables on the left and the right sides of the cell interface. As mentioned before, these flow variables are usually reconstructed by the upwind scheme.

For supersonic flows (i.e., $|Ma| \geq 1$), the two parts of inviscid fluxes are given by

$$\begin{aligned} &\mathbf{F}_c^+ = \mathbf{F}_L,\ \mathbf{F}_c^- = \mathbf{0} \quad \text{if } Ma \geq +1, \\ &\mathbf{F}_c^+ = \mathbf{0},\ \mathbf{F}_c^- = \mathbf{F}_R \quad \text{if } Ma \leq -1. \end{aligned} \tag{1.37}$$

Here, $\mathbf{F}_L$ and $\mathbf{F}_R$ both represent the inviscid fluxes given by Eq. (1.13), which can be calculated from the left and the right states, respectively. In supersonic flow regimes, all eigenvalues of the Jacobian matrix for inviscid fluxes are either larger than zero for $Ma \geq +1$ or less than zero for $Ma \leq -1$. Adopting Eq. (1.37) in such flow condition leads the van Leer's scheme to the standard upwind differencing scheme.

For subsonic flow (i.e., $|Ma| < 1$), the forward and the backward inviscid fluxes are evaluated from

$$\mathbf{F}_c^{\pm} = \begin{bmatrix} f_{\text{mass}}^{\pm} \\ f_{\text{mass}}^{\pm}\left[n_x\left(-u_n \pm 2c\right)/\gamma + u\right] \\ f_{\text{mass}}^{\pm}\left[n_y\left(-u_n \pm 2c\right)/\gamma + v\right] \\ f_{\text{mass}}^{\pm}\left[n_z\left(-u_n \pm 2c\right)/\gamma + w\right] \\ f_{\text{energy}}^{\pm} \end{bmatrix}, \tag{1.38}$$

where $c = \sqrt{\gamma p/\rho}$ is the sound speed; components of the mass and energy flux are defined as

$$\begin{aligned} f_{\text{mass}}^{+} &= +\rho_L c_L \frac{\left(Ma_L + 1\right)^2}{4}, \\ f_{\text{mass}}^{-} &= -\rho_R c_R \frac{\left(Ma_R - 1\right)^2}{4}, \\ f_{\text{energy}}^{\pm} &= f_{\text{mass}}^{\pm}\left\{\frac{\left[\left(\gamma-1\right)u_n \pm 2c\right]^2}{2\left(\gamma^2-1\right)} + \frac{u^2+v^2+w^2-u_n^2}{2}\right\}_{L/R}. \end{aligned} \tag{1.39}$$

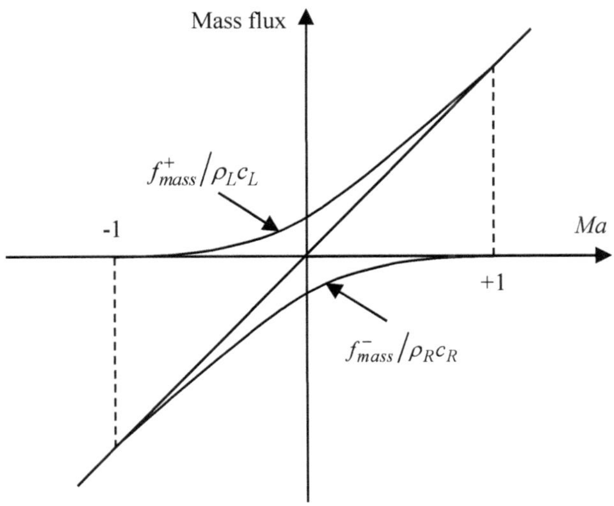

Figure 1.5. Variation of mass flux with Mach number for van Leer's scheme.

In subsonic flow regimes, the Jacobian matrix for inviscid fluxes has both positive and negative eigenvalues, which means that both sides of the cell interface contribute to the calculation of numerical fluxes. Besides, it can be seen from Eq. (1.38) and Figure 1.5 that the split mass flux of the van Leer's scheme is continuously differentiable at the sonic and the stagnation points. Such smoothing nature favors the numerical approximation of inviscid fluxes. However, when solving the Navier-Stokes equations in which the viscous effect is encountered, the van Leer's scheme may smear the boundary layers and lead to inaccurate stagnation and wall temperatures [Hänel *et al.*, 1987; Van Leer *et al.*, 1987]. Some remedial measures have been suggested in the work of Hänel and Schwane [1989].

1.5.2.2 *AUSM*

The AUSM follows a similar splitting strategy as the van Leer's scheme but employs an extra decomposition step before performing the splitting [Liou, 1993; Liou and Steffen, 1993; Liou, 2001]. Specifically, the inviscid flux vector $\mathbf{F}_c$ is first decomposed into the convective part $\mathbf{F}_c^{(c)}$ and the pressure part $\mathbf{F}_c^{(p)}$, namely

$$\mathbf{F}_c = \begin{bmatrix} \rho u_n \\ \rho u u_n \\ \rho v u_n \\ \rho w u_n \\ \rho H u_n \end{bmatrix} + \begin{bmatrix} 0 \\ n_x p \\ n_y p \\ n_z p \\ 0 \end{bmatrix} = \mathbf{F}_c^{(c)} + \mathbf{F}_c^{(p)}. \tag{1.40}$$

As a result, the inviscid fluxes on the cell interface can be written as

$$\mathbf{F}_{c,i+1/2} = \mathbf{F}_{c,i+1/2}^{(c)} + \mathbf{F}_{c,i+1/2}^{(p)}. \tag{1.41}$$

According to Liou and Steffen [1993], the convective part is given by

$$\mathbf{F}_{c,i+1/2}^{(c)} = \frac{Ma}{2}\left(\mathbf{F}_L^{(c)} + \mathbf{F}_R^{(c)}\right) + \frac{|Ma|}{2}\left(\mathbf{F}_L^{(c)} - \mathbf{F}_R^{(c)}\right), \tag{1.42}$$

where

$$\mathbf{F}_L^{(c)} = \begin{bmatrix} \rho c \\ \rho c u \\ \rho c v \\ \rho c w \\ \rho c H \end{bmatrix}_L , \quad \mathbf{F}_R^{(c)} = \begin{bmatrix} \rho c \\ \rho c u \\ \rho c v \\ \rho c w \\ \rho c H \end{bmatrix}_R .$$

Here, the Mach number normal to the face of the control volume is evaluated by Eq. (1.34). In addition, the flow variables ρ, u, v, w, H and c on the left and the right sides of the cell interface are interpolated from the properties at cell centers by the upwind scheme. As shown in Eq. (1.42), for $Ma \geq 0$, the convective flux component is fully determined by the left state, while for $Ma < 0$, it is completely evaluated from the right state.

The splitting of the pressure flux component is essentially the splitting of the pressure itself, which gives

$$p_{i+1/2} = p_L^+ + p_R^- . \tag{1.43}$$

The following two splitting algorithms have been suggested by Liou and Steffen [1993]:

$$p_L^+ = \begin{cases} \dfrac{p_L}{4}(Ma_L + 1)^2 (2 - Ma_L) & \text{if } |Ma_L| < 1, \\ \dfrac{p_L}{2}\dfrac{(Ma_L + |Ma_L|)}{Ma_L} & \text{if } |Ma_L| \geq 1. \end{cases} \tag{1.44}$$

$$p_R^- = \begin{cases} \dfrac{p_R}{4}(Ma_R - 1)^2 (2 + Ma_R) & \text{if } |Ma_R| < 1, \\ \dfrac{p_R}{2}\dfrac{(Ma_R - |Ma_R|)}{Ma_R} & \text{if } |Ma_R| \geq 1. \end{cases} \tag{1.45}$$

and

$$p_L^+ = \begin{cases} \dfrac{p_L}{2}(1+Ma_L) & \text{if } |Ma_L|<1, \\ \dfrac{p_L}{2}\dfrac{(Ma_L+|Ma_L|)}{Ma_L} & \text{if } |Ma_L|\geq 1. \end{cases} \tag{1.46}$$

$$p_R^- = \begin{cases} \dfrac{p_R}{2}(1-Ma_R) & \text{if } |Ma_R|<1, \\ \dfrac{p_R}{2}\dfrac{(Ma_R-|Ma_R|)}{Ma_R} & \text{if } |Ma_R|\geq 1. \end{cases} \tag{1.47}$$

It is indicated in Eqs. (1.43)-(1.47) that the splitting of the pressure flux component is fully upwind only in the case of supersonic flow. In the subsonic flow regime, both the flow variables of the left and the right states contribute to the calculation of the pressure on the cell interface.

In practice, the AUSM could provide a sharp resolution of strong shocks and accurate results for boundary layers. However, the original AUSM might present spurious pressure oscillations around the shocks for the case where the flow is aligned with the grid [Radespiel and Kroll, 1995]. To alleviate such issues, various improved schemes, such as AUSM+ [Liou, 1996] and AUSMDV [Wada and Liou, 1997], have been developed.

1.5.3 *Riemann solvers*

Different from the flux-vector splitting schemes which consider the direction of the wave propagation in the reconstruction of inviscid fluxes, Riemann solvers aim to approach exact/approximate solutions to the Riemann problem established over the interface to reconstruct the inviscid fluxes. Pioneering work in this area was carried out by Godunov [1959] who was pursuing the exact solution of the local Riemann problem. However, the process of the analytical solution is very complicated and involves tedious iterative procedures. Thus, some approximate Riemann solvers have been proposed in the literature to improve computational efficiency [Roe, 1981; Harten *et al.*, 1983; Toro and Titarev, 2006]. In the following text, the Godunov scheme and three representative approximate Riemann solvers, namely the Harten-Lax-van Leer (HLL) scheme [Harten *et al.*, 1983; Einfeldt, 1988], the Harten-Lax-van Leer with contact discontinuities (HLLC) scheme [Toro *et al.*, 1994; Batten *et al.*, 1997] and the Roe scheme [Roe, 1981; Roe and Pike, 1985; Kim *et al.*, 2002], will be briefly introduced.

1.5.3.1 *Godunov scheme*

The Riemann problem is essentially an initial-value problem (IVP). It resolves the one-dimensional, time-dependent Euler equation with a piece-wise constant initial condition consisting of the left and the right states, which gives

$$\left.\begin{aligned} &\mathbf{W}_t + \mathbf{F}(\mathbf{W})_x = \mathbf{0} \\ &\mathbf{W}(x,0) = \mathbf{W}^{(0)}(x) = \begin{cases} \mathbf{W}_L & \text{if } x<0 \\ \mathbf{W}_R & \text{if } x>0 \end{cases} \end{aligned}\right\}, \tag{1.48}$$

where $\mathbf{F}(\mathbf{W})$ is the vector of inviscid fluxes. $\mathbf{W}_L$ and $\mathbf{W}_R$ are the initial states on the left and the right sides of the cell interface, respectively. The physical analogue of the Riemann problem is the shock tube

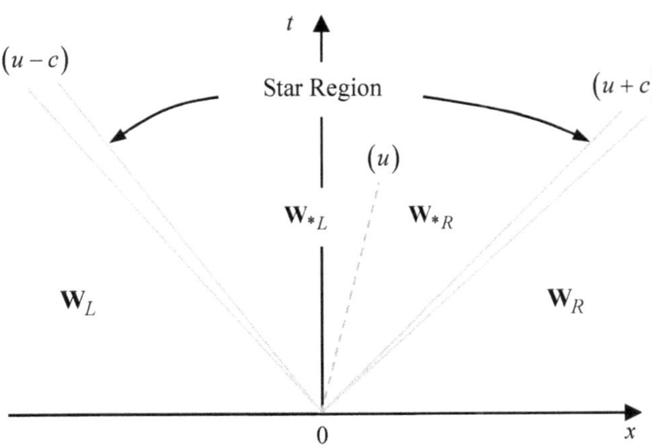

Figure 1.6. Solution of the Riemann problem on the x-t plane for one-dimensional Euler equation. The yellow solid line represents either shock wave or rarefaction wave and the green dashed line denotes contact discontinuity.

problem, in which the velocities u_L and u_R on either side of the interface are idealized by an initial discontinuity.

For the cases without vacuum, the similarity solution $\mathbf{W}(x/t)$ can be decomposed into three wave families associated with the eigenvalues of the Jacobian matrix of the one-dimensional Euler equation, $\lambda_1 = u - c$, $\lambda_2 = u$ and $\lambda_3 = u + c$, where c is the local sound speed, as shown in Figure 1.6. Furthermore, the wave associated with the characteristic curve λ_2 is a contact discontinuity and those associated with the characteristics λ_1 and λ_3 could be either the rarefaction wave (also known as the expansion wave) or the shock wave. Specifically, across the contact discontinuity, both the pressure p and the velocity u are constant but the density is discontinuous. The rarefaction wave spans over a fan-shaped area which is bounded by the characteristic lines corresponding to the head and the tail of the wave. In this area, the density ρ, velocity u and pressure p vary, while the Generalized

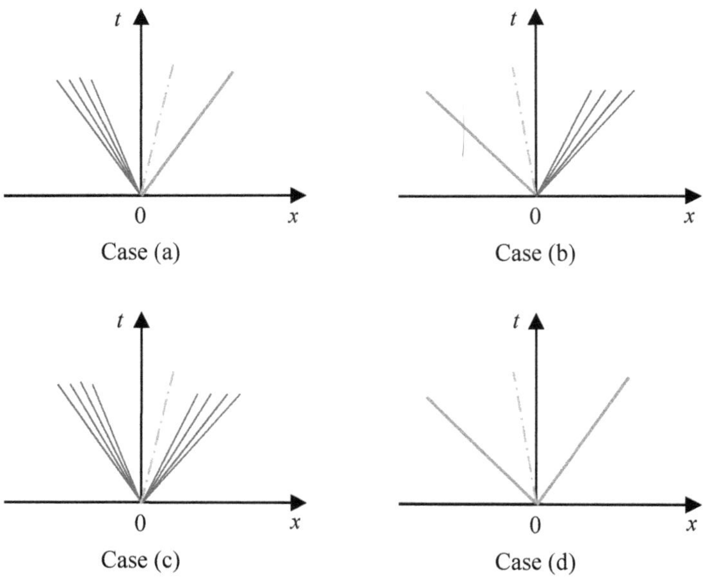

Figure 1.7. Possible wave patterns of the solution for Riemann problem: (a) left rarefaction, contact, right shock; (b) left shock, contact, right rarefaction; (c) left rarefaction, contact, right rarefaction; (d) left shock, contact, right shock. Red solid line denotes shock wave, blue solid line represents rarefaction wave and green dashdot line stands for the contact discontinuity.

Riemann Invariants and Isentropic Law are applicable. For the shock wave, all three quantities ρ, u and p change according to the Rankine-Hugoniot Conditions.

As shown in Figure 1.6, the unknown region between the left and the right waves is named the Star Region. The Star Region is divided by the contact discontinuity into two subregions named as the Star Left ($\mathbf{W}_{*L}$) and the Star Right ($\mathbf{W}_{*R}$). According to the characteristics of the contact discontinuity, the pressure p_* and the velocity u_* remain constant within the Star Region, while the density has distinct values of ρ_{*L} and ρ_{*R} across the contact discontinuity. So, the unknowns to be resolved in the Star Region include p_*, u_*, ρ_{*L} and ρ_{*R}, which requires specifying the wave patterns on the left and the right sides of the contact discontinuity. As illustrated in Figure 1.7, there are four possible wave patterns.

First, we consider the case with the left wave being a shock wave moving with the speed S_L. The pre-shock flow quantities are ρ_L, u_L and p_L, and the post-shock flow quantities are ρ_{*L}, u_* and p_*. In the frame of reference that is moving with the shock, the shock speed is zero and the relative pre-shock and post-shock velocities are

$$\hat{u}_L = u_L - S_L, \quad \hat{u}_* = u_* - S_L. \tag{1.49}$$

Then the Rankine-Hugoniot conditions yield

$$\rho_L \hat{u}_L = \rho_{*L} \hat{u}_*, \tag{1.50a}$$

$$\rho_L \hat{u}_L^2 + p_L = \rho_{*L} \hat{u}_*^2 + p_*, \tag{1.50b}$$

$$\hat{u}_L \left(\rho_L \hat{E}_L + p_L \right) = \hat{u}_* \left(\rho_{*L} \hat{E}_{*L} + p_* \right), \tag{1.50c}$$

where $\hat{E} = \hat{u}^2/2 + p/[\rho(\gamma - 1)]$ is the total energy. Substituting Eq. (1.49) into Eqs. (1.50a)-(1.50c), the unknowns in Eqs. (1.50a)-(1.50c) turn into ρ_{*L}, u_*, p_* and S_L. It is noted that the equation system in Eqs. (1.50a)-(1.50c) is not closed. However, after some manipulations, the following relationship between u_* and p_* can be established:

$$u_* = u_L - f_L(p_*, \mathbf{W}_L), \tag{1.51}$$

where

$$f_L(p_*,\mathbf{W}_L)=(p_*-u_L)\left[\frac{A_L}{p_*+B_L}\right]^{\frac{1}{2}}, \tag{1.52}$$

and

$$A_L=\frac{2}{(\gamma+1)\rho_L},\quad B_L=\frac{(\gamma-1)p_L}{(\gamma+1)}. \tag{1.53}$$

Eq. (1.51) has two unknowns of u_* and p_*, which should be coupled with another equation from the right state to close the system.

In the second case, we assume that the right wave is a shock with the moving speed of S_R. After a similar derivation as the first case, the relation between u_* and p_* can be expressed as

$$u_*=u_R+f_R(p_*,\mathbf{W}_R), \tag{1.54}$$

where

$$f_R(p_*,\mathbf{W}_R)=(p_*-u_R)\left[\frac{A_R}{p_*+B_R}\right]^{\frac{1}{2}}, \tag{1.55}$$

and

$$A_R=\frac{2}{(\gamma+1)\rho_R},\quad B_R=\frac{(\gamma-1)p_R}{(\gamma+1)}. \tag{1.56}$$

The third scenario is that the left wave is a rarefaction wave. With the notion that the Generalized Riemann Invariant $I_L(u,c)$ is constant across the rarefaction wave, we can derive

$$I_L(u,c)=u_L+\frac{2c_L}{\gamma-1}=u_*+\frac{2c_{*L}}{\gamma-1}. \tag{1.57}$$

Furthermore, according to the Isentropic Law, we have

$$\frac{p_L}{\rho_L^\gamma} = \frac{p_*}{\rho_{*L}^\gamma}. \tag{1.58}$$

Substituting Eq. (1.58) into Eq. (1.57) and using the definition of the sound speed, we can derive the relationship between u_* and p_* as follows:

$$u_* = u_L - f_L\left(p_*, \mathbf{W}_L\right), \tag{1.59}$$

where

$$f_L\left(p_*, \mathbf{W}_L\right) = \frac{2c_L}{(\gamma - 1)}\left[\left(\frac{p_*}{p_L}\right)^{\frac{\gamma-1}{2\gamma}} - 1\right]. \tag{1.60}$$

Finally, in the condition that the right wave is a rarefaction wave, adopting the Generalized Riemann Invariant $I_R(u,c)$ and the Isentropic Law for the right rarefaction wave yields

$$I_R(u,c) = u_* - \frac{2c_{*R}}{\gamma - 1} = u_R - \frac{2c_R}{\gamma - 1}, \tag{1.61}$$

$$\frac{p_*}{\rho_{*R}^\gamma} = \frac{p_R}{\rho_R^\gamma}. \tag{1.62}$$

After some manipulations, we have

$$u_* = u_R + f_R\left(p_*, \mathbf{W}_R\right), \tag{1.63}$$

where

$$f_R\left(p_*, \mathbf{W}_R\right) = \frac{2c_R}{(\gamma - 1)}\left[\left(\frac{p_*}{p_R}\right)^{\frac{\gamma-1}{2\gamma}} - 1\right]. \tag{1.64}$$

Note that Eqs. (1.51), (1.54), (1.59) and (1.63) provide the relationships between u_* and p_* for all four possible wave patterns shown in Figure 1.7. The pressure p_* in the Star Region can then be evaluated by eliminating the velocity u_* from the derived relationships, which gives

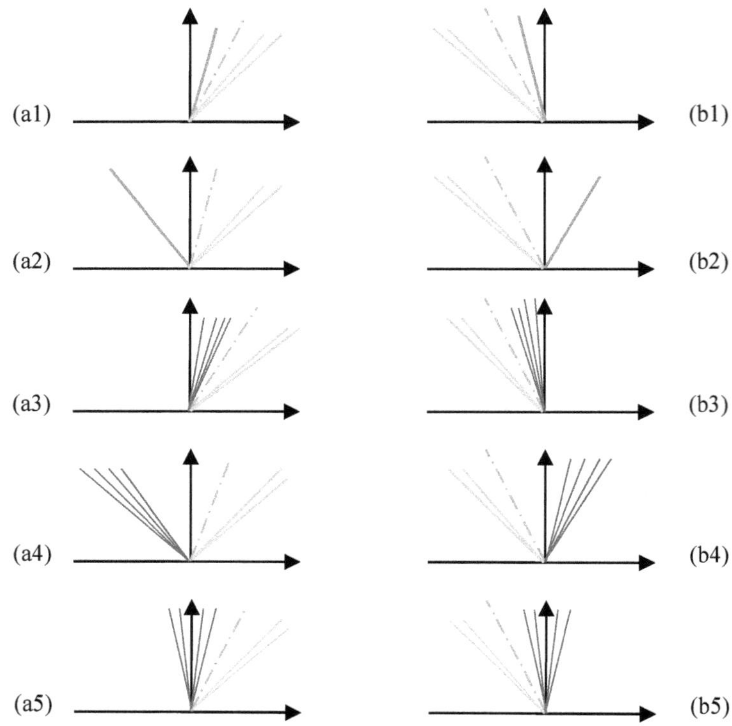

Figure 1.8. Possible wave patterns in evaluating the Godunov fluxes for Euler equation: (a) positive velocity in the Star Region; (b) negative velocity in the Star Region.

Table 1.1. Values of $\mathbf{W}(x/t=0)$ for all ten possible wave patterns in the solution of the Riemann problem.

Sub-case	Case (a)	Case (b)
1	$\mathbf{W}_L$	$\mathbf{W}_R$
2	$\mathbf{W}_{*L}$	$\mathbf{W}_{*R}$
3	$\mathbf{W}_L$	$\mathbf{W}_R$
4	$\mathbf{W}_{*L}$	$\mathbf{W}_{*R}$
5	$\mathbf{W}_{Lfan}$	$\mathbf{W}_{Rfan}$

$$f\left(p_*, \mathbf{W}_L, \mathbf{W}_R\right) = f_L\left(p_*, \mathbf{W}_L\right) + f_R\left(p_*, \mathbf{W}_R\right) + \left(u_R - u_L\right) = 0\,. \quad (1.65)$$

Eq. (1.65) reveals an equation with a single unknown p_*. It can be solved by common iterative methods such as the Newton-Raphson iteration. After that, the velocity u_* can be computed by

$$u_* = \frac{1}{2}\left(u_L + u_R\right) + \frac{1}{2}\left[f_R\left(p_*, \mathbf{W}_R\right) - f_L\left(p_*, \mathbf{W}_L\right)\right]. \quad (1.66)$$

And the densities ρ_{*L} and ρ_{*R} in the Star Region can then be calculated straightforwardly according to the wave patterns.

Eqs. (1.65) and (1.66) provide the exact solutions of the Riemann problem defined in Eq. (1.48). This exact solution can be adopted in the reconstruction of inviscid fluxes on the cell interface, and the obtained fluxes are called the Godunov fluxes. Given the discontinuity nature of the initial condition, this reconstruction is based on the flow variables resolved on the cell interface after an infinitesimal time interval, namely, $\mathbf{W}\left(x/t = 0\right)$. As shown in Figure 1.8, there are ten possible wave patterns in evaluating the Godunov fluxes for the Euler equation. For the cases of (a1)-(a5), the velocity in the Star Region u_* has a positive value, while for the cases of (b1)-(b5), u_* is negative. In addition, in the cases of (a1) and (a3), all eigenvalues are positive, which correspond to the supersonic flow at $Ma > 1$. So, all the flow variables on the cell interface $\mathbf{W}\left(x/t = 0\right)$ are evaluated from the left state. On the contrary, in the cases of (b1) and (b3), all eigenvalues are negative and $\mathbf{W}\left(x/t = 0\right)$ are fully determined by the right state. And in the cases of (a2), (a4), (b2) and (b4), $\mathbf{W}\left(x/t = 0\right)$ can be calculated from the exact solution of the Riemann problem. Finally, in the cases of (a5) and (b5), the location of the cell interface is inside the rarefaction wave, which calls for special treatments.

The values of $\mathbf{W}\left(x/t = 0\right)$ needed for evaluating the Godunov fluxes are summarized in Table 1.1 for all ten possible wave patterns. For the cases of (a5) and (b5), the flow variables in the vectors of $\mathbf{W}_{Lfan}$ and $\mathbf{W}_{Rfan}$ can be calculated by setting $x/t = 0$ in the following equations [Munz, 1994; Toro, 2009]:

$$\mathbf{W}_{Lfan} = \begin{cases} \rho = \rho_L \left[\dfrac{2}{(\gamma+1)} + \dfrac{(\gamma-1)}{(\gamma+1)c_L} \left(u_L - \dfrac{x}{t} \right) \right]^{\frac{2}{\gamma-1}}, \\ u = \dfrac{2}{(\gamma+1)} \left[c_L + \dfrac{(\gamma-1)}{2} u_L + \dfrac{x}{t} \right], \\ p = p_L \left[\dfrac{2}{(\gamma+1)} + \dfrac{(\gamma-1)}{(\gamma+1)c_L} \left(u_L - \dfrac{x}{t} \right) \right]^{\frac{2\gamma}{\gamma-1}}. \end{cases} \tag{1.67}$$

and

$$\mathbf{W}_{Rfan} = \begin{cases} \rho = \rho_R \left[\dfrac{2}{(\gamma+1)} - \dfrac{(\gamma-1)}{(\gamma+1)c_R} \left(u_R - \dfrac{x}{t} \right) \right]^{\frac{2}{\gamma-1}}, \\ u = \dfrac{2}{(\gamma+1)} \left[-c_R + \dfrac{(\gamma-1)}{2} u_R + \dfrac{x}{t} \right], \\ p = p_R \left[\dfrac{2}{(\gamma+1)} - \dfrac{(\gamma-1)}{(\gamma+1)c_R} \left(u_R - \dfrac{x}{t} \right) \right]^{\frac{2\gamma}{\gamma-1}}. \end{cases} \tag{1.68}$$

Once $\mathbf{W}(x/t=0)$ is obtained, the inviscid flux $\mathbf{F}_{c,i+1/2}$ can be calculated directly by its definition as shown in Eq. (1.13), i.e., $\mathbf{F}_{c,i+1/2} = \mathbf{F}(\mathbf{W}(x/t=0))$. It should be noted that this reconstruction strategy originates from the exact solution of the one-dimensional Euler equation. Thus, in multi-dimensional cases, the reconstruction should be carried out along the normal direction of the cell interface and the tangential velocities must be treated as the passive scalar quantities. In the cases where the normal velocity on the cell interface is positive, the tangential velocities on the cell interface take the values from the left state. Otherwise, they are evaluated from the right state.

1.5.3.2 *HLL scheme*

It has been noted that, the exact solution to the Riemann problem defined in Eq. (1.48) encounters tedious iterations to determine the flow variables in the Star Region. To reduce computational efforts, various approximate solutions have been proposed. Harten, Lax and van Leer [Harten *et al.*, 1983] utilized a two-wave model to approximate the solutions of the Riemann problem and proposed the so-called HLL scheme. The configuration of the two-wave model in the HLL scheme is shown in Figure 1.9. The exact solution in the Star Region is uniformly approximated by its integral average $\mathbf{W}^{hll}$. To calculate $\mathbf{W}^{hll}$, a control volume $[TS_L,\ TS_R]\times[0,\ T]$ on the $x-t$ plane is considered, where S_L and S_R are velocities of the two shock waves generated by the initial discontinuity, and T is a specific time. Applying conservation laws within the control volume $[TS_L,\ TS_R]\times[0,\ T]$ gives

$$\begin{aligned}\int_{TS_L}^{TS_R}\mathbf{W}(x,T)\,dx &= \int_{TS_L}^{TS_R}\mathbf{W}(x,0)\,dx \\ &+\int_0^T \mathbf{F}(\mathbf{W}(TS_L,t))\,dt-\int_0^T \mathbf{F}(\mathbf{W}(TS_R,t))\,dt.\end{aligned} \tag{1.69}$$

The right-hand side of Eq. (1.69) can be evaluated by

$$\int_{TS_L}^{TS_R}\mathbf{W}(x,T)\,dx = T\left(S_R\mathbf{W}_R - S_L\mathbf{W}_L + \mathbf{F}_L - \mathbf{F}_R\right), \tag{1.70}$$

where $\mathbf{F}_L=\mathbf{F}(\mathbf{W}_L)$ and $\mathbf{F}_R=\mathbf{F}(\mathbf{W}_R)$. The average of the solution in the Star Region yields

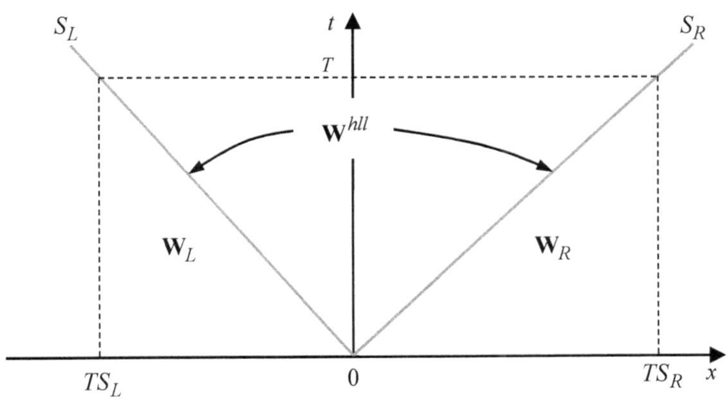

Figure 1.9. Configuration of the two-wave model for the HLL scheme.

$$\mathbf{W}^{hll} = \frac{1}{T\left(S_R - S_L\right)} \int_{TS_L}^{TS_R} \mathbf{W}\left(x,T\right) dx = \frac{S_R \mathbf{W}_R - S_L \mathbf{W}_L + \mathbf{F}_L - \mathbf{F}_R}{S_R - S_L}. \quad (1.71)$$

As shown in Eq. (1.71), $\mathbf{W}^{hll}$ can be explicitly evaluated once the shockwave speeds S_L and S_R are provided. To calculate the inviscid fluxes at $x/t = 0$ in the Star Region, the control volume $[TS_L,\ TS_R]\times[0,\ T]$ is divided into the left portion $[TS_L,\ 0]\times[0,\ T]$ and the right portion $[0,\ TS_R]\times[0,\ T]$, and the following conservation laws can be established:

$$\int_{TS_L}^{0} \mathbf{W}\left(x,T\right) dx = -TS_L \mathbf{W}_L + T\left(\mathbf{F}_L - \mathbf{F}\left(x/t = 0\right)\right), \quad (1.72)$$

$$\int_{0}^{TS_R} \mathbf{W}\left(x,T\right) dx = TS_R \mathbf{W}_R + T\left(\mathbf{F}\left(x/t = 0\right) - \mathbf{F}_R\right). \quad (1.73)$$

Simple manipulations of the above equations yield

$$\mathbf{F}\left(x/t = 0\right) = \mathbf{F}_L - S_L \mathbf{W}_L - \frac{1}{T}\int_{TS_L}^{0} \mathbf{W}\left(x,T\right) dx, \quad (1.74)$$

$$\mathbf{F}\left(x/t = 0\right) = \mathbf{F}_R - S_R \mathbf{W}_R + \frac{1}{T}\int_{0}^{TS_R} \mathbf{W}\left(x,T\right) dx. \quad (1.75)$$

In fact, the calculation of inviscid fluxes by Eqs. (1.74) and (1.75) is accurate if the exact solution to the Riemann problem is obtained. However, $\mathbf{W}\left(x,T\right)$ is calculated in an approximate manner in the HLL scheme. Therefore, the HLL scheme is an approximate Riemann solver.

In summary, the approximate similarity solution to the Riemann problem in the HLL scheme can be expressed as

$$\widetilde{\mathbf{W}}\left(x/t\right) = \begin{cases} \mathbf{W}_L & \text{if} \quad x/t \le S_L, \\ \mathbf{W}^{hll} & \text{if} \ S_L < x/t < S_R, \\ \mathbf{W}_R & \text{if} \quad x/t \ge S_R. \end{cases} \quad (1.76)$$

It is noted that this approximate solution consists of three constant states that are separated by two waves. The essence here is that a constant state is assumed in the Star Region, which is denoted as $\mathbf{W}^{hll}$. Once the approximate solution is obtained, the corresponding inviscid fluxes can be calculated from Eqs. (1.74) and (1.75), which gives

$$\mathbf{F}_{c,i+1/2}^{hll} = \mathbf{F}(x/t=0)$$
$$= \begin{cases} \mathbf{F}_L & \text{if} \quad 0 \leq S_L, \\ \dfrac{S_R \mathbf{F}_L - S_L \mathbf{F}_R + S_L S_R (\mathbf{W}_R - \mathbf{W}_L)}{S_R - S_L} & \text{if } S_L < 0 < S_R, \\ \mathbf{F}_R & \text{if} \quad 0 \geq S_R. \end{cases} \tag{1.77}$$

By now, the remaining unknowns in the flux reconstruction of the HLL scheme are the shockwave speeds S_L and S_R. As suggested by Einfeldt [1988] and Davis [1988], S_L and S_R can be approximated by eigenvalues evaluated from the Roe averages, which give

$$S_L = \tilde{u} - \tilde{c}, \quad S_R = \tilde{u} + \tilde{c}, \tag{1.78}$$

where $\tilde{u}$ and $\tilde{c}$ are respectively the Roe average velocity and sound speed

$$\tilde{u} = \frac{u_L \sqrt{\rho_L} + u_R \sqrt{\rho_R}}{\sqrt{\rho_L} + \sqrt{\rho_R}}, \quad \tilde{c} = \sqrt{(\gamma - 1)\left(\tilde{H} - \frac{1}{2}\tilde{u}^2\right)}. \tag{1.79}$$

In Eq. (1.79), the Roe average enthalpy is computed by

$$\tilde{H} = \frac{H_L \sqrt{\rho_L} + H_R \sqrt{\rho_R}}{\sqrt{\rho_L} + \sqrt{\rho_R}}. \tag{1.80}$$

Similar to the Godunov solver, the HLL scheme is based on the one-dimensional Riemann problem. Therefore, for multi-dimensional cases, the velocity u should be taken as the normal velocity on the cell interface and the tangential velocities should be treated as the passive scalar quantities.

1.5.3.3 *HLLC scheme*

By using the average solution in the whole Star Region, the HLL scheme naturally omits the contact discontinuities, shear waves, material interfaces, or any type of intermediate waves. To overcome this defect, Toro *et al.* [1994] proposed the HLLC scheme, where the extra "C" stands for contact discontinuities. The HLLC scheme is essentially a three-wave model as shown in Figure 1.10. With this model, it approximates the solutions to the Riemann problem as

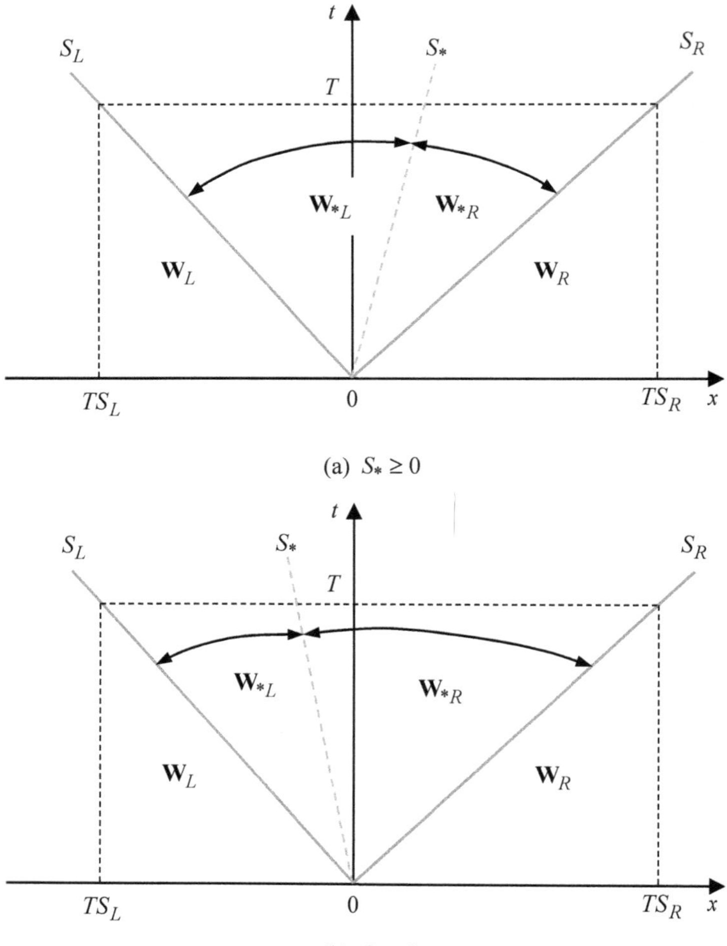

(b) $S_* < 0$

Figure 1.10. Configuration of the three-wave model for the HLLC scheme.

$$\widetilde{\mathbf{W}}(x/t) = \begin{cases} \mathbf{W}_L & \text{if} \quad x/t \le S_L, \\ \mathbf{W}_{*L} & \text{if } S_L < x/t \le S_*, \\ \mathbf{W}_{*R} & \text{if } S_* < x/t \le S_R, \\ \mathbf{W}_R & \text{if} \quad x/t > S_R. \end{cases} \tag{1.81}$$

And the corresponding inviscid fluxes can be reconstructed as

$$\mathbf{F}_{c,i+1/2}^{hllc} = \begin{cases} \mathbf{F}_L & \text{if} \quad 0 \le S_L, \\ \mathbf{F}_{*L} & \text{if } S_L < 0 \le S_*, \\ \mathbf{F}_{*R} & \text{if } S_* < 0 \le S_R, \\ \mathbf{F}_R & \text{if} \quad 0 > S_R. \end{cases} \tag{1.82}$$

The principle of the HLLC scheme is still to select appropriate control volume and apply conservation laws. Specifically, for the case of $S_* \ge 0$ (see Figure 1.10a), by applying the integral form of conservation laws in the control volume $[TS_L,\ 0] \times [0,\ T]$, we have

$$\int_{TS_L}^{0} \mathbf{W}(x,T)\,dx = -TS_L \mathbf{W}_L + T(\mathbf{F}_L - \mathbf{F}_{*L}). \tag{1.83}$$

Substituting Eq. (1.81) into Eq. (1.83), we can obtain

$$\mathbf{F}_{*L} = \mathbf{F}_L + S_L(\mathbf{W}_{*L} - \mathbf{W}_L). \tag{1.84}$$

Similarly, for the case of $S_* < 0$ (see Figure 1.10b), by applying the integral form of conservation laws in the control volume $[0,\ TS_R] \times [0,\ T]$, we have

$$\int_{0}^{TS_R} \mathbf{W}(x,T)\,dx = TS_R \mathbf{W}_R + T(\mathbf{F}_{*R} - \mathbf{F}_R). \tag{1.85}$$

Substituting Eq. (1.81) into Eq. (1.85) yields

$$\mathbf{F}_{*R} = \mathbf{F}_R + S_R(\mathbf{W}_{*R} - \mathbf{W}_R). \tag{1.86}$$

As shown in Eqs. (1.82), (1.84) and (1.86), to obtain the explicit expression of the numerical flux for the HLLC scheme, the flow variables in the Star Region (i.e., $\mathbf{W}_{*L}$ and $\mathbf{W}_{*R}$) have to be calculated in advance. One evaluation strategy is to directly use the characteristic of the contact discontinuity. According to the analysis shown in the Godunov solver, the velocity and the pressure maintain constant across the contact discontinuity. Thus, there are four unknowns in the Star Region, i.e., ρ_{*L}, ρ_{*R}, u_* and p_*. Note that in Figure 1.10, the notation of $S_* = u_*$ has been used for convenience.

To calculate ρ_{*L}, ρ_{*R}, u_* and p_* in the Star Region, Eqs. (1.84) and (1.86) can be reformulated as

$$\mathbf{F}_L - S_L \mathbf{W}_L = \mathbf{F}_{*L} - S_L \mathbf{W}_{*L}, \tag{1.87}$$

$$\mathbf{F}_R - S_R \mathbf{W}_R = \mathbf{F}_{*\mathrm{R}} - S_R \mathbf{W}_{*R}. \tag{1.88}$$

The vectors on the left-hand side of the above equations are known constants. Eqs. (1.87) and (1.88) contain six independent relations for the one-dimensional Euler equation, while there only exist four unknowns. To make the solution well-posed, a common strategy is to ignore the last components in Eqs. (1.87) and (1.88), which leads to

$$\begin{aligned} \rho_L(u_L - S_L) &= \rho_{*L}(S_* - S_L), \\ \rho_L u_L(u_L - S_L) + p_L &= \rho_{*L} S_*(S_* - S_L) + p_*, \end{aligned} \tag{1.89}$$

$$\begin{aligned} \rho_R(u_R - S_R) &= \rho_{*R}(S_* - S_R), \\ \rho_R u_R(u_R - S_R) + p_R &= \rho_{*R} S_*(S_* - S_R) + p_*. \end{aligned} \tag{1.90}$$

The solutions to Eqs. (1.89) and (1.90) yield

$$S_* = \frac{p_R - p_L + \rho_L u_L(S_L - u_L) - \rho_R u_R(S_R - u_R)}{\rho_L(S_L - u_L) - \rho_R(S_R - u_R)}, \tag{1.91}$$

$$\rho_{*K} = \rho_K \frac{(S_K - u_K)}{(S_K - S_*)}, \tag{1.92}$$

where the subscript "K" represents either "L" or "R". By substituting Eqs. (1.91) and (1.92) into Eqs. (1.84) and (1.86), $\mathbf{F}_{*L}$ and $\mathbf{F}_{*R}$ can be expressed as

$$\mathbf{F}_{*K} = \mathbf{F}_K + S_K\left(\mathbf{W}_{*K} - \mathbf{W}_K\right), \tag{1.93}$$

with

$$\mathbf{W}_{*K} = \rho_{*K}\begin{bmatrix} 1 \\ S_* \\ E_K + \left(S_* - u_K\right)\left[S_* + \dfrac{p_K}{\rho_K\left(S_K - u_K\right)}\right] \end{bmatrix}. \tag{1.94}$$

The shockwave speeds S_L and S_R can be approximated by the eigenvalues evaluated from the Roe averages, which are essentially the same as shown in Eqs. (1.78)-(1.80). Once $\mathbf{F}_{*K}$ is obtained, the final expressions of the HLLC flux in Eq. (1.82) are available. For the multi-dimensional case, the reconstruction of the HLLC flux is carried out along the normal direction of the cell interface, and the tangential velocities are treated as the passive scalar quantities.

1.5.3.4 *Roe scheme*

The Roe scheme is one of the most well-known approximate Riemann solvers for the inviscid fluxes. Originally proposed by Professor Philip L. Roe in 1981, this method has been continuously refined and applied to various physical problems. Unlike the Godunov scheme and the HLL/HLLC schemes which exactly or approximately resolve the Riemann problem from the perspective of wave patterns, the Roe scheme constructs the inviscid fluxes by solving the Riemann problem with an approximate Jacobian matrix that represents the average behavior between the left and the right states. To begin with, the Roe scheme replaces the original Riemann problem in Eq. (1.48) by the approximate Riemann problem as follows

$$\left.\begin{aligned} &\mathbf{W}_t + \tilde{\mathbf{A}}\mathbf{W}_x = \mathbf{0} \\ &\mathbf{W}(x,0) = \mathbf{W}^{(0)}(x) = \begin{cases} \mathbf{W}_L & \text{if } x < 0 \\ \mathbf{W}_R & \text{if } x > 0 \end{cases} \end{aligned}\right\}, \tag{1.95}$$

where $\tilde{\mathbf{A}}$ is a constant Jacobian matrix given by

$$\tilde{\mathbf{A}} = \left.\frac{\partial \mathbf{F}}{\partial \mathbf{W}}\right|_{(\mathbf{W}_L,\mathbf{W}_R)} = \tilde{\mathbf{A}}(\mathbf{W}_L, \mathbf{W}_R). \tag{1.96}$$

With the constant Jacobian matrix, exact solutions to the approximate Riemann problem in Eq. (1.95) can be pursued, based on which the inviscid fluxes can be reconstructed.

The key of the Roe scheme is to construct an approximate Jacobian matrix $\tilde{\mathbf{A}}$ that fulfills the following criteria [Roe, 1981]:

(1) Hyperbolicity of the system, i.e., $\tilde{\mathbf{A}}$ is diagonalizable with real eigenvalues and has a complete set of linearly independent right eigenvectors.

(2) Consistency with the exact Jacobian, i.e., $\tilde{\mathbf{A}}(\mathbf{W},\mathbf{W}) = \mathbf{A}$.

(3) Conservation across discontinuities, i.e., $\mathbf{F}(\mathbf{W}_R) - \mathbf{F}(\mathbf{W}_L) = \tilde{\mathbf{A}}(\mathbf{W}_R - \mathbf{W}_L)$.

Conditions (1) and (2) are the obvious requirements, which preserves the mathematical characteristics of the original nonlinear system and ensures consistency with the original governing equations. Condition (3) can be interpreted as that the instantaneous growth rate $\mathbf{A}$ in the original Riemann problem is replaced by its average value $\tilde{\mathbf{A}}$ between the left and the right states in the approximate Riemann problem, as shown in Figure 1.11, where $\widetilde{\mathbf{W}}$ is the vector of so-called Roe averages at the Roe point for determining the approximate Jacobian matrix $\tilde{\mathbf{A}}$.

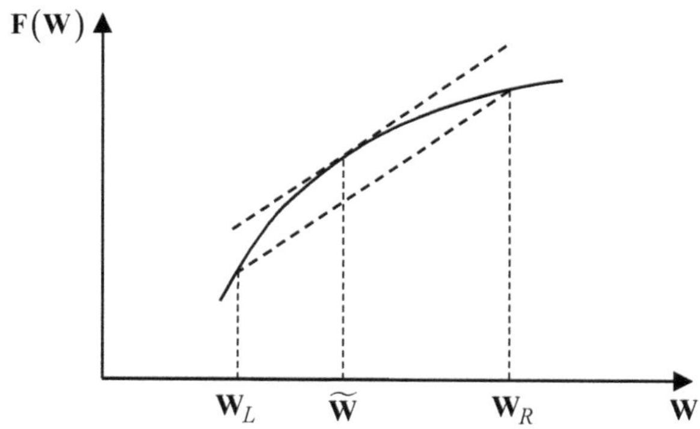

Figure 1.11. Schematic of Roe scheme.

The existence of the Roe point can be validated by the Lagrange's mean value theorem; and a general strategy for finding the Roe point is to construct a quadratic homogeneous function through coordinate transformation, so that the average slope would be equal to the local slope of the mid-point in the transformed coordinate. Performing such strategy on the Jacobian matrix of the Euler equation, the following Roe averages can be derived:

$$\tilde{\rho} = \sqrt{\rho_L \rho_R}\,, \tag{1.97}$$

$$\tilde{u} = \frac{u_L\sqrt{\rho_L} + u_R\sqrt{\rho_R}}{\sqrt{\rho_L} + \sqrt{\rho_R}}\,, \tag{1.98}$$

$$\tilde{v} = \frac{v_L\sqrt{\rho_L} + v_R\sqrt{\rho_R}}{\sqrt{\rho_L} + \sqrt{\rho_R}}\,, \tag{1.99}$$

$$\tilde{w} = \frac{w_L\sqrt{\rho_L} + w_R\sqrt{\rho_R}}{\sqrt{\rho_L} + \sqrt{\rho_R}}\,, \tag{1.100}$$

$$\tilde{H} = \frac{H_L\sqrt{\rho_L} + H_R\sqrt{\rho_R}}{\sqrt{\rho_L} + \sqrt{\rho_R}}\,, \tag{1.101}$$

$$\tilde{c} = \sqrt{(\gamma - 1)\left[\tilde{H} - \frac{\left(\tilde{u}^2 + \tilde{v}^2 + \tilde{w}^2\right)}{2}\right]}\,. \tag{1.102}$$

By knowing the Roe averages, the resolution of the approximate Riemann problem in Eq. (1.95) starts from the diagonalization of the approximate Jacobian matrix $\tilde{\mathbf{A}}$, which gives

$$\tilde{\mathbf{A}} = \tilde{\mathbf{S}}^{-1}\tilde{\mathbf{\Lambda}}\tilde{\mathbf{S}}\,, \tag{1.103}$$

where $\tilde{\mathbf{S}}^{-1}$ and $\tilde{\mathbf{S}}$ are the matrices composed by the right and the left eigenvectors, respectively. $\tilde{\mathbf{\Lambda}}$ is a diagonal matrix composed by the eigenvalues of $\tilde{\mathbf{A}}$. Substituting Eq. (1.103) into Eq. (1.95), we have

$$\frac{\partial \mathbf{W}}{\partial t} + \tilde{\mathbf{S}}^{-1}\tilde{\mathbf{\Lambda}}\tilde{\mathbf{S}}\frac{\partial \mathbf{W}}{\partial x} = \mathbf{0}. \tag{1.104}$$

Eq. (1.104) can also be expressed as

$$\frac{\partial \tilde{\mathbf{S}}\mathbf{W}}{\partial t} + \tilde{\mathbf{\Lambda}}\frac{\partial \tilde{\mathbf{S}}\mathbf{W}}{\partial x} = \mathbf{0}. \tag{1.105}$$

By introducing an intermediate variable $\overline{\mathbf{W}} = \tilde{\mathbf{S}}\mathbf{W}$, Eq. (1.105) can be rewritten as

$$\frac{\partial \overline{\mathbf{W}}}{\partial t} + \tilde{\mathbf{\Lambda}}\frac{\partial \overline{\mathbf{W}}}{\partial x} = \mathbf{0}. \tag{1.106}$$

It can be seen that Eq. (1.106) is a system of linear equations with constant coefficients. The solution of Eq. (1.106) yields

$$\overline{\mathbf{W}}_{i+1/2} = \frac{1}{2}\left(\overline{\mathbf{W}}_L + \overline{\mathbf{W}}_R\right) - \frac{1}{2} sign\left(\tilde{\mathbf{\Lambda}}\right)\left(\overline{\mathbf{W}}_R - \overline{\mathbf{W}}_L\right), \tag{1.107}$$

where $sign(\bullet)$ is a sign function. By incorporating $\tilde{\mathbf{S}}^{-1}$ into the above equation, we have

$$\mathbf{W}_{i+1/2} = \frac{1}{2}\left(\mathbf{W}_L + \mathbf{W}_R\right) - \frac{1}{2}\tilde{\mathbf{S}}^{-1} sign\left(\tilde{\mathbf{\Lambda}}\right)\tilde{\mathbf{S}}\left(\mathbf{W}_R - \mathbf{W}_L\right). \tag{1.108}$$

Finally, the corresponding inviscid fluxes on the cell interface can be expressed by

$$\mathbf{F}_{c,i+1/2} = \frac{1}{2}\left(\mathbf{F}_L + \mathbf{F}_R\right) - \frac{1}{2}\left|\tilde{\mathbf{A}}\right|\left(\mathbf{W}_R - \mathbf{W}_L\right), \tag{1.109}$$

where $\left|\tilde{\mathbf{A}}\right| = \tilde{\mathbf{S}}^{-1}\left|\tilde{\mathbf{\Lambda}}\right|\tilde{\mathbf{S}}$. After some manipulations, the inviscid fluxes for the three-dimensional case given by the Roe scheme can be written as

$$\mathbf{F}_{c,i+1/2} = \frac{1}{2}\left(\mathbf{F}_L + \mathbf{F}_R\right) - \frac{1}{2}\left(\left|\Delta\mathbf{F}_1\right| + \left|\Delta\mathbf{F}_{2,3,4}\right| + \left|\Delta\mathbf{F}_5\right|\right), \tag{1.110}$$

where

$$\left|\Delta\mathbf{F}_1\right| = \left|\tilde{u}_n - \tilde{c}\right| \left(\frac{\Delta p - \tilde{\rho}\tilde{c}\Delta u_n}{2\tilde{c}^2}\right) \begin{bmatrix} 1 \\ \tilde{u} - \tilde{c}n_x \\ \tilde{v} - \tilde{c}n_y \\ \tilde{w} - \tilde{c}n_z \\ \tilde{H} - \tilde{c}\tilde{u}_n \end{bmatrix}, \tag{1.111}$$

$$\begin{aligned} \left|\Delta\mathbf{F}_{2,3,4}\right| = \left|\tilde{u}_n\right| \Bigg\{ & \left(\Delta\rho - \frac{\Delta p}{\tilde{c}^2}\right) \begin{bmatrix} 1 \\ \tilde{u} \\ \tilde{v} \\ \tilde{w} \\ \tilde{q}^2/2 \end{bmatrix} \\ & + \tilde{\rho} \begin{bmatrix} 0 \\ \Delta u - \Delta u_n n_x \\ \Delta v - \Delta u_n n_y \\ \Delta w - \Delta u_n n_z \\ \tilde{u}\Delta u + \tilde{v}\Delta v + \tilde{w}\Delta w - \tilde{u}_n \Delta u_n \end{bmatrix} \Bigg\}, \end{aligned} \tag{1.112}$$

$$\left|\Delta\mathbf{F}_5\right| = \left|\tilde{u}_n + \tilde{c}\right| \left(\frac{\Delta p + \tilde{\rho}\tilde{c}\Delta u_n}{2\tilde{c}^2}\right) \begin{bmatrix} 1 \\ \tilde{u} + \tilde{c}n_x \\ \tilde{v} + \tilde{c}n_y \\ \tilde{w} + \tilde{c}n_z \\ \tilde{H} + \tilde{c}\tilde{u}_n \end{bmatrix}. \tag{1.113}$$

Here, $\tilde{u}_n = \tilde{u}n_x + \tilde{v}n_y + \tilde{w}n_z$ and $\tilde{q}^2 = \tilde{u}^2 + \tilde{v}^2 + \tilde{w}^2$. In addition, the jump condition is defined as $\Delta(\bullet) = (\bullet)_L - (\bullet)_R$.

As indicated in [Peery and Imlay, 1988; Lin, 1995], the original Roe scheme may generate an unphysical expansion shock in the case of stationary expansion and could lead to the "carbuncle phenomenon" when a perturbation grows ahead of a strong bow shock along the stagnation line. To fix these flaws, the entropy correction is commonly introduced to the Roe scheme to modify the modulus of eigenvalues. The details can be found in the work of Harten and Hyman [1983].

1.6 Motivation to Develop Lattice Boltzmann and Gas Kinetic Flux Solvers

The fluid flow problems are governed by conservation laws of the mass, momentum and energy. Together with the continuum assumption, the Stokes hypothesis and Fourier's law, the Navier-Stokes equations can be established for the viscous flows within this framework. And the simplification under the inviscid assumption leads to Euler equations. These equations are fully coupled nonlinear partial differential equations, due to which pursuing their analytical solutions would be extremely difficult, if not impossible. Given such fact, various numerical approaches have been proposed to resolve these equations, among which FVM is an attractive one due to its conservative nature and flexibility on unstructured mesh. The FVM updates the flow variables defined at cell centers, and the key issue is to reconstruct the numerical fluxes on the cell interface, which could be realized through either the mathematical or the physical reconstruction strategies. Different flux solvers have thus been developed for calculating both the inviscid and the viscous fluxes.

As a typical representative of the mathematical reconstruction, the JST scheme evaluates the inviscid fluxes on the cell interface by introducing additional numerical dissipations into the central schemes. The purpose of the extra numerical dissipation is to smooth the solution and stabilize the computation in discontinuous regions. Different from the mathematical reconstruction, the physical reconstruction calculates the inviscid fluxes on the cell interface according to the direction of the wave propagation or from the exact/approximate solutions to the Riemann problem established over the interface. In particular, the van Leer's scheme and the ASUM compute the inviscid fluxes based on the sign of the local Mach number; the Godunov scheme evaluates the inviscid fluxes by the exact solution to the Riemann problem; the HLL and HLLC schemes calculate the inviscid fluxes from the approximate solutions to the Riemann problem; and the Roe scheme reconstructs the inviscid fluxes by the solution to the approximate Riemann problem. Apart from these schemes, there are many other approaches of flux reconstruction in the CFD community.

In most conventional flux solvers, no matter whether the mathematical reconstruction or the physical reconstruction is used, the inviscid fluxes and viscous fluxes are evaluated separately. The inviscid fluxes can be computed by various flux solvers, such as the aforementioned JST scheme, AUSM, HLLC scheme and Roe scheme etc., while the viscous fluxes are usually evaluated by the mathematical manner (smooth function approximation). Even for the inviscid flux solvers, the physical solution is based on the one-dimensional Euler equation. Currently, there are still some difficulties to reconstruct local solutions to multi-dimensional Euler equations. Therefore, for multi-dimensional problems, the one-dimensional flux solvers are applied along the normal direction, and some approximations are made to tackle the tangential parts. In this sense, these flux solvers cannot be considered as truly physical reconstruction of the solution on the cell interface. One may ask whether it is possible to develop flux solvers which can truly satisfy the physical governing equations on the cell interface. The answer to this question motivates the present book.

It is well known that apart from the Euler/Navier-Stokes equations, the fluid behaviors can also be interpreted by the Boltzmann equation established at the mesoscopic scale. Under the continuum assumption, consistency between these two systems can be established via the Chapman-Enskog expansion analysis. In other words, it guarantees that the solutions from the Boltzmann equation can also satisfy Euler/Navier-Stokes equations. This gives us a new way to reconstruct the truly physical solution on the cell interface. To be specific, for multi-dimensional problems, although we encounter difficulties to find local solutions to the multi-dimensional Euler/Navier-Stokes equations, we may be able to find local solutions to the multi-dimensional Boltzmann equation or its discrete form. This idea leads to the development of various versions of the lattice Boltzmann flux solver (LBFS) [Shu *et al.*, 2014; Wang *et al.*, 2014b; Wang *et al.*, 2017a] and the gas kinetic scheme/gas kinetic flux solver (GKS/GKFS) [Xu, 2001; Xu *et al.*, 2005; Sun *et al.*, 2015; Yang *et al.*, 2015]. These novel flux reconstruction strategies can eliminate limitations of the conventional flux reconstruction approaches. In particular, both LBFS and GKFS can physically reconstruct the whole numerical fluxes without separating them into the inviscid and the viscous parts. Due to these distinct merits, LBFS and GKFS could be promising solution candidates for engineering problems. In the remaining chapters of this book, detailed derivations and implementations of LBFS and GKFS will be introduced.

Chapter 2

Kinetic Equations

Under their macroscopic behaviors, the fluid flows actually consist of a large number of particles (atoms or molecules). The motions of these particles are random and accompanied by collisions. The macroscopic continuum assumption fails in describing such random behaviors at small scale. To fully understand the fluid system, other numerical models should be utilized to interpret the moving disciplines of these fluid particles. Direct modelling of such microscopic motions yields the well-known molecular dynamics method. Although being successfully adopted in some categories of problems like the nanoflows, this method is intrinsically restricted to small-scale problems due to its high demand in computational resources. An alternative and more efficient solution is to describe the probability of particle motions rather than to directly track individual particles, which leads to the kinetic theory and the associated statistical mechanics. In this chapter, the kinetic theory and its connection to the macroscopic governing equations of the fluid system will be described in detail.

2.1 Kinetic Theory

The kinetic theory is established at the mesoscopic scale. It essentially describes the collective behavior of N-body fluid system with the probability distribution function $f_N(\mathbf{q}, \mathbf{p}, t)$ in the phase space. The phase space is a high-dimensional space that identifies all possible position and momentum variables. Here, $(\boldsymbol{q}, \boldsymbol{p})$ are the generalized coordinates in the $6N$ dimensional phase space, with $\mathbf{q}=(\mathbf{q}_1, \mathbf{q}_2,..., \mathbf{q}_N)$ being the $3N$ spatial coordinates of N molecules and $\mathbf{p}=(\mathbf{p}_1, \mathbf{p}_2,..., \mathbf{p}_N)$ representing the $3N$

conjugate momenta. With the distribution function f_N, the probability of the state in the infinitesimal volume [**p**, **p**+d**p**]×[**q**, **q**+d**q**] reads f_Nd**q**d**p**. The evolution of the distribution function follows Liouville's theorem which asserts that the distribution function is constant along any trajectory in the phase space. Thus, the following Liouville equation can be derived:

$$\frac{Df_N}{Dt} = \frac{\partial f_N}{\partial t} + \sum_{i=1}^{N}\left[\frac{\partial f_N}{\partial \mathbf{p}_i}\cdot\dot{\mathbf{p}}_i + \frac{\partial f_N}{\partial \mathbf{q}_i}\cdot\dot{\mathbf{q}}_i\right] = 0, \tag{2.1}$$

where i is the index of the particle, and the dot on the hat denotes the time derivative. With the notions that the time derivatives of the position and momenta are respectively related to the velocity and the force, the above formulation can be reduced to

$$\frac{\partial f_N}{\partial t} + \sum_{i=1}^{N}\left[\frac{\partial f_N}{\partial \mathbf{q}_i}\cdot\frac{\mathbf{p}_i}{m} + \frac{\partial f_N}{\partial \mathbf{p}_i}\cdot\mathbf{F}_i\right] = 0, \tag{2.2}$$

where m is the particle mass, and **F** means the net force exerted on the particle. To facilitate the resolution, the reduced distribution functions for the s-particle system are defined as

$$\begin{aligned} &f_s\left(\mathbf{q}_1 \cdots \mathbf{q}_s, \mathbf{p}_1 \cdots \mathbf{p}_s, t\right) \\ &= \int f_N\left(\mathbf{q}_1 \cdots \mathbf{q}_N, \mathbf{p}_1 \cdots \mathbf{p}_N, t\right) \mathrm{d}\mathbf{q}_{s+1}\mathrm{d}\mathbf{p}_{s+1} \ldots \mathrm{d}\mathbf{q}_N\mathrm{d}\mathbf{p}_N. \end{aligned} \tag{2.3}$$

The Liouville equation can also be integrated over part of the variables to obtain a chain of equations, which is known as the BBGKY (Bogoliubov-Born-Green-Kirkwood-Yvon) hierarchy. In this chain of equations, the sth equation associates the s-particle probability distribution function f_s with the (s+1)-particle probability distribution function f_{s+1}, which reads

$$\frac{\partial f_s}{\partial t} + \sum_{i=1}^{s}\left[\frac{\partial f_s}{\partial \mathbf{q}_i}\cdot\frac{\mathbf{p}_i}{m} + \frac{\partial f_s}{\partial \mathbf{p}_i}\cdot\mathbf{F}_i\right] = H\left(f_{s+1}\right), \tag{2.4}$$

where the operator H represents the influence from f_{s+1}.

It is noted that the set of equations (2.4) are fully coupled, which is equivalent to resolving the full Liouville equation. A simpler alternative is to specify f_{s+1} and then resolve f_s from Eq. (2.4). This is a common strategy of deriving kinetic equations within the framework of the kinetic theory. One such case is the Boltzmann equation, in which the single-particle probability distribution function f_1 is evolved, and the two-particle probability distribution function f_2 is modeled as the collision integral based on the molecular chaos hypothesis.

2.2 BGK-Boltzmann Equation and Maxwellian Distribution Function

2.2.1 *Continuous Boltzmann equation*

Under some assumptions [Grad, 1958], the BBGKY hierarchy can be truncated at certain orders so that a smaller set of equations can be obtained to approximate the original chain. Actually, this approach has been a common strategy for developing kinetic equations in many applications of the kinetic theory. Particularly, the approximation equation truncated at the first order leads to the well-known Boltzmann equation for the velocity distribution function f, which is defined as

$$f(\mathbf{x},\xi,t)=mNf_s(\mathbf{q}_1,\mathbf{p}_1,t), \tag{2.5}$$

where $\mathbf{x}=\mathbf{q}_1$ is the position of a particle and $\xi=\mathbf{p}_1/m$ is its velocity. As a result, the original phase space ($\mathbf{q}_1,\mathbf{p}_1$) is now expressed by the physical space $\mathbf{x}$ and the particle velocity space ξ.

With the definition of the velocity distribution function (2.5) and by ignoring the external force term acting on the particle, the Liouville equation (2.4) can be reduced to the Boltzmann equation as follows

$$\frac{\partial f}{\partial t}+\xi\cdot\nabla f=\Omega(f), \tag{2.6}$$

where $\Omega(f)$ is the collision integral describing the binary molecular collisions with the following form:

$$\Omega(f) = \iint (f'f_1' - ff_1)|\mathbf{g}|\cos(\theta) d_D^2 d\omega d\xi_1 . \tag{2.7}$$

Here, $f = f(\mathbf{x}, \xi, t)$ and $f_1 = f(\mathbf{x}, \xi_1, t)$. f' and f_1' are the velocity distribution functions after the collision. $\mathbf{g} = \xi - \xi_1$ is the relative particle velocity, θ is the angle of refraction, d_D is the diameter of the elastic sphere of the particle, and $d\omega$ is the area element on the unit sphere. After the collision, the particle velocities ξ and ξ_1 (incoming) are transformed into ξ' and ξ_1' (outgoing).

With the velocity distribution function, the fluid density ρ, velocity $\mathbf{u}$, and internal energy e of the monatomic gas can be determined from the associated moment relationships of the distribution function:

$$\rho = \langle f \rangle = \langle f^{eq} \rangle, \tag{2.8}$$

$$\rho\mathbf{u} = \langle \xi f \rangle = \langle \xi f^{eq} \rangle, \tag{2.9}$$

$$\rho e = \left\langle \frac{1}{2} c^2 f \right\rangle = \left\langle \frac{1}{2} c^2 f^{eq} \right\rangle, \tag{2.10}$$

and the heat flux $\mathbf{q}$ and the stress tensor $\boldsymbol{\tau}$ can be computed by

$$\mathbf{q} = \left\langle \frac{1}{2} \mathbf{c} c^2 f \right\rangle, \tag{2.11}$$

$$\boldsymbol{\tau} = \langle \mathbf{c}\mathbf{c} f \rangle, \tag{2.12}$$

where $\mathbf{c} = \xi - \mathbf{u}$ is the peculiar velocity of the particle and $c = |\mathbf{c}|$ denotes its magnitude. Note that $c^2 = \mathbf{c} \cdot \mathbf{c}$ is used in Eqs. (2.10)-(2.11) and the following text. f^{eq} is the velocity distribution function at the equilibrium state. The symbol $\langle \bullet \rangle$ denotes the integral over the particle velocity space, which is given by

$$\langle \phi \rangle = \int_{-\infty}^{+\infty} \phi d\Xi . \tag{2.13}$$

Here, ϕ is an arbitrary function and $d\Xi$ is the volume element in the particle velocity space.

2.2.2 *Maxwellian distribution function and BGK collision model*

One key issue in recovering hydrodynamic behaviors from the continuous Boltzmann equation (2.6) is the definition of local equilibrium. The *H*-theorem defines the *H*-function $H = \langle f \ln f \rangle$ from the Boltzmann equation which is monotonically decreasing with time [Grad, 1958], i.e.,

$$\frac{DH}{Dt} = \frac{\partial H}{\partial t} + \boldsymbol{\xi} \cdot \nabla H \le 0 . \tag{2.14}$$

The equality holds if and only if the system reaches the thermodynamic equilibrium state at which the so-called balance condition is fulfilled

$$f'f_1' = ff_1 . \tag{2.15}$$

This condition states that any direct/inverse collision is dynamically balanced by an inverse/direct partner. Taking logarithms of Eq. (2.15) yields

$$\ln f' + \ln f_1' = \ln f + \ln f_1 . \tag{2.16}$$

Eq. (2.16) shows that, at the thermodynamic equilibrium, the microscopic property of a system does not vary with the collision effect. Thus, the conservation laws of the mass, momentum and energy always hold. To fulfill that requirement, $\ln f$ must be a function of dynamic elementary collision invariants which include

$$\boldsymbol{\varphi}_\alpha = \left(1, \boldsymbol{\xi}, \frac{\xi^2}{2} \right)^T . \tag{2.17}$$

Here, $\xi^2 = \boldsymbol{\xi} \cdot \boldsymbol{\xi}$. The three components of $\boldsymbol{\varphi}_\alpha$ are respectively proportional to mass, momentum and kinetic energy. As a result, the general form of the equilibrium distribution function can be established as

$$\ln f^{eq} = A\xi^2 + \mathbf{B} \cdot \boldsymbol{\xi} + C . \tag{2.18}$$

By defining $\mathbf{c} = \boldsymbol{\xi} - \mathbf{u}$, Eq. (2.18) can be rewritten as

$$\ln f^{eq} = Ac^2 + \left(2A\mathbf{u} + \mathbf{B}\right)\cdot\mathbf{c} + \mathbf{B}\cdot\mathbf{u} + Au^2 + C\,, \tag{2.19}$$

where $u^2 = \mathbf{u}\cdot\mathbf{u}$. To fulfill the thermodynamic equilibrium, the distribution function should be isotropic. Thus, the second term on the right-hand side of Eq. (2.19) is zero, which yields

$$\mathbf{B} = -2A\mathbf{u}\,. \tag{2.20}$$

Substituting Eq. (2.20) into Eq. (2.19), we have

$$f^{eq} = \exp\left(Ac^2 - Au^2 + C\right). \tag{2.21}$$

Let us define $\alpha = \exp\left(-Au^2 + C\right)$ and $\lambda = -A$. Eq. (2.21) can then be rewritten as

$$f^{eq} = \alpha\exp\left(-\lambda c^2\right). \tag{2.22}$$

Substituting Eq. (2.22) into Eqs. (2.8) and (2.10), we can get

$$\rho = \alpha\left(\frac{\pi}{\lambda}\right)^{\frac{3}{2}}, \tag{2.23}$$

$$\frac{3}{2}\rho RT = \frac{3\alpha}{4}\sqrt{\frac{\pi^3}{\lambda^5}}, \tag{2.24}$$

where R is the gas constant and T is the temperature. The solutions of Eqs. (2.23)-(2.24) yield

$$\alpha = \rho\left(\frac{\lambda}{\pi}\right)^{\frac{3}{2}}, \tag{2.25}$$

$$\lambda = \frac{1}{2RT}. \tag{2.26}$$

As a result, the equilibrium distribution function for the monatomic gas can be written as

$$f^{eq} = \rho \frac{1}{(2\pi RT)^{3/2}} \exp\left[-\frac{c^2}{2RT}\right]. \tag{2.27}$$

The above equation is referred as the Maxwellian distribution function.

In a more general form, the Maxwellian distribution function can be written as [Xu, 1998]

$$f^{eq} = \rho \left(\frac{\lambda}{\pi}\right)^{\frac{D+K}{2}} e^{-\lambda\left[\sum_{i=1}^{D}(\xi_i - u_i)^2 + \sum_{j=1}^{K}\zeta_j^2\right]}, \tag{2.28}$$

where u_i is the macroscopic flow velocity in the i th-direction of the D-dimensional space. $\boldsymbol{\xi} = (\xi_1, \xi_2, ..., \xi_D)$ is the translational particle velocity in the D-dimensional space and $\boldsymbol{\zeta} = (\zeta_1, \zeta_2, ..., \zeta_K)$ represents the rest components of the translational particle velocity (ξ_1, ξ_2, ξ_3) in the 3-dimensional space and the rotational particle velocity, i.e., $K = 3 - D + N$, where N denotes the number of rotational degrees of freedom. The specific heat ratio γ is defined by

$$\gamma = \frac{b+2}{b} = \frac{K+D+2}{K+D}, \tag{2.29}$$

where b is the total number of degrees of freedom of molecules. For monatomic gas and diatomic gas, b takes the values of 3 and 5, respectively.

One of the major challenges in dealing with the Boltzmann equation (2.6) is the complicated nature of the collision term $\Omega(f)$. To facilitate numerical and analytical solutions of the Boltzmann equation, the integral form of the collision term is often replaced by a simpler expression. The rationality behind this replacement is that the collision does not take net effects on the thermodynamic equilibrium, namely $\Omega(f^{eq}) = 0$, and a large amount of detailed two-body interactions is not likely to significantly alter the macroscopic physical behaviors. Based on these facts, the Bhatnagar-Gross-Krook (BGK) collision model was proposed by Bhatnagar, Gross and Krook [1954], which assumes

$$\Omega(f) = \frac{f^{eq} - f}{\tau}, \tag{2.30}$$

where τ is the relaxation time. This model reflects the overall effect of intermolecular collisions, i.e., the distribution function relaxes to the equilibrium state after the collision process. It can be easily verified that the BGK collision model conserves mass, momentum, and energy, and satisfies the *H*-theorem. Due to its simplicity, the BGK model has been widely used in practical engineering problems. However, this model also suffers from some limitations. For example, the Prandtl number (*Pr*) of the BGK model is fixed at 1, while the full Boltzmann collision operator gives *Pr*=2/3. Some modified models have been proposed to overcome this defect, such as the Ellipsoidal Statistical BGK model [Holway, 1966] and the Shakhov model [Shakhov, 1968].

2.2.3 *Chapman-Enskog expansion analysis*

At the macroscopic scale, we investigate the collective behavior of a large number of individual particles without detailing with molecular structures and interactions. It implies that even an infinitesimally small (in the sense of differential calculus) element of the fluid still contains a sufficiently large number of particles, for which we can specify mean velocity and mean kinetic energy while reflecting the molecular properties by the transport coefficients of the fluid such as the viscosity and the thermal conductivity. This assumption is usually referred as the continuum assumption. An important parameter to identify the applicability of the continuum assumption is the Knudsen (*Kn*) number, which is defined as the ratio of the mean free path λ_∞ to the characteristic length L_∞:

$$Kn = \frac{\lambda_\infty}{L_\infty}. \tag{2.31}$$

The mean free path is the average distance of fluid molecules between two successive collisions. For an ideal gas composed of hard-sphere molecules, λ_∞ can be linked to the dynamic viscosity through [Grad, 1949; Yang and Huang, 1995; Struchtrup, 2005]

$$\lambda_\infty = \frac{\mu_\infty}{\rho_\infty} \frac{16}{5\sqrt{2\pi RT_\infty}}, \tag{2.32}$$

where ρ_∞ and T_∞ are the reference density and the reference temperature, respectively. μ_∞ is the dynamic viscosity at the reference state. Substituting Eq. (2.32) into Eq. (2.31), we have

$$Kn = \frac{Ma}{Re} \frac{16}{5} \sqrt{\frac{\gamma}{2\pi}}. \tag{2.33}$$

Here, $Ma = u_\infty / \sqrt{\gamma RT_\infty}$ is the Mach number and $Re = \rho_\infty u_\infty L_\infty / \mu_\infty$ is the Reynolds number, where u_∞ is the reference velocity. It can be seen that the Knudsen number is proportional to the Mach number and inversely proportional to the Reynolds number. Usually, the continuum assumption holds in the regime of $Kn \le 0.01$. When $Kn > 0.01$, the fluid flow is considered to be the rarefied flow for which the continuum assumption may no longer be valid.

The macroscopic governing equations (Navier-Stokes equations) and the Boltzmann equation are both applicable in the continuum flow regime. Their consistency within this flow regime can be established via the Chapman-Enskog expansion analysis. Utilizing the Maxwellian distribution function (2.28), the following 7 moment relationships (from the zeroth order to the fourth order) can be derived:

$$\left\langle f^{eq} \right\rangle = \rho, \tag{2.34}$$

$$\left\langle \xi_\alpha f^{eq} \right\rangle = \rho u_\alpha, \tag{2.35}$$

$$\left\langle \left(\xi_\alpha \xi_\alpha + \sum_{j=1}^{K} \zeta_j^2 \right) f^{eq} \right\rangle = \rho \left(u_\alpha u_\alpha + bRT \right), \tag{2.36}$$

$$\left\langle \xi_\alpha \xi_\beta f^{eq} \right\rangle = \rho u_\alpha u_\beta + p\delta_{\alpha\beta}, \tag{2.37}$$

$$\left\langle \left(\xi_\alpha \xi_\alpha + \sum_{j=1}^{K} \zeta_j^2 \right) \xi_\beta f^{eq} \right\rangle = \rho \left[u_\alpha u_\alpha + (b+2) RT \right] u_\beta, \tag{2.38}$$

$$\left\langle \xi_\alpha \xi_\beta \xi_\chi f^{eq} \right\rangle = p\left(u_\alpha \delta_{\beta\chi} + u_\beta \delta_{\chi\alpha} + u_\chi \delta_{\alpha\beta} \right) + \rho u_\alpha u_\beta u_\chi, \tag{2.39}$$

$$\begin{aligned} &\left\langle \left(\xi_\chi \xi_\chi + \sum_{j=1}^{K} \zeta_j^2 \right) \xi_\alpha \xi_\beta f^{eq} \right\rangle \\ &= \rho \left\{ (b+2) R^2 T^2 \delta_{\alpha\beta} + \left[(b+4) u_\alpha u_\beta + u_\chi u_\chi \delta_{\alpha\beta} \right] RT \right. \\ &\quad \left. + u_\chi u_\chi u_\alpha u_\beta \right\}, \end{aligned} \tag{2.40}$$

where the subscripts α, β and χ denote the directions in the velocity space; ξ and u represent the particle velocity and the macroscopic flow velocity, respectively. In fact, these moment relationships can easily be derived from the recurrence relationship shown in **Appendix A**. Eqs. (2.34)-(2.40) would be fundamental relationships in recovering the Navier-Stokes equations from the Boltzmann equation.

The Boltzmann equation with BGK model can be rewritten as

$$f = f^{eq} - \tau Df, \tag{2.41}$$

where $D = (\partial_t + \boldsymbol{\xi} \cdot \nabla)$ represents the substantial derivative. According to the Chapman-Enskog expansion analysis, the first-order expansion of Eq. (2.41) in terms of the collision time scale can be written as

$$f = f^{eq} - \tau D f^{eq} + O(\tau^2). \tag{2.42}$$

To fulfill the compatibility condition, the non-equilibrium part of the distribution function should not contribute to the calculation of conservative flow variables, which yields

$$\left\langle \boldsymbol{\varphi}_\alpha \left(f - f^{eq} \right) \right\rangle = 0, \tag{2.43}$$

where $\boldsymbol{\varphi}_\alpha$ stands for the moment vector, which is given by

$$\boldsymbol{\varphi}_\alpha = \left(1, \boldsymbol{\xi}, \frac{1}{2}\left(\sum_{i=1}^{D} \xi_i^2 + \sum_{j=1}^{K} \zeta_j^2\right)\right)^T. \tag{2.44}$$

Performing integration over the particle velocity space for Eq. (2.42) and utilizing the compatibility condition, we have

$$\partial_t \left\langle \boldsymbol{\varphi}_a f^{eq} \right\rangle + \nabla \cdot \left\langle \boldsymbol{\xi} \boldsymbol{\varphi}_a f^{eq} \right\rangle = O(\tau). \tag{2.45}$$

With the conservation relationships in Eqs. (2.34)-(2.38), Eq. (2.45) can be rewritten as

$$\partial_t \rho + \nabla \cdot (\rho \mathbf{u}) = O(\tau), \tag{2.46}$$

$$\partial_t (\rho \mathbf{u}) + \nabla \cdot (\rho \mathbf{u}\mathbf{u} + p\mathbf{I}) = O(\tau), \tag{2.47}$$

$$\partial_t (\rho E) + \nabla \cdot \left[(\rho E + p) \mathbf{u} \right] = O(\tau), \tag{2.48}$$

where $E = e + \mathbf{u} \cdot \mathbf{u}/2$ is the total energy and $\mathbf{I}$ is the unit tensor. It can be seen that, when truncated at $O(\tau)$, the Boltzmann equation can recover the Euler equations.

In order to derive the Navier-Stokes equations from the Boltzmann equation, higher-order terms should be considered. To begin with, let us multiply the operator D on both sides of Eq. (2.42)

$$Df = Df^{eq} - D\left(\tau D f^{eq}\right) + O\left(\tau^2\right). \tag{2.49}$$

From Eq. (2.41), we have $Df = \left(f^{eq} - f\right)/\tau$. Furthermore, multiplying $\boldsymbol{\varphi}_a$ on both sides of Eq. (2.49) and then performing integration over the particle velocity space yield

$$\begin{aligned} &\partial_t \left\langle \boldsymbol{\varphi}_a f^{eq} \right\rangle + \nabla \cdot \left\langle \boldsymbol{\xi} \boldsymbol{\varphi}_a f^{eq} \right\rangle \\ &= \partial_t \left\langle \boldsymbol{\varphi}_a \left(\tau D f^{eq}\right) \right\rangle + \nabla \cdot \left\langle \boldsymbol{\xi} \boldsymbol{\varphi}_a \left(\tau D f^{eq}\right) \right\rangle + O\left(\tau^2\right). \end{aligned} \tag{2.50}$$

Note that the compatibility condition has been used in the derivation of Eq. (2.50). Substituting Eq. (2.42) into the first term on the right-hand side of Eq. (2.50) and utilizing the compatibility condition, we can get

$$\partial_t \left\langle \boldsymbol{\varphi}_a f^{eq} \right\rangle + \nabla \cdot \left\langle \boldsymbol{\xi} \boldsymbol{\varphi}_a f^{eq} \right\rangle = \nabla \cdot \left\langle \boldsymbol{\xi} \boldsymbol{\varphi}_a \left(\tau D f^{eq} \right) \right\rangle + O\left(\tau^2\right). \quad (2.51)$$

Substituting different components of the moment vector $\boldsymbol{\varphi}_a$ into Eq. (2.51) can derive the continuity, the momentum and the energy equations at the macroscopic scale.

The first component of $\boldsymbol{\varphi}_a$ is 1, which gives the following relationship when being substituted into Eq. (2.51):

$$\partial_t \left\langle f^{eq} \right\rangle + \nabla \cdot \left\langle \boldsymbol{\xi} f^{eq} \right\rangle = \nabla \cdot \left\langle \boldsymbol{\xi} \left(\tau D f^{eq} \right) \right\rangle + O\left(\tau^2\right). \quad (2.52)$$

Utilizing Eq. (2.42) and the compatibility condition (2.43) allows further simplification of the above formula into

$$\partial_t \left\langle f^{eq} \right\rangle + \nabla \cdot \left\langle \boldsymbol{\xi} f^{eq} \right\rangle = O\left(\tau^2\right). \quad (2.53)$$

Applying the conservation relationships in Eqs. (2.34) and (2.35), Eq. (2.53) can be rewritten as

$$\partial_t \rho + \nabla \cdot \left(\rho \mathbf{u} \right) = O\left(\tau^2\right), \quad (2.54)$$

which recovers the continuity equation in the order of $O\left(\tau^2\right)$.

Considering the second component of $\boldsymbol{\varphi}_a$ in Eq. (2.51), which gives

$$\begin{aligned} &\partial_t \left\langle \boldsymbol{\xi} f^{eq} \right\rangle + \nabla \cdot \left\langle \boldsymbol{\xi}\boldsymbol{\xi} f^{eq} \right\rangle \\ &= \nabla \cdot \left[\tau \left(\partial_t \left\langle \boldsymbol{\xi}\boldsymbol{\xi} f^{eq} \right\rangle + \nabla \cdot \left\langle \boldsymbol{\xi}\boldsymbol{\xi}\boldsymbol{\xi} f^{eq} \right\rangle \right) \right] + O\left(\tau^2\right). \end{aligned} \quad (2.55)$$

Applying the relationships (2.35), (2.37) and (2.39) and noting that

$$\begin{aligned} \partial_t \left\langle \boldsymbol{\xi}\boldsymbol{\xi} f^{eq} \right\rangle &= \partial_t \left(\rho \mathbf{u}\mathbf{u} + p\mathbf{I} \right) \\ &= \partial_t \left(\rho \mathbf{u} \right) \mathbf{u} + \mathbf{u} \partial_t \left(\rho \mathbf{u} \right) - \mathbf{u}\mathbf{u} \partial_t \rho + \left(\gamma - 1 \right) \partial_t \left(\rho e \right) \mathbf{I} \\ &= -\nabla \cdot \left(\rho \mathbf{u}\mathbf{u}\mathbf{u} \right) - \left[\left(\nabla p \right) \mathbf{u} + \mathbf{u} \nabla p \right] - \nabla \cdot \left(p \mathbf{u} \right) \mathbf{I} \\ &\quad - \left(\gamma - 1 \right) p (\nabla \cdot \mathbf{u}) \mathbf{I} + O(\tau), \end{aligned}$$

$$\begin{aligned}\nabla\cdot\left\langle \boldsymbol{\xi\xi\xi} f^{eq}\right\rangle &= \partial_\chi\left\langle \xi_\alpha\xi_\beta\xi_\chi f^{eq}\right\rangle\\ &= \partial_\chi\left[p(u_\alpha\delta_{\beta\chi}+u_\beta\delta_{\chi\alpha}+u_\chi\delta_{\alpha\beta})+\rho u_\alpha u_\beta u_\chi\right]\\ &= \nabla\cdot(\rho\mathbf{uuu})+\nabla\cdot(p\mathbf{u})\mathbf{I}+\left[(\nabla p)\mathbf{u}+\mathbf{u}\nabla p\right]+p\left[\nabla\mathbf{u}+(\nabla\mathbf{u})^T\right],\end{aligned}$$

we have

$$\partial_t(\rho\mathbf{u})+\nabla\cdot(\rho\mathbf{uu}+p\mathbf{I})=\nabla\cdot\mathbf{\Pi}+O(\tau^2), \tag{2.56}$$

where

$$\mathbf{\Pi}=\mu\left[\left(\nabla\mathbf{u}+(\nabla\mathbf{u})^T\right)-\frac{2}{3}(\nabla\cdot\mathbf{u})\mathbf{I}\right]+\mu_b(\nabla\cdot\mathbf{u})\mathbf{I},$$

$$\mu=\tau p,$$

$$\mu_b=\left(\frac{5}{3}-\gamma\right)\mu.$$

Note that Eqs. (2.46)-(2.48) have been used in the calculation of the term $\partial_t(\rho\mathbf{uu}+p\mathbf{I})$. Here, μ is the dynamic viscosity and μ_b is the bulk viscosity. Nevertheless, even if the momentum equation is recovered in the order of $O(\tau^2)$, the bulk viscosity only vanishes in the case of monatomic gas with the specific heat ratio of $\gamma=5/3$.

Finally, we consider the last component of $\boldsymbol{\varphi}_a$ in Eq. (2.51), which gives

$$\begin{aligned}&\partial_t\left\langle\frac{1}{2}\left(\sum_{i=1}^{D}\xi_i^2+\sum_{j=1}^{K}\zeta_j^2\right)f^{eq}\right\rangle+\nabla\cdot\left\langle\frac{1}{2}\left(\sum_{i=1}^{D}\xi_i^2+\sum_{j=1}^{K}\zeta_j^2\right)\boldsymbol{\xi}f^{eq}\right\rangle\\ &=\nabla\cdot\left[\tau\left(\partial_t\left\langle\frac{1}{2}\left(\sum_{i=1}^{D}\xi_i^2+\sum_{j=1}^{K}\zeta_j^2\right)\boldsymbol{\xi}f^{eq}\right\rangle\right.\right.\\ &\left.\left.\quad+\nabla\cdot\left\langle\frac{1}{2}\left(\sum_{i=1}^{D}\xi_i^2+\sum_{j=1}^{K}\zeta_j^2\right)\boldsymbol{\xi\xi}f^{eq}\right\rangle\right)\right]+O(\tau^2).\end{aligned} \tag{2.57}$$

Applying the relationships (2.36), (2.38), (2.40) and noting that

$$\begin{aligned}
&\partial_t \left\langle \frac{1}{2}\left(\sum_{i=1}^{D} \xi_i^2 + \sum_{j=1}^{K} \varsigma_j^2 \right) \boldsymbol{\xi} f^{eq} \right\rangle \\
&= \partial_t \left[(\rho E + p)\mathbf{u} \right] \\
&= -\nabla \cdot (\rho E \mathbf{u}\mathbf{u}) - 2\mathbf{u}\nabla \cdot (p\mathbf{u}) - E\nabla p - p\mathbf{u} \cdot \nabla \mathbf{u} \\
&\quad -(\gamma - 1)e\nabla p - (\gamma - 1)p\mathbf{u}\nabla \cdot \mathbf{u} + O(\tau),
\end{aligned}$$

$$\begin{aligned}
&\nabla \cdot \left\langle \frac{1}{2}\left(\sum_{i=1}^{D} \xi_i^2 + \sum_{j=1}^{K} \varsigma_j^2 \right) \boldsymbol{\xi}\boldsymbol{\xi} f^{eq} \right\rangle \\
&= \partial_\beta \left\langle \frac{1}{2}\left(\sum_{i=1}^{D} \xi_i^2 + \sum_{j=1}^{K} \varsigma_j^2 \right) \xi_\alpha \xi_\beta f^{eq} \right\rangle \\
&= \partial_\beta \left\{ \frac{\rho}{2}\left[(b+2)R^2T^2\delta_{\alpha\beta} + \left[(b+4)u_\alpha u_\beta + u_\chi u_\chi \delta_{\alpha\beta} \right] RT \right.\right. \\
&\quad \left.\left. + u_\chi u_\chi u_\alpha u_\beta \right] \right\} \\
&= \nabla(pE) + \nabla \cdot (\rho E \mathbf{u}\mathbf{u}) + 2\mathbf{u}\nabla \cdot (p\mathbf{u}) + p\mathbf{u} \cdot \left[\nabla \mathbf{u} + (\nabla \mathbf{u})^T \right] \\
&\quad + (\gamma - 1)e\nabla p + Rp\nabla T,
\end{aligned}$$

we have

$$\partial_t(\rho E) + \nabla \cdot \left[(\rho E + p)\mathbf{u} \right] = \nabla \cdot (\kappa \nabla T) + \nabla \cdot (\mathbf{u} \cdot \boldsymbol{\Pi}) + O(\tau^2). \tag{2.58}$$

Here, κ denotes the thermal conductivity, which is related to the collision time scale τ through

$$\kappa = \tau p(c_V + R) = \mu c_p. \tag{2.59}$$

And the Prandtl number thus yields

$$Pr = \frac{\mu c_p}{\kappa} = 1. \tag{2.60}$$

Note that Eqs. (2.46)-(2.48) have been used in the calculation of the term $\partial_t[(\rho E + p)\mathbf{u}]$. The above derivations suggest that the energy equation can be recovered from the Boltzmann equation after omitting higher order terms in $O(\tau^2)$, although the resultant Prandtl number is fixed at 1.

From Eq. (2.56), the relationship between the dynamic viscosity and the collision time scale is given by

$$\tau = \frac{\mu}{p}. \tag{2.61}$$

At the reference state, $\mu = \mu_\infty$ and $p = \rho_\infty RT_\infty$, the collision time scale can thus be written as

$$\tau = \frac{Ma^2}{Re}\frac{\gamma L_\infty}{u_\infty} = Kn \times Ma \frac{5\sqrt{2\pi\gamma}}{16}\frac{L_\infty}{u_\infty}. \tag{2.62}$$

The above formulation indicates that, for given Mach number and reference velocity, the collision time scale is proportional to the Knudsen number. Since the Knudsen number in the continuum flow regime is a small quantity, the truncation errors in $O(\tau^2)$ can be omitted, which ensures recovering the Navier-Stokes equations from the Boltzmann equation.

2.3 Discrete Velocity Boltzmann Equation and Its Solution

2.3.1 *From continuous Boltzmann equation to DVBE*

To numerically resolve the Boltzmann equation, numerical discretization should be performed on the computational domains. The computational domains in resolving the Boltzmann equation involve both the physical space and the particle velocity space. Given the infinite nature of the particle velocity space, numerical resolution of the Boltzmann equation requires rational truncation of the particle velocity space and appropriate discretization strategies, which yields the discrete velocity Boltzmann equation (DVBE)

$$\frac{\partial f_\alpha}{\partial t} + \xi_\alpha \cdot \nabla f_\alpha = \frac{f_\alpha^{eq} - f_\alpha}{\tau}, \quad \alpha = 1, \ldots, N_V, \tag{2.63}$$

where N_V is the number of discrete velocity points. If the Maxwellian distribution function is used as the equilibrium state, all moment relationships of the distribution function (more than the moment relationships shown in Eqs. (2.34)-(2.40)) can be approximately satisfied by the numerical quadrature. This strategy has been widely used in the discrete velocity method (DVM) for simulation of rarefied flows [Mieussens, 2000; Li and Zhang, 2009; Xu and Huang, 2010; Guo *et al.*, 2013; Yang *et al.*, 2016c]. To guarantee the accuracy of the numerical quadrature, DVM normally utilizes a large number of points to discretize the particle velocity space, which leads to heavy computational effort and memory requirement [Zhu *et al.*, 2016; Yang *et al.*, 2018b].

The above problem can be greatly simplified if being restricted to the continuum flow regime in which only the 7 moment relationships shown in Eqs. (2.34)-(2.40) should be fulfilled to recover the correct macroscopic equations. It thus indicates that the Maxwellian distribution function can be replaced by a discrete equilibrium state for simulation of continuous flows as long as Eqs. (2.34)-(2.40) are satisfied. This kind of discrete equilibrium state is also named as the lattice Boltzmann model. Up to now, massive lattice Boltzmann models have been developed for either incompressible flows [Chen *et al.*, 1991; He *et al.*, 1999; Luo,

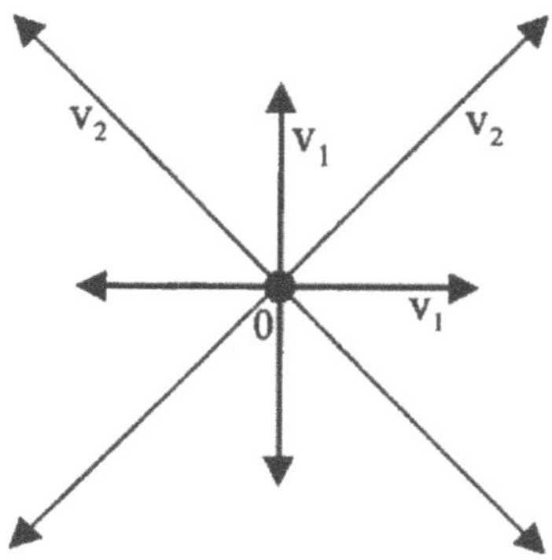

Figure 2.1. Distribution of the D2Q9 model of Kataoka and Tsutahara in the particle velocity space.

2000; Inamuro *et al.*, 2004; Yan and Zu, 2007; Zhang *et al.*, 2012] or compressible flows [Hu *et al.*, 1997; Sun, 1998; Qu *et al.*, 2007a; Dellar, 2008]. For example, the D2Q9 model proposed by Kataoka and Tsutahara [2004a] for compressible inviscid flows, which only needs to satisfy the moment relationships of Eqs. (2.34)-(2.38), has the following form:

$$f_\alpha^{eq} = \rho\left(A_\alpha + B_\alpha u_i e_{\alpha i} + D_\alpha u_i e_{\alpha i} u_j e_{\alpha j}\right) \quad \text{for } \alpha = 1, 2, \ldots, 9. \tag{2.64}$$

Note that the particle velocity $\boldsymbol{\xi}_\alpha$ is denoted by $\mathbf{e}_\alpha = \left(e_{\alpha i}, e_{\alpha j}\right)$ in Eq. (2.64) and the following text. The distribution of the discrete particle velocity is shown in Figure 2.1 and given by

$$\mathbf{e}_\alpha = \begin{cases} (0,0), & \text{for } \alpha = 1, \\ v_1\left(\cos(\pi\alpha/2), \sin(\pi\alpha/2)\right), & \text{for } \alpha = 2,3,4,5, \\ v_2\left[\cos\left(\pi(\alpha/2 + 1/4)\right), \sin\left(\pi(\alpha/2 + 1/4)\right)\right], & \text{for } \alpha = 6,7,8,9. \end{cases} \tag{2.65}$$

$$\eta_\alpha = \begin{cases} \eta_0 & \text{for } \alpha = 1, \\ 0 & \text{for } \alpha = 2,3,\ldots,9. \end{cases} \tag{2.66}$$

The coefficients A_α, B_α and D_α in Eq. (2.64) can be determined by substituting Eq. (2.64) into Eqs. (2.34)-(2.38), which yield

$$A_\alpha = \begin{cases} \dfrac{b-2}{\eta_0^2} T, & \text{for } \alpha = 1, \\ \dfrac{-v_2^2 + \left[(b-2)\dfrac{v_2^2}{\eta_0^2} + 2\right] T + \dfrac{v_2^2}{v_1^2} u_i^2}{4\left(v_1^2 - v_2^2\right)}, & \text{for } \alpha = 2,3,4,5, \\ \dfrac{-v_1^2 + \left[(b-2)\dfrac{v_1^2}{\eta_0^2} + 2\right] T + \dfrac{v_1^2}{v_2^2} u_i^2}{4\left(v_2^2 - v_1^2\right)}, & \text{for } \alpha = 6,7,8,9. \end{cases} \tag{2.67}$$

$$B_\alpha = \begin{cases} \dfrac{-v_2^2 + (b+2)T + u_j^2}{2v_1^2\left(v_1^2 - v_2^2\right)}, & \text{for } \alpha = 2,3,4,5, \\ \dfrac{-v_1^2 + (b+2)T + u_j^2}{2v_2^2\left(v_2^2 - v_1^2\right)}, & \text{for } \alpha = 6,7,8,9. \end{cases} \tag{2.68}$$

$$D_\alpha = \begin{cases} \dfrac{1}{2v_1^4}, & \text{for } \alpha = 2,3,4,5, \\ \dfrac{1}{2v_2^4}, & \text{for } \alpha = 6,7,8,9. \end{cases} \tag{2.69}$$

where v_1, $v_2\left(\neq v_1\right)$ and η_0 are nonzero constants. Note that, in the above lattice Boltzmann model, the integral of the potential energy of particles is approximated by

$$\left\langle \sum_{j=1}^{K} \zeta_j^2 f^{eq} \right\rangle = \sum_{\alpha=1}^{9} \eta_\alpha^2 f_\alpha^{eq}. \tag{2.70}$$

2.3.2 *Finite volume solution of DVBE*

The DVBE is in fact a set of partial differential equations, which can be solved directly by the commonly-used finite difference method (FDM) [Mei and Shyy, 1998; Watari and Tsutahara, 2003; Mezrhab *et al.*, 2004; Gan *et al.*, 2008] or finite volume method (FVM) [Xi *et al.*, 1999; Peng *et al.*, 1999; Chew *et al.*, 2002; Stiebler *et al.*, 2006; Patil and Lakshmisha, 2009]. For problems with complex geometries, the FVM seems to be a more flexible alternative. Within the FVM framework, various flux solvers such as the upwind schemes can be embedded. For illustration, the 2D DVBE can be rewritten into the following conservative form [Qu *et al.*, 2007b]:

$$\frac{\partial f_\alpha}{\partial t} + \frac{\partial F_\alpha}{\partial x} + \frac{\partial G_\alpha}{\partial y} = \Omega_\alpha, \tag{2.71}$$

where

$$F_\alpha = f_\alpha e_{\alpha x},\ G_\alpha = f_\alpha e_{\alpha y},\ \Omega_\alpha = \frac{f_\alpha^{eq} - f_\alpha}{\tau}. \tag{2.72}$$

On the Cartesian grid, the FVM discretization of Eq. (2.71) yields

$$\begin{aligned}\frac{f_\alpha^{n+1} - f_\alpha^n}{\Delta t} &= -\frac{1}{\Delta x}\left[F^n_{\alpha,i+\frac{1}{2},j} - F^n_{\alpha,i-\frac{1}{2},j}\right] \\ &- \frac{1}{\Delta y}\left[G^n_{\alpha,i,j+\frac{1}{2}} - G^n_{\alpha,i,j-\frac{1}{2}}\right] + \Omega^n_{\alpha,i,j}.\end{aligned} \tag{2.73}$$

Note that the explicit Euler scheme has been used in Eq. (2.73) for temporal discretization. $F^n_{\alpha,i\pm 1/2,j}$ and $G^n_{\alpha,i,j\pm 1/2}$ are numerical fluxes on the faces of the cell (i,j), which are evaluated by resolving the local Riemann problem on the cell interface. Taking $F^n_{\alpha,i+1/2,j}$ as an example, the local Riemann problem is established as

$$\left.\begin{aligned}&\frac{\partial f_\alpha}{\partial t} + \frac{\partial F_\alpha}{\partial x} = 0 \\ &f_\alpha(x,0) = f_\alpha^{(0)}(x) = \begin{cases} f_\alpha^L & \text{if } x < x_{i+\frac{1}{2}} \\ f_\alpha^R & \text{if } x > x_{i+\frac{1}{2}} \end{cases}\end{aligned}\right\}, \tag{2.74}$$

where f_α^L and f_α^R are the initial distribution functions on the left and the right sides of the cell interface, respectively. The solution of the above Riemann problem yields

$$f_{\alpha,i+\frac{1}{2},j} = \begin{cases} f_\alpha^L & \text{if } x < x_{i+\frac{1}{2}}, \\ f_\alpha^R & \text{if } x > x_{i+\frac{1}{2}}. \end{cases} \tag{2.75}$$

As a result, $F^n_{\alpha,i+1/2,j}$ can be computed by

$$F^n_{\alpha,i+\frac{1}{2},j} = f_{\alpha,i+\frac{1}{2},j} e_{\alpha x} = \begin{cases} f_\alpha^L e_{\alpha x} & \text{if } e_{\alpha x} > 0, \\ f_\alpha^R e_{\alpha x} & \text{if } e_{\alpha x} < 0. \end{cases} \tag{2.76}$$

$F^n_{\alpha,i-1/2,j}$ and $G^n_{\alpha,i,j\pm 1/2}$ can be evaluated in the same way.

To ensure high accuracy while maintaining smoothness of the solution, the third-order MUSCL (Monotone Upstream-centered Scheme for Conservation Laws) scheme [Van Leer, 1997] with van Albada limiter [Van Albada *et al.*, 1997] can be applied to reconstruct the initial distribution functions on two sides of an interface, which gives

$$\begin{cases} f^L = f_i + \left\{\frac{s}{4}\left[\left(1-\kappa s\right)\Delta_- + \left(1+\kappa s\right)\Delta_+\right]\right\}_i, \\ f^R = f_{i+1} - \left\{\frac{s}{4}\left[\left(1-\kappa s\right)\Delta_+ + \left(1+\kappa s\right)\Delta_-\right]\right\}_{i+1}, \end{cases} \tag{2.77}$$

where $\kappa = 1/3$ and s is van Albada limiter given by

$$s = \frac{2\Delta_+\Delta_- + \varepsilon}{\Delta_+^2 + \Delta_-^2 + \varepsilon}. \tag{2.78}$$

Here ε is a small number (usually set as $\varepsilon = 10^{-6}$) to avoid zero denominators, and

$$\left(\Delta_+\right)_i = f_{i+1} - f_i, \quad \left(\Delta_-\right)_i = f_i - f_{i-1}. \tag{2.79}$$

Note that the subscript "α" has been omitted in Eqs. (2.77)-(2.79).

2.3.3 *Advantages and limitations of DVBE solver*

The DVBE is actually the discrete form of the continuous Boltzmann equation in the particle velocity space. It can be used for simulation of either rarefied flows or continuous flows by choosing appropriate equilibrium distribution functions. Specifically, to ensure general applicability in the whole flow regime, the discrete Maxwellian distribution function associated with a large number of discrete points in the particle velocity space should be used to recover all moment relationships. Such strategy is redundant if the scope of study is restricted within continuous flows. With the continuum assumption, simplifications can be made by maintaining a few moment relationships that are necessary for the recovery of Euler/Navier-Stokes equations. The lattice

Boltzmann model with much fewer discrete points in the particle velocity space can thus be derived. The DVBE with the lattice Boltzmann model is more efficient than the DVBE with the discrete Maxwellian distribution function. In the meantime, the DVBE can be resolved within the FVM framework on both structured and unstructured grids, which enriches its flexibility in problems with complex geometries. Furthermore, the flux limiters used in the conventional computational fluid dynamics (CFD) can be introduced into the DVBE solver for capturing shock waves in compressible flows.

However, the DVBE solver also suffers from some drawbacks, especially in simulating continuous flows. The first one is its tediousness. The number of unknowns in DVBE is equal to the number of discrete points in the particle velocity space, which is much larger than the number of variables in the Euler/Navier-Stokes equations. The second one is the critical stability condition of the DVBE solver. As commented by Kataoka and Tsutahara [2004b], the time step Δt and the mesh spacing Δx used for solving DVBE are usually constrained by the collision time scale τ and $\tau\sqrt{RT}$, respectively. For continuous flows at high Reynolds numbers, τ is very small, which leads to very small time step and very fine mesh spacing, and thus low computational efficiency. The last limitation is related to the interpretation of physical boundary conditions. Since the distribution function is evolved in the DVBE solver, the physical boundary conditions for macroscopic variables need to be transformed into conditions for distribution functions. This process is very complicated, specifically for a curved boundary.

2.4 Lattice Boltzmann Equation and Lattice Boltzmann Method

2.4.1 *From DVBE to lattice Boltzmann equation*

The finite volume solution for the DVBE requires the evaluation of fluxes on the cell interface, which inevitably introduces interpolation or reconstruction algorithms and deteriorates the compactness of the scheme. An alternative strategy is to integrate DVBE from t to $t+\delta_t$ along the characteristic line and approximate the collision term by trapezoidal rule, which yields

$$f_\alpha(\mathbf{x}+\mathbf{e}_\alpha\delta_t, t+\delta_t) - f_\alpha(\mathbf{x}, t) = \frac{\delta_t}{2}\left[\frac{f_\alpha^{eq}(\mathbf{x}, t) - f_\alpha(\mathbf{x}, t)}{\tau} + \frac{f_\alpha^{eq}(\mathbf{x}+\mathbf{e}_\alpha\delta_t, t+\delta_t) - f_\alpha(\mathbf{x}+\mathbf{e}_\alpha\delta_t, t+\delta_t)}{\tau}\right]. \tag{2.80}$$

This approximation has the second order of accuracy in time. To remove the implicitness in the above evolution, the following transformation can be introduced:

$$\overline{f}_\alpha = f_\alpha - \frac{\delta_t}{2\tau}\left(f_\alpha^{eq} - f_\alpha\right). \tag{2.81}$$

Then Eq. (2.80) can be rewritten as

$$\overline{f}_\alpha(\mathbf{x}+\mathbf{e}_\alpha\delta_t, t+\delta_t) - \overline{f}_\alpha(\mathbf{x}, t) = \frac{1}{\tau/\delta_t + 0.5}\left[f_\alpha^{eq}(\mathbf{x}, t) - \overline{f}_\alpha(\mathbf{x}, t)\right]. \tag{2.82}$$

Defining $\tau' = \tau/\delta_t + 0.5$, the following lattice Boltzmann equation (LBE) is obtained:

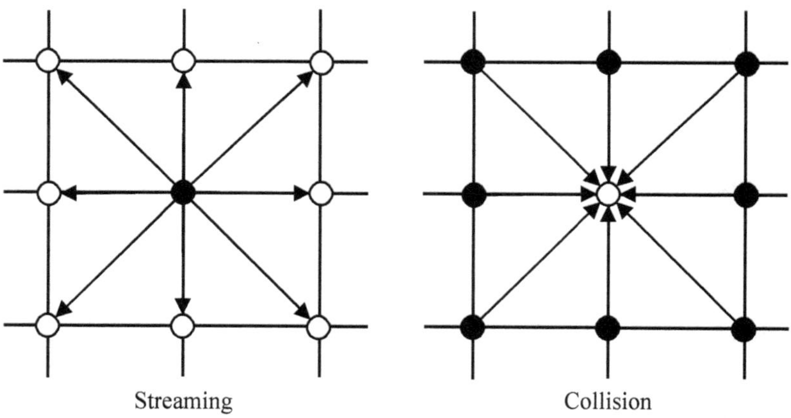

Figure 2.2. Solution process of the lattice Boltzmann equation with D2Q9 model.

$$\bar{f}_\alpha(\mathbf{x}+\mathbf{e}_\alpha\delta_t, t+\delta_t)-\bar{f}_\alpha(\mathbf{x},t) \\ =\frac{f_\alpha^{eq}(\mathbf{x},t)-\bar{f}_\alpha(\mathbf{x},t)}{\tau'}, \quad \alpha=0,1,\dots,N_V. \tag{2.83}$$

According to the Chapman-Enskog expansion analysis shown in **Section 2.2.3**, the dimensionless relaxation parameter τ' can be expressed as

$$\tau'=\frac{\mu}{p\delta_t}+\frac{1}{2}=\frac{\mu}{\rho RT\delta_t}+\frac{1}{2}. \tag{2.84}$$

Noted that the conservative variables calculated by moments of f_α are consistent with those of $\bar{f}_\alpha$ due to the compatibility condition, i.e,

$$\rho=\sum_\alpha f_\alpha=\sum_\alpha \bar{f}_\alpha, \tag{2.85}$$

$$\rho\mathbf{u}=\sum_\alpha \mathbf{e}_\alpha f_\alpha=\sum_\alpha \mathbf{e}_\alpha \bar{f}_\alpha. \tag{2.86}$$

For simplicity, $\bar{f}_\alpha$ and τ' are denoted by f_α and τ in the following text.

The lattice Boltzmann equation (2.83) is essentially a linear algebraic equation of the distribution function and its equilibria. The physical interpretation of its solution process is illustrated in Figure 2.2. The left-hand side of Eq. (2.83) represents the streaming process in which the particles are streamed from the central mesh point to the neighboring ones (or vice versa) within a time interval, and its left-hand side describes the collision process which happens instantaneously at the central mesh point.

2.4.2 *Lattice Boltzmann models*

Numerical strategies adopted to solve the LBE are generally denoted as the lattice Boltzmann method (LBM) or lattice Boltzmann models. The lattice Boltzmann models with BGK collision operator, or simply the LBGK models, utilize a single relaxation time in the collision process. It means a constant relaxation rate towards the equilibria for all distribution functions. Although such an assumption may sometimes over-simplify the physics, it brings compactness to the mathematical formulations. Until now, the LBGK models are perhaps the most popular models in the LBM community, and they have been extensively applied in various flow problems. The LBGK models are associated with a specific lattice velocity model, and the most successful ones are the set of DnQb models proposed by Qian *et al.* [1992]. Here, n and b denote the dimension and the number of lattice velocity vectors, respectively. In these models, the equilibrium distribution function can be obtained by expanding the Maxwellian distribution function in terms of the Mach number and truncating to $O(Ma^3)$ as follows

$$\begin{aligned} f_\alpha^{eq} &= \rho(2\pi RT)^{-D/2}\exp\left[-\frac{(\mathbf{e}_\alpha-\mathbf{u})\cdot(\mathbf{e}_\alpha-\mathbf{u})}{2RT}\right] \\ &= \rho(2\pi RT)^{-D/2}\exp\left[-\frac{\mathbf{e}_\alpha\cdot\mathbf{e}_\alpha-2\mathbf{e}_\alpha\cdot\mathbf{u}+\mathbf{u}\cdot\mathbf{u}}{2RT}\right] \\ &= \rho(2\pi RT)^{-D/2}\exp\left[-\frac{\mathbf{e}_\alpha\cdot\mathbf{e}_\alpha}{2RT}\right]\exp\left[\frac{2\mathbf{e}_\alpha\cdot\mathbf{u}-\mathbf{u}\cdot\mathbf{u}}{2RT}\right] \\ &= \omega_\alpha\rho\left(1+\frac{\mathbf{e}_\alpha\cdot\mathbf{u}}{c_s^2}+\frac{(\mathbf{e}_\alpha\cdot\mathbf{u})^2}{2c_s^4}-\frac{\mathbf{u}\cdot\mathbf{u}}{2c_s^2}\right)+O(Ma^3), \end{aligned} \tag{2.87}$$

where

Table 2.1. Parameters of typical DnQb models.

Model	$\mathbf{e}_\alpha$	ω_α
D1Q3	0, ±1	2/3, 1/6
D2Q9	(0,0), (±1,0), (0,±1), (±1,±1)	4/9, 1/9, 1/36
D3Q15	(0,0,0), (±1,0,0), (0,±1,0), (0,0,±1), (±1, ±1, ±1)	2/9, 1/9, 1/72
D3Q19	(0,0,0), (±1,0,0), (0,±1,0), (0,0,±1), (±1, ±1, 0), (±1, 0, ±1), (0, ±1, ±1)	1/3, 1/18, 1/36

$$\omega_\alpha = (2\pi RT)^{-D/2} \exp\left[-\frac{\mathbf{e}_\alpha \cdot \mathbf{e}_\alpha}{2RT}\right]. \tag{2.88}$$

Here, ω_α is the weight coefficient and $c_s = \sqrt{RT}$ is the sound speed. For typical D*n*Q*b* models, the lattice velocity vectors $\mathbf{e}_\alpha$ and the values of ω_α are specified to ensure the isotropy of up to the fourth-order lattice tensors; and the sound speed c_s will be adjusted accordingly. The parameters of some typical D*n*Q*b* models are illustrated in Table 2.1. It should be noted that the truncation error of $O(Ma^3)$ imposes low Mach number limit to these models and thus restricts their applicability within incompressible flows.

With the isotropic properties of lattice tensors, the relationships between the moments of the equilibrium distribution function and the macroscopic properties can be established as follows:

$$\sum_\alpha f_\alpha^{(eq)} = \rho, \tag{2.89a}$$

$$\sum_\alpha e_{\alpha i} f_\alpha^{(eq)} = \rho u_i, \tag{2.89b}$$

$$\sum_\alpha e_{\alpha i} e_{\alpha j} f_\alpha^{(eq)} = \rho u_i u_j + p\delta_{ij}, \tag{2.89c}$$

$$\sum_\alpha e_{\alpha i} e_{\alpha j} e_{\alpha k} f_\alpha^{(eq)} = c_s^2 \rho \left(u_i \delta_{jk} + u_j \delta_{ki} + u_k \delta_{ij}\right). \tag{2.89d}$$

2.4.3 *Chapman-Enskog expansion analysis*

The physical consistency requires that the collective behavior of microscopic particles should recover the macroscopic fluid system. Therefore, in mathematics, the LBE should be consistent with the Navier-Stokes equations at the macroscopic scale. Such a connection can be established through the Chapman-Enskog expansion analysis [Frisch *et al.*, 1987; Benzi *et al.*, 1992].

The essence of the Chapman-Enskog expansion analysis is the multi-scale expansion. To begin with, the density distribution function, the temporal derivative and the spatial derivative can be expanded as

$$f_\alpha = f_\alpha^{(0)} + \varepsilon f_\alpha^{(1)} + \varepsilon^2 f_\alpha^{(2)}, \tag{2.90a}$$

$$\frac{\partial}{\partial t} = \varepsilon \frac{\partial}{\partial t_0} + \varepsilon^2 \frac{\partial}{\partial t_1}, \tag{2.90b}$$

$$\nabla = \varepsilon \nabla_0, \tag{2.90c}$$

where ε is a small parameter proportional to the Knudsen number [Inamuro *et al.*, 1997]; t_0 and t_1 represent the convective time scale and the diffusive time scale, respectively. By using the truncated Taylor series expansion, the lattice Boltzmann equation (2.83) can be reduced to the following formulation with second-order accuracy both in time and in space:

$$\left(\frac{\partial}{\partial t} + \mathbf{e}_\alpha \cdot \nabla\right) f_\alpha + \frac{\delta_t}{2}\left(\frac{\partial}{\partial t} + \mathbf{e}_\alpha \cdot \nabla\right)^2 f_\alpha + \frac{1}{\tau \delta_t}\left(f_\alpha - f_\alpha^{eq}\right) + O\left(\delta_t^2\right) = 0. \tag{2.91}$$

Substituting the multi-scale expansions (Eqs. (2.90a)-(2.90c)) into the above equation (2.91) gives the following formulations at different orders of ε:

$$O\left(\varepsilon^0\right): \quad \left(f_\alpha^{(0)} - f_\alpha^{eq}\right) \big/ \left(\tau \delta_t\right) = 0, \tag{2.92}$$

$$O\left(\varepsilon^1\right): \quad \left(\frac{\partial}{\partial t_0} + \mathbf{e}_\alpha \cdot \nabla_0\right) f_\alpha^{(0)} + \frac{1}{\tau \delta_t} f_\alpha^{(1)} = 0, \tag{2.93}$$

$$O\left(\varepsilon^2\right): \quad \frac{\partial f_\alpha^{(0)}}{\partial t_1} + \left(1 - \frac{1}{2\tau}\right)\left(\frac{\partial}{\partial t_0} + \mathbf{e}_\alpha \cdot \nabla_0\right) f_\alpha^{(1)} + \frac{1}{\tau \delta_t} f_\alpha^{(2)} = 0. \tag{2.94}$$

From Eq. (2.92), it is obtained that

$$f_\alpha^{(0)} = f_\alpha^{eq}. \tag{2.95}$$

With the conservation laws and the expansion formulation in Eq. (2.90a), we have

$$\sum_{\alpha} f_{\alpha}^{(0)} = \rho, \quad \sum_{\alpha} \mathbf{e}_{\alpha} f_{\alpha}^{(0)} = \rho \mathbf{u}, \tag{2.96}$$

$$\sum_{\alpha} f_{\alpha}^{(n)} = 0 \text{ and } \sum_{\alpha} \mathbf{e}_{\alpha} f_{\alpha}^{(n)} = \mathbf{0} \quad \text{for} \quad n > 0. \tag{2.97}$$

And Eq. (2.93) gives

$$\varepsilon f_{\alpha}^{(1)} = -\tau \delta_t \varepsilon \left(\frac{\partial}{\partial t_0} + \mathbf{e}_{\alpha} \cdot \nabla_0 \right) f_{\alpha}^{eq}. \tag{2.98}$$

The above formulation indicates that, if being truncated at the order of $O(\varepsilon^2)$, the equilibrium and the non-equilibrium parts of the distribution function can be respectively represented by $f_{\alpha}^{(0)}$ and $\varepsilon f_{\alpha}^{(1)}$, and these two parts are correlated through

$$\begin{aligned} f_{\alpha}^{neq} &= \varepsilon f_{\alpha}^{(1)} \\ &= -\tau \delta_t \left(\frac{\partial}{\partial t} + \mathbf{e}_{\alpha} \cdot \nabla \right) f_{\alpha}^{eq}. \end{aligned} \tag{2.99}$$

To shed light on the equivalent macroscopic equations, we sum up Eqs. (2.93) and (2.94) in all lattice velocity directions and thus derive

$$\frac{\partial \rho}{\partial t_0} + \nabla_0 \cdot \left(\sum_{\alpha} \mathbf{e}_{\alpha} f_{\alpha}^{(0)} \right) = 0, \tag{2.100}$$

$$\frac{\partial \rho}{\partial t_1} = 0. \tag{2.101}$$

Consolidating the above two formulations yields

$$\frac{\partial \rho}{\partial t} + \nabla \cdot (\rho \mathbf{u}) = 0, \tag{2.102}$$

which clearly recovers the continuity equation.

Similarly, by multiplying Eqs. (2.93) and (2.94) with the lattice velocity $\mathbf{e}_\alpha$ and making summation over all lattice velocity directions, the following formulations in time scales t_0 and t_1 are obtained:

$$\frac{\partial \rho \mathbf{u}}{\partial t_0} + \nabla_0 \cdot \mathbf{\Pi}^{(0)} = \mathbf{0}, \tag{2.103}$$

$$\frac{\partial \rho \mathbf{u}}{\partial t_1} + \left(1 - \frac{1}{2\tau}\right) \nabla_0 \cdot \mathbf{\Pi}^{(1)} = \mathbf{0}, \tag{2.104}$$

where $\mathbf{\Pi}^{(0)}$ and $\mathbf{\Pi}^{(1)}$ are respectively the zeroth order and the first-order momentum flux tensors in the following forms:

$$\Pi_{ij}^{(0)} = \sum_\alpha e_{\alpha i} e_{\alpha j} f_\alpha^{(0)}, \tag{2.105}$$

$$\Pi_{ij}^{(1)} = \sum_\alpha e_{\alpha i} e_{\alpha j} f_\alpha^{(1)}. \tag{2.106}$$

With the moment relationships in Eqs. (2.89a)-(2.89d), the above momentum fluxes can be related to the macroscopic properties by

$$\Pi_{ij}^{(0)} = \sum_\alpha e_{\alpha i} e_{\alpha j} f_\alpha^{(0)} = \rho u_i u_j + p\delta_{ij}, \tag{2.107}$$

$$\begin{aligned}
-\frac{1}{\tau \delta_t} \Pi_{ij}^{(1)} &= -\frac{1}{\tau \delta_t} \sum_\alpha e_{\alpha i} e_{\alpha j} f_\alpha^{(1)} \\
&= \frac{\partial}{\partial t_0} \sum_\alpha e_{\alpha i} e_{\alpha j} f_\alpha^{(0)} + \partial_{0k} \sum_\alpha e_{\alpha i} e_{\alpha j} e_{\alpha k} f_\alpha^{(0)} \\
&= \frac{\partial}{\partial t_0} \left(\rho u_i u_j + c_s^2 \rho \delta_{ij} \right) + \partial_{0k} \left[c_s^2 \rho \left(u_i \delta_{jk} + u_j \delta_{ki} + u_k \delta_{ij} \right) \right] \\
&= c_s^2 \left[\frac{\partial \rho}{\partial t_0} + \partial_{0k} (\rho u_k) \right] \delta_{ij} + u_j \left[\frac{\partial (\rho u_i)}{\partial t_0} + \partial_{0i} p \right] \\
&\quad + u_i \left[\frac{\rho \partial u_j}{\partial t_0} + \partial_{0j} p \right] + c_s^2 \rho \left[\partial_{0i} u_j + \partial_{0j} u_i \right].
\end{aligned} \tag{2.108}$$

With the notions that

$$\frac{\partial \rho}{\partial t_0} + \partial_{0i}(\rho u_i) = 0, \tag{2.109}$$

$$\frac{\partial(\rho u_i)}{\partial t_0} = -\partial_{0j}\Pi_{ij}^{(0)} = -\partial_{0i}p - \partial_{0j}(\rho u_i u_j), \tag{2.110}$$

$$\frac{\rho \partial u_i}{\partial t_0} = -\partial_{0i}p - \rho u_j \partial_{0j} u_i. \tag{2.111}$$

Eq. (2.108) can thus be reduced to

$$\begin{aligned} -\frac{1}{\tau\delta_t}\Pi_{ij}^{(1)} &= c_s^2 \rho\left[\partial_{0i}u_j + \partial_{0j}u_i\right] - \partial_{0k}(\rho u_i u_j u_k) \\ &= c_s^2 \rho\left[\partial_{0i}u_j + \partial_{0j}u_i\right] + O(Ma^3), \end{aligned} \tag{2.112}$$

where Ma is the Mach number, and minor terms in the order of (Ma^3) are omitted under the incompressible limit.

Combining the relationships in Eqs. (2.107) and (2.112), the following momentum equation is recovered at the macroscopic scale:

$$\frac{\partial(\rho \mathbf{u})}{\partial t} + \nabla\cdot(\rho \mathbf{u}\mathbf{u}) = -\nabla p + \nabla\cdot\left[\rho\nu\left(\nabla\mathbf{u} + \nabla\mathbf{u}^T\right)\right], \tag{2.113}$$

where the kinematic viscosity ν is associated with the relaxation parameter τ through

$$\nu = c_s^2\left(\tau - \frac{1}{2}\right)\delta_t. \tag{2.114}$$

It can be seen that the relationship (2.114) is consistent with Eq. (2.84) since $\nu = \mu/\rho$ and $c_s = \sqrt{RT}$.

2.4.4 *Advantages and limitations of LBM*

After decades of development, the LBM has grown up to form an active community with many researchers [Guo and Zhao, 2002; Krüger *et al.*,

2016; Montessori and Falcucci, 2018; Valero-Lara, 2018; Chen *et al.*, 2018] working on it. Compared with the conventional Navier-Stokes equations, LBM bears some appealing characteristics. The most obvious one is its kinetic nature. Originating from the kinetic theory, LBM is not constrained by the mathematical continuum assumption as required in macroscopic methods. It is thus a more general model and has the potential in solving more complex fluid problems even beyond the capability of macroscopic Navier-Stokes solvers. One typical example is the application of LBM in microflows [Shu *et al.*, 2005; Ansumali *et al.*, 2006; Verhaeghe *et al.*, 2009] which would be difficult for Navier-Stokes solvers under the continuum assumption.

LBM is also endorsed by its simplicity. The evolution formulations of LBM are essentially algebraic equations. It only requires the linear evolution formulation to recover the fluid field with complex nonlinearities. In contrast, the Navier-Stokes equations are partial differential equations with strong nonlinearities. Special numerical strategies are usually needed to ensure accurate and stable approximations of the nonlinear terms involved.

The explicitness is another advantage of LBM. One clear obstacle in implementing the incompressible Navier-Stokes solvers is the solution of the pressure Poisson equation (PPE). The solution process encounters an equation system in the order of the number of total grid points, which is implicit and extremely time-consuming. In LBM, on the other hand, the weak compressibility of fluid is allowed under the incompressible limit. It adopts the equation of state (EOS) to explicitly connect the pressure with the density. The explicitness of LBM not only removes the tediousness and favors its implementation, but also facilitates the parallelization of the code.

The limitations of LBM are as obvious as its advantages. Firstly, LBM demands large amounts of virtual memory due to its direct evolution of the distribution functions. For instance, when using the typical D3Q15 lattice velocity model, at least 15 variables must be stored at each grid point. Performing 3D simulations with the macroscopic solvers, however, only requires storing 5 variables (density, velocity vector with 3 components, and pressure) in theory, which is far fewer than LBM.

The second drawback is the tedious implementation of physical boundary conditions. Since the LBM evolves distribution function, the physical boundary conditions for macroscopic variables need to be transformed into conditions for distribution functions so as to be implemented [Chen *et al.*, 1996; Zou and He, 1997; Guo *et al.*, 2002b; Guo *et al.*, 2002c]. Such transformation is tedious and sometimes challenging for problems with complex boundaries.

The symmetricity of lattice velocity vectors yields the third limitation: lattice uniformity. It constrains the direct implementation of LBM to uniform meshes. Extra fitting or interpolation algorithms are required to enable its application on non-uniform or unstructured meshes which are preferable for engineering problems [He *et al.*, 1996; Shu *et al.*, 2002; Dixit and Babu, 2006].

Another significant limitation of LBM, or more specifically, LBGK models, is its poor numerical stability. Von Neumann stability performed on the conventional LBGK model indicates that, even by using different lattice velocity models, LBGK tends to be highly unstable as the single relaxation parameter τ approaches 0.5 which is equivalent to the high Reynolds number scenarios [Sterling and Chen, 1996; Niu *et al.*, 2004]. As a result, LBE would require very fine meshes to avoid numerical instability when simulating cases at higher Reynolds numbers, which further exaggerates its high cost in virtual memory.

Various endeavors have been made to eliminate the above drawbacks of the conventional LBM, especially to improve its numerical stability [Ladd, 1994]. The first solution was proposed by using multiple-relaxation-time (MRT) tuned with the linear stability analysis [D'Humières, 1994]. The entropic LBGK method appeared as the second solution and was conceived based on the maximization of the entropy, by locally tuning the BGK single relaxation time [Karlin *et al.*, 1999; Gorban and Packwood, 2014]. The third alternative is to work with regularized LB equations [Latt and Chopard, 2006; Mattila *et al.*, 2015; Mattila *et al.*, 2017]. In this approach, the LBE is rewritten in a manner as to filter the undesirable ghost moments from the numerical scheme, which could effectively suppress numerical instability in the conventional LBGK model.

2.5 Summary

In this chapter, the kinetic theory established at the mesoscopic scale is introduced. A typical example in this theory, the Boltzmann equation, is presented and correlated with the macroscopic flow behaviors. The Boltzmann equation evolves the distribution function which is essentially a probability function in the particle velocity space. At the thermodynamic equilibrium, the Maxwellian distribution function can be derived to represent the equilibrium state of the distribution function. Numerical resolution of the Boltzmann equation normally starts from the truncation and the discretization of the particle velocity space, which yields the DVBE. And the integration of the DVBE along the characteristic line leads to the LBE. In principle, both DVBE and LBE can be used to simulate flows from the free molecular regime to the continuum regime as long as the discrete Maxwellian distribution function is adopted to describe the equilibrium state. If the flow is restricted to the continuum regime, the equilibrium state can be simplified into the lattice Boltzmann model which contains much fewer numbers of discrete particle velocity points as compared with the discrete Maxwellian distribution function, and LBM can be established.

Comparing with conventional Navier-Stokes solvers, the most appealing characteristic of the solvers based on DVBE and LBE is the kinetic nature, which enriches their physical robustness, preserves positive pressure, and naturally fulfills the entropy condition. It was reported in the literature that the solvers of DVBE can effectively prevent unphysical solutions such as the "Carbuncle phenomenon" in the application of Riemann solvers. This phenomenon is essentially a numerical instability that arises from the capturing of shock waves, which originates from inaccurate numerical interpretations of physics. In addition, the LBE is essentially a set of algebraic equations, which facilitates the implementation and parallelization. However, the solvers of DVBE and LBE directly evolve the distribution functions instead of the conservative variables, which leads to high cost in virtual memory and complicates the implementation of physical boundary conditions.

According to the Chapman-Enskog expansion analysis shown in **Section 2.2.3**, the consistency between the Euler/Navier-Stokes equations and the Boltzmann equation in the continuum flow regime relies on two important factors. One is the truncation error of the expanded distribution function in terms of the collision time scale (see Eq. (2.42)). The other is that key moment relationships of the equilibrium state (see Eqs. (2.34)-(2.40)) should be satisfied. With the recognition of such consistency, one may find an alternative strategy to simulate continuous flows with the help of the Boltzmann equation. The idea is to update the conservative variables by resolving the Euler/Navier-Stokes equations with the FVM discretization while reconstructing numerical fluxes on the cell interface with local solutions to the Boltzmann equation. This strategy consolidates the advantages of both the conventional Navier-Stokes solvers and the solvers of DVBE and LBE. On one hand, the direct evolution of macroscopic variables reduces the memory cost and facilitates the implementation of physical boundary conditions. On the other hand, the physical reconstruction of numerical fluxes inherits the kinetic nature from the Boltzmann equation. Successful applications of this strategy include the lattice Boltzmann flux solver (LBFS) [Shu *et al.*, 2014], the gas kinetic scheme [Xu, 2001; Xu *et al.*, 2005], the Maxwellian function-based gas kinetic flux solver (M-GKFS) [Sun *et al.*, 2015], the circular function-based gas kinetic flux solver (C-GKFS) [Yang *et al.*, 2014b] and the sphere function-based gas kinetic flux solver (S-GKFS) [Yang *et al.*, 2015], which will be elaborated in the following chapters.

Chapter 3

Lattice Boltzmann Flux Solver for Isothermal and Thermal Flows

In this chapter, the idea of using local solutions of the lattice Boltzmann equation (LBE) in the flux reconstruction of Navier-Stokes solvers will be firstly implemented in the incompressible isothermal/thermal flows, which leads to the lattice Boltzmann flux solver (LBFS) and thermal lattice Boltzmann flux solver (TLBFS). The derivations are rooted in the recognition that Navier-Stokes equations and the LBE are consistent mathematical models to describe the same physical problem in the continuum flow regime. This consistency can be established through the Chapman-Enskog expansion analysis, and both the macroscopic variables and fluxes in the Navier-Stokes equations can be equivalently recovered by the lattice moments of density distribution functions. The Chapman-Enskog expansion analysis is applicable in the whole flow domain and at any time level, which establishes the validity of using solutions to the LBE in the flux reconstruction on the cell interface.

Progresses have been made in the flux reconstruction using the solutions to either the continuous Boltzmann equation [Tang and Xu, 2000; Xu, 2001; Yuan *et al.*, 2015] or the LBE [Shu *et al.*, 2014; Wang *et al.*, 2014a; Wang *et al.*, 2014c; Yang *et al.*, 2019]. The latter one leads to the LBFS and TLBFS to be discussed in this chapter, which is applicable to incompressible flows. In LBFS and TLBFS, the macroscopic equations recovered by the LBE are discretized by the finite volume method (FVM). The inviscid and viscous fluxes on the cell interface are reconstructed simultaneously by the local solution to the LBE. And the macroscopic variables defined at the cell center are updated by implementing proper time marching schemes.

As compared to lattice Boltzmann method (LBM) which applies the LBE globally in the whole computational domain [Succi, 2001; Guo and Shu, 2013; Mohamad, 2019], the LBFS only applies the solution to the LBE locally at each cell interface, and the flow variables at cell centers are given from the solution of Navier-Stokes equations. Due to the local application of the solution to the LBE in LBFS, the drawbacks of LBM, such as the limitation to uniform mesh, tie-up of mesh spacing and time step and inconvenient implementation of physical boundary conditions, are completely removed. In the meantime, the weak compressibility assumption brought by the LBM allows an explicit evaluation of the pressure with the equation of state, which avoids the tedious resolution of the pressure Poisson equation in incompressible Navier-Stokes solvers.

3.1 Macroscopic Governing Equations

For isothermal and thermal incompressible fluid flows, the macroscopic governing equations of mass, momentum and energy conservation laws, i.e., the Navier-Stokes equations, can be written in a weakly compressible form [Chen and Doolen, 1998]:

$$\frac{\partial \rho}{\partial t} + \nabla \cdot (\rho \mathbf{u}) = 0, \tag{3.1}$$

$$\frac{\partial \rho \mathbf{u}}{\partial t} + \nabla \cdot (\rho \mathbf{u}\mathbf{u}) + \nabla p - \nabla \cdot \left[\mu \left(\nabla \mathbf{u} + (\nabla \mathbf{u})^T \right) \right] = \mathbf{f}, \tag{3.2}$$

$$\frac{\partial \rho e}{\partial t} + \nabla \cdot (\rho \mathbf{u} e) = \chi \nabla^2 (\rho e), \tag{3.3}$$

where ρ, $\mathbf{u}$, p, e, μ and χ represent the density, the velocity vector, the pressure, the internal energy, the dynamic viscosity and the thermal diffusivity, respectively; $\mathbf{f}$ is the body force. For simulation of isothermal flows, only Eqs. (3.1) and (3.2) are needed to resolve, while for thermal flows, the energy equation (3.3) should also be considered. In the following sections, the LBFS for isothermal flows will be introduced first and the counterpart for thermal flows will be presented thereafter.

3.2 Lattice Boltzmann Flux Solver for Isothermal Flows

The LBFS for isothermal flows has been proposed based on the standard LBM. It directly solves the governing equations of the mass and momentum derived from the LBE using the Chapman-Enskog expansion analysis in which the fluxes are lattice moments of particle distribution functions. In LBFS, the FVM is applied to discretize the Navier-Stokes equations and the fluxes are evaluated numerically at the cell interface by reconstructing the LBE solution locally. For solving isothermal flows by the LBFS, the governing equations (3.1) and (3.2) are rewritten in the following forms [Shu *et al.*, 2014]:

$$\frac{\partial \mathbf{W}}{\partial t} + \nabla \cdot \mathbf{F} = \mathbf{F}_E, \tag{3.4}$$

where

$$\mathbf{W} = \begin{bmatrix} \rho \\ \rho \mathbf{u} \end{bmatrix}, \quad \mathbf{F} = \begin{bmatrix} \mathbf{P} \\ \mathbf{\Pi} \end{bmatrix}, \tag{3.5}$$

$$\mathbf{P} = \sum_{\alpha=0}^{N} \mathbf{e}_\alpha f_\alpha^{eq}, \tag{3.6}$$

$$\Pi_{\beta\gamma} = \sum_{\alpha=0}^{N} (\mathbf{e}_\alpha)_\beta (\mathbf{e}_\alpha)_\gamma f_\alpha^{\wedge}, \tag{3.7}$$

$$f_\alpha^{\wedge} = f_\alpha^{eq} + \left(1 - \frac{1}{2\tau_v}\right) f_\alpha^{neq}, \tag{3.8}$$

$$f_\alpha^{neq} = f_\alpha - f_\alpha^{eq} = \varepsilon f_\alpha^{(1)} = -\tau_v \delta_t \left(\frac{\partial}{\partial t} + \mathbf{e}_\alpha \cdot \nabla\right) f_\alpha^{eq}. \tag{3.9}$$

Here, the external forcing term $\mathbf{F}_E$ is included and treated as a source term to consider the additional actions such as gravitational, buoyancy and electromagnetic forces. f_α^{eq} and f_α^{neq} are respectively the equilibrium and non-equilibrium density distribution functions along the

α^{th} lattice velocity direction; τ_v is the single relaxation parameter which is related to the dynamic viscosity; δ_t is the streaming time step; N is the total number of discrete particles in the lattice velocity model; $\mathbf{e}_\alpha$ is the lattice velocity vector in the αth direction and is determined by the lattice velocity model; $(\mathbf{e}_\alpha)_\beta$ is the component of the lattice velocity vector $\mathbf{e}_\alpha$ in the βth coordinate direction. As shown in Figure 3.1, for simulation of two-dimensional (2D) incompressible flows, the commonly-used 9-bit lattice velocity model (D2Q9) can be adopted

$$\mathbf{e}_\alpha = \begin{cases} 0, & \alpha = 0, \\ \left(\cos\left[(\alpha-1)\pi/2\right], \sin\left[(\alpha-1)\pi/2\right]\right)c, & \alpha = 1,2,3,4, \\ \sqrt{2}\left(\cos\left[\frac{(\alpha-5)\pi}{2}+\frac{\pi}{4}\right], \sin\left[\frac{(\alpha-5)\pi}{2}+\frac{\pi}{4}\right]\right)c, & \alpha = 5,6,7,8. \end{cases} \quad (3.10)$$

and for simulation of three-dimensional (3D) incompressible flows, the 15-bit lattice velocity model (D3Q15) can be applied

$$\mathbf{e}_\alpha = \begin{cases} 0, & \alpha = 0, \\ (\pm1,0,0),(0,\pm1,0),(0,0,\pm1)c, & \alpha = 1 \sim 6, \\ (\pm1,\pm1,\pm1)c, & \alpha = 7 \sim 14. \end{cases} \quad (3.11)$$

Here $c = \delta_x/\delta_t$, where δ_x is the streaming distance.

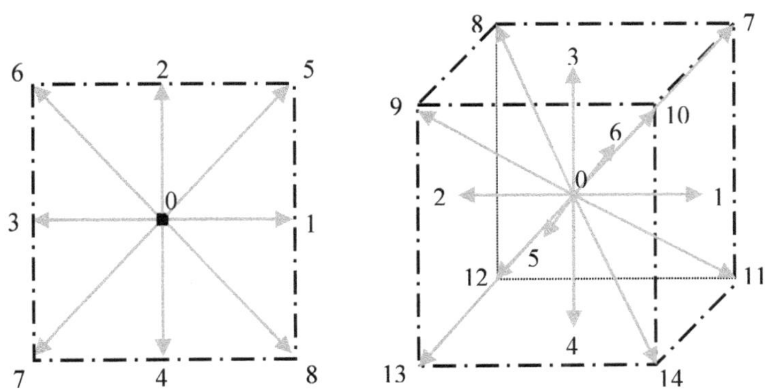

Figure 3.1. D2Q9 (left) and D3Q15 (right) lattice velocity models.

The equilibrium density distribution function for the flow field can be given by [Qian *et al.*, 1992]

$$f_\alpha^{eq} = \rho w_\alpha \left[1 + \frac{\mathbf{e}_\alpha \cdot \mathbf{u}}{c_s^2} + \frac{(\mathbf{e}_\alpha \cdot \mathbf{u})^2 - (c_s |\mathbf{u}|)^2}{2c_s^4} \right]. \tag{3.12}$$

For the D2Q9 model, the coefficients w_α and the sound speed c_s are given as $w_0 = 4/9$, $w_{1\sim4} = 1/9$, $w_{5\sim8} = 1/36$ and $c_s = c/\sqrt{3}$. For the D3Q15 model, $w_0 = 2/9$, $w_{1\sim6} = 1/9$, $w_{7\sim14} = 1/72$ and $c_s = c/\sqrt{3}$. Other models can refer to Table 2.1.

3.2.1 *Governing equations for isothermal LBFS*

Under the incompressible limit, the macroscopic governing equations (3.1) and (3.2) for isothermal flows are consistent with the LBE established at the mesoscopic scale [Chen *et al.*, 1991, Qian *et al.*, 1992]:

$$f_\alpha(\mathbf{r} + \mathbf{e}_\alpha \delta_t, t + \delta_t) = f_\alpha(\mathbf{r}, t) + \frac{f_\alpha^{eq}(\mathbf{r}, t) - f_\alpha(\mathbf{r}, t)}{\tau_v}, \quad \alpha = 0, 1, \ldots, N, \tag{3.13}$$

where f_α is the density distribution function along the αth direction. The macroscopic properties of the density ρ and the velocity vector $\mathbf{u}$ are computed by the lattice moments in the following forms:

$$\rho = \sum_{\alpha=0}^{N} f_\alpha, \quad \rho\mathbf{u} = \sum_{\alpha=0}^{N} \mathbf{e}_\alpha f_\alpha. \tag{3.14}$$

And τ_v is associated with the dynamic viscosity of the fluid through

$$\mu = \left(\tau_v - \frac{1}{2} \right) \rho c_s^2 \delta_t. \tag{3.15}$$

The Chapman-Enskog expansion analysis showed in **Section 2.4.3** establishes the correlation between the LBE in Eq. (3.13) and the macroscopic mass and momentum equations for incompressible flows [Chen and Doolen, 1998; Aidun and Clausen, 2010], which provides a solid foundation for the LBFS. The macroscopic equations recovered from LBE read

$$\frac{\partial \rho}{\partial t} + \nabla \cdot \mathbf{P} = 0, \tag{3.16}$$

$$\frac{\partial \rho \mathbf{u}}{\partial t} + \nabla \cdot \mathbf{\Pi} = \mathbf{0}, \tag{3.17}$$

where, $\mathbf{P}$ and $\mathbf{\Pi}$ are respectively the mass flux vector and the momentum flux tensor reproduced by the lattice Boltzmann model. By looking into the moment relationships of equilibrium distribution functions, it can be shown that $\mathbf{P}$ and $\mathbf{\Pi}$ are equivalent to the macroscopic fluxes given in Eqs. (3.1) and (3.2), that is,

$$\mathbf{P} = \rho \mathbf{u} = \sum_{\alpha=0}^{N} \mathbf{e}_\alpha f_\alpha^{eq}, \tag{3.18}$$

$$\mathbf{\Pi} = \rho \mathbf{u}\mathbf{u} + p\mathbf{I} - \mu\left(\nabla \mathbf{u} + \left(\nabla \mathbf{u}\right)^T\right) = \sum_{\alpha=0}^{N} \mathbf{e}_\alpha \mathbf{e}_\alpha f_\alpha^{\wedge}, \tag{3.19}$$

where $f_\alpha^{\wedge}$ is defined in Eq. (3.8).

Conventional Navier-Stokes solvers apply well-established schemes to discretize Eqs. (3.1) and (3.2) for simulating isothermal flows.

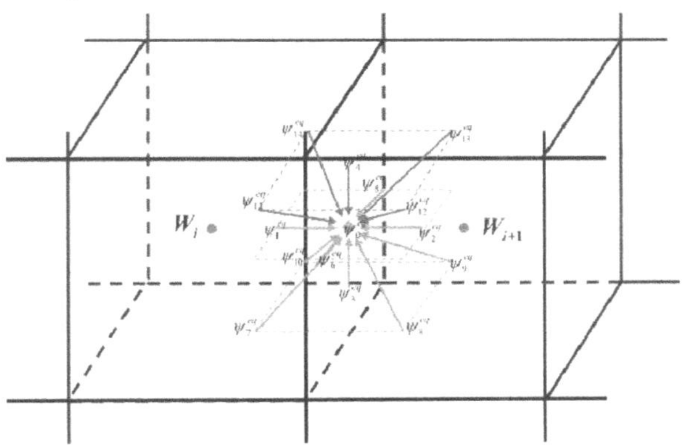

Figure 3.2. Local reconstruction of the LBE solution at an interface between two control cells.

Particularly, the FVM updates conservative flow variables in a control volume by evaluating the macroscopic fluxes at the cell interfaces. By correlating the macroscopic and the mesoscopic expressions of fluxes through Eqs. (3.18) and (3.19), alternative flux reconstruction strategy has been explored. The LBFS proposed in [Shu *et al.*, 2014] aims to solve Eqs. (3.1) and (3.2) by applying the FVM discretization while reconstructing the fluxes **P** and **Π** from the solutions to the mesoscopic LBE. The implementation details of LBFS for solving isothermal flow problems will be detailed in the next section.

3.2.2 *Numerical algorithms of isothermal LBFS*

When applying the FVM to simulate incompressible isothermal flows, the mass and the momentum fluxes should be reconstructed on the cell interface. In general conditions, the second-order of accuracy should be fulfilled in the reconstruction process. Figure 3.2 depicts the schematic diagram of LBFS for reconstructing numerical fluxes on the cell interface. The macroscopic flow variables ρ and $\rho\mathbf{u}$ are updated at the cell center in each control volume while the inviscid and the viscous fluxes are calculated simultaneously on the cell interface through the local reconstruction of the LBE solution. The derivation starts from integrating Eq. (3.4) within a control volume, which yields the following discretized form of governing equations:

$$\frac{\partial \mathbf{W}_i}{\partial t} = -\frac{1}{\Omega_i}\sum_{j=1}^{N_f}\mathbf{R}_j S_j + \mathbf{F}_{E,i}, \text{ with } \mathbf{R}_j = (\mathbf{n}\cdot\mathbf{F})_j, \tag{3.20}$$

where Ω_i is the volume of the ith control cell, N_f represents the number of surfaces for a given cell, S_j denotes the area of the jth control surface enclosed the cell i, $\mathbf{n} = (n_x, n_y, n_z)$ is the unit outer normal vector of the jth control interface. The general applicability of FVM and its flexibility on various types of meshes have broaden the application of LBFS to various flow problems [Wang *et al.*, 2015d; Wang *et al.*, 2016b; Pellerin *et al.*, 2017; Zhang *et al.*, 2018; Zhang *et al.*, 2019; Wang *et al.*, 2019], such as isothermal flows, thermal flows, porous flows and turbulent flows with complex geometries. At the same time, the LBFS retains the simplicity of LBM in reconstructing fluxes at the cell interface.

The key of LBFS is to evaluate the flux vector $\mathbf{R}_j$ at each cell interface which includes both the inviscid and the viscous parts. With the lattice velocity models given in Eq. (3.10), the expressions of $\mathbf{R}_j$ can be given explicitly via lattice moments of distribution functions. In the 2D case and with the D2Q9 model, $\mathbf{R}_j$ is given by

$$\mathbf{R}_j^{2D} = \begin{bmatrix} n_x\psi_1 + n_y\psi_2 \\ n_x\left(f_1^{\wedge} + f_3^{\wedge} + f_5^{\wedge} + f_6^{\wedge} + f_7^{\wedge} + f_8^{\wedge}\right) + n_y\left(f_5^{\wedge} - f_6^{\wedge} + f_7^{\wedge} - f_8^{\wedge}\right) \\ n_x\left(f_5^{\wedge} - f_6^{\wedge} + f_7^{\wedge} - f_8^{\wedge}\right) + n_y\left(f_2^{\wedge} + f_4^{\wedge} + f_5^{\wedge} + f_6^{\wedge} + f_7^{\wedge} + f_8^{\wedge}\right) \end{bmatrix}_j, \quad (3.21)$$

where

$$\psi_1 = f_1^{eq} - f_3^{eq} + f_5^{eq} - f_6^{eq} - f_7^{eq} + f_8^{eq}, \quad (3.22a)$$

$$\psi_2 = f_2^{eq} - f_4^{eq} + f_5^{eq} + f_6^{eq} - f_7^{eq} - f_8^{eq}. \quad (3.22b)$$

In the 3D case and with the D3Q15 model, $\mathbf{R}_j$ is given by

$$\mathbf{R}_j^{3D} = \begin{bmatrix} n_x q_1 + n_y q_2 + n_z q_3 \\ n_x\phi_{11} + n_y\phi_{12} + n_z\phi_{13} \\ n_x\phi_{21} + n_y\phi_{22} + n_z\phi_{23} \\ n_x\phi_{31} + n_y\phi_{32} + n_z\phi_{33} \end{bmatrix}_j, \quad (3.23)$$

where q_i and ϕ_{ij} are simple algebraic combinations of particle distribution functions,

$$q_1 = f_1^{eq} - f_2^{eq} + f_7^{eq} - f_8^{eq} - f_9^{eq} + f_{10}^{eq} + f_{11}^{eq} - f_{12}^{eq} - f_{13}^{eq} + f_{14}^{eq}, \quad (3.24a)$$

$$q_2 = f_3^{eq} - f_4^{eq} + f_7^{eq} + f_8^{eq} + f_9^{eq} + f_{10}^{eq} - f_{11}^{eq} - f_{12}^{eq} - f_{13}^{eq} - f_{14}^{eq}, \quad (3.24b)$$

$$q_3 = f_5^{eq} - f_6^{eq} - f_7^{eq} - f_8^{eq} + f_9^{eq} + f_{10}^{eq} - f_{11}^{eq} - f_{12}^{eq} + f_{13}^{eq} + f_{14}^{eq}, \quad (3.24c)$$

$$\phi_{11} = f_1^{\wedge} + f_2^{\wedge} + f_7^{\wedge} + f_8^{\wedge} + f_9^{\wedge} + f_{10}^{\wedge} + f_{11}^{\wedge} + f_{12}^{\wedge} + f_{13}^{\wedge} + f_{14}^{\wedge}, \quad (3.24d)$$

$$\phi_{12} = \phi_{21} = f_7^{\wedge} - f_8^{\wedge} - f_9^{\wedge} + f_{10}^{\wedge} - f_{11}^{\wedge} + f_{12}^{\wedge} + f_{13}^{\wedge} - f_{14}^{\wedge}, \tag{3.24e}$$

$$\phi_{13} = \phi_{31} = -f_7^{\wedge} + f_8^{\wedge} - f_9^{\wedge} + f_{10}^{\wedge} - f_{11}^{\wedge} + f_{12}^{\wedge} - f_{13}^{\wedge} + f_{14}^{\wedge}, \tag{3.24f}$$

$$\phi_{22} = f_3^{\wedge} + f_4^{\wedge} + f_7^{\wedge} + f_8^{\wedge} + f_9^{\wedge} + f_{10}^{\wedge} + f_{11}^{\wedge} + f_{12}^{\wedge} + f_{13}^{\wedge} + f_{14}^{\wedge}, \tag{3.24g}$$

$$\phi_{23} = \phi_{32} = -f_7^{\wedge} - f_8^{\wedge} + f_9^{\wedge} + f_{10}^{\wedge} + f_{11}^{\wedge} + f_{12}^{\wedge} - f_{13}^{\wedge} - f_{14}^{\wedge}, \tag{3.24h}$$

$$\phi_{33} = f_5^{\wedge} + f_6^{\wedge} + f_7^{\wedge} + f_8^{\wedge} + f_9^{\wedge} + f_{10}^{\wedge} + f_{11}^{\wedge} + f_{12}^{\wedge} + f_{13}^{\wedge} + f_{14}^{\wedge}. \tag{3.24i}$$

In the above equations, q_i only includes the equilibrium density distribution functions and ϕ_{ij} is symmetric which is consistent with its macroscopic counterpart. Note that, for computations on the Cartesian mesh, normally only one of n_x, n_y and n_z is non-zero. This feature can be used to effectively reduce computational efforts. To obtain $\mathbf{R}_j$, the key is to evaluate f_α^{eq} and $f_\alpha^{\wedge}$ at each cell interface. The calculation procedures of fluxes $\mathbf{R}_j$, which have been shown in [Shu *et al.*, 2014; Wang *et al.*, 2015e], are described next.

3.2.2.1 *Evaluation of f_α^{eq} and $f_\alpha^{\wedge}$ at a cell interface*

The evaluation of numerical fluxes $\mathbf{R}_j$ relies on the calculation of f_α^{eq} and $f_\alpha^{\wedge}$ on the cell interface as shown in Figure 3.2. This can be performed by reconstructing the local solution to the LBE through the application of LBM in a dummy streaming time step. Note that $f_\alpha^{\wedge}$ includes both the equilibrium distribution function f_α^{eq} and the non-equilibrium distribution function f_α^{neq}.

Consider the non-equilibrium part f_α^{neq} first. According to the Chapman-Enskog expansion analysis shown in **Section 2.4.3**, f_α^{neq} can be approximated by Eq. (3.9). Assume that the physical locations of cell centers of two adjacent control cells and the center of their interface are respectively $\mathbf{r}_i$, $\mathbf{r}_{i+1}$ and $\mathbf{r}$. Performing the Taylor-series expansion on $f_\alpha^{eq}\left(\mathbf{r} - \mathbf{e}_\alpha \delta_t, t - \delta_t\right)$ with respect to the space and time $(\mathbf{r}, t)$ gives

$$f_\alpha^{eq}(\mathbf{r}, t) - f_\alpha^{eq}\left(\mathbf{r} - \mathbf{e}_\alpha \delta_t, t - \delta_t\right) = \delta_t \left(\frac{\partial}{\partial t} + \mathbf{e}_\alpha \cdot \nabla\right) f_\alpha^{eq} + O\left(\delta_t^2\right). \tag{3.25}$$

Then the following second-order approximation to the non-equilibrium distribution function f_α^{neq} can be established:

$$\begin{aligned} f_\alpha^{neq}(\mathbf{r},t) &= -\tau_\nu \delta_t \left(\frac{\partial}{\partial t} + \mathbf{e}_\alpha \cdot \nabla \right) f_\alpha^{eq} \\ &= -\tau_\nu \left[f_\alpha^{eq}(\mathbf{r},t) - f_\alpha^{eq}(\mathbf{r} - \mathbf{e}_\alpha \delta_t, t - \delta_t) \right] + O\left(\delta_t^2\right). \end{aligned} \tag{3.26}$$

Eq. (3.26) shows that the non-equilibrium distribution function f_α^{neq} can be simply approximated by the difference of equilibrium density distribution functions at the cell interface and its surrounding points with the second-order of accuracy in terms of the streaming time step δ_t. According to Eq. (3.12), f_α^{eq} can be expressed by simple explicit functions of the density ρ and the velocity $\mathbf{u}$ which are available at the cell centers. The density and the velocity at the location $\mathbf{r} - \mathbf{e}_\alpha \delta_t$ can be easily obtained through second-order interpolation, which gives

$$\phi(\mathbf{r} - \mathbf{e}_\alpha \delta_t) = \begin{cases} \phi(\mathbf{r}_i) + (\mathbf{r} - \mathbf{e}_\alpha \delta_t - \mathbf{r}_i) \cdot \nabla \phi(\mathbf{r}_i), & \text{if } \mathbf{r} - \mathbf{e}_\alpha \delta_t \in \Omega_i, \\ \phi(\mathbf{r}_{i+1}) + (\mathbf{r} - \mathbf{e}_\alpha \delta_t - \mathbf{r}_{i+1}) \cdot \nabla \phi(\mathbf{r}_{i+1}), & \text{if } \mathbf{r} - \mathbf{e}_\alpha \delta_t \in \Omega_{i+1}. \end{cases} \tag{3.27}$$

where ϕ represents either ρ or $\mathbf{u}$. The gradients in Eq. (3.27) can be computed through the Gaussian theorem when the arbitrary unstructured mesh is used. If the Cartesian mesh is applied for simulation, they can also be calculated in a much simpler way by using the one-dimensional finite difference scheme (FDM) with the second-order of accuracy

$$\begin{aligned} \left.\frac{\partial \phi}{\partial x}\right|_{x=x_i} &= \frac{\Delta x_{i-1,i}}{\left(\Delta x_{i-1,i} \Delta x_{i,i+1} + \Delta x_{i,i+1}^2\right)} (\phi_{i+1} - \phi_i) \\ &+ \frac{\Delta x_{i,i+1}}{\left(\Delta x_{i-1,i}^2 + \Delta x_{i-1,i} \Delta x_{i,i+1}\right)} (\phi_i - \phi_{i-1}), \end{aligned} \tag{3.28}$$

$$\begin{aligned} \left.\frac{\partial \phi}{\partial y}\right|_{y=y_i} &= \frac{\Delta y_{i-1,i}}{\left(\Delta y_{i-1,i} \Delta y_{i,i+1} + \Delta y_{i,i+1}^2\right)} (\phi_{i+1} - \phi_i) \\ &+ \frac{\Delta y_{i,i+1}}{\left(\Delta y_{i-1,i}^2 + \Delta y_{i-1,i} \Delta y_{i,i+1}\right)} (\phi_i - \phi_{i-1}), \end{aligned} \tag{3.29}$$

$$\left.\frac{\partial \phi}{\partial z}\right|_{z=z_i} = \frac{\Delta z_{i-1,i}}{\left(\Delta z_{i-1,i}\Delta z_{i,i+1} + \Delta z_{i,i+1}^2\right)}\left(\phi_{i+1} - \phi_i\right) + \frac{\Delta z_{i,i+1}}{\left(\Delta z_{i-1,i}^2 + \Delta z_{i-1,i}\Delta z_{i,i+1}\right)}\left(\phi_i - \phi_{i-1}\right), \tag{3.30}$$

where $\Delta x_{i,j}$, $\Delta y_{i,j}$ and $\Delta z_{i,j}$ denote the distances between cell centers of cell i and cell j along x-, y- and z- directions, respectively. With ρ and $\mathbf{u}$ at $\mathbf{r} - \mathbf{e}_\alpha \delta_t$, $f_\alpha^{eq}\left(\mathbf{r} - \mathbf{e}_\alpha \delta_t, t - \delta_t\right)$ can be calculated straightforwardly.

The only undetermined variable is $f_\alpha^{eq}(\mathbf{r},t)$. The computation of $f_\alpha^{eq}(\mathbf{r},t)$ requires $\rho(\mathbf{r},t)$ and $\mathbf{u}(\mathbf{r},t)$ according to Eq. (3.12). These physical properties are correlated to the distribution functions through the relationships shown in Eq. (3.14):

$$\rho(\mathbf{r},t) = \sum_{\alpha=0}^{N} f_\alpha(\mathbf{r},t), \tag{3.31}$$

$$\rho(\mathbf{r},t)\mathbf{u}(\mathbf{r},t) = \sum_{\alpha=0}^{N} \mathbf{e}_\alpha f_\alpha(\mathbf{r},t). \tag{3.32}$$

$f_\alpha(\mathbf{r},t)$ consists of two parts: $f_\alpha^{eq}(\mathbf{r},t)$ and $f_\alpha^{neq}(\mathbf{r},t)$, i.e.,

$$f_\alpha(\mathbf{r},t) = f_\alpha^{eq}(\mathbf{r},t) + f_\alpha^{neq}(\mathbf{r},t). \tag{3.33}$$

Substituting Eq. (3.26) into Eq. (3.33) gives

$$f_\alpha(\mathbf{r},t) = f_\alpha^{eq}(\mathbf{r},t) - \tau_v\left[f_\alpha^{eq}(\mathbf{r},t) - f_\alpha^{eq}\left(\mathbf{r} - \mathbf{e}_\alpha\delta_t, t - \delta_t\right)\right]. \tag{3.34}$$

According to the compatibility condition, the zeroth and first lattice moments of Eq. (3.34) yield

$$\rho(\mathbf{r},t) = \sum_{\alpha=0}^{N} f_\alpha^{eq}\left(\mathbf{r} - \mathbf{e}_\alpha\delta_t, t - \delta_t\right), \tag{3.35}$$

$$\rho(\mathbf{r},t)\mathbf{u}(\mathbf{r},t) = \sum_{\alpha=0}^{N} \mathbf{e}_\alpha f_\alpha^{eq}\left(\mathbf{r} - \mathbf{e}_\alpha\delta_t, t - \delta_t\right). \tag{3.36}$$

Eqs. (3.35) and (3.36) show that, at each cell interface, the conservative flow variables can be evaluated from the equilibrium distribution functions $f_\alpha^{eq}(\mathbf{r}-\mathbf{e}_\alpha\delta_t, t-\delta_t)$ at the streaming nodes around the cell interface. Since these equilibrium distribution functions are explicit functions of macroscopic flow variables, the storage of density distribution functions is not required, which saves considerable virtual memory. In practical calculations, the LBE solution is reconstructed at the cell interface with the available conservative flow variables and their derivatives at cell centers, from which the numerical fluxes can be evaluated. By applying Eqs. (3.35) and (3.36), $f_\alpha^{eq}(\mathbf{r},t)$ can be determined. Once $f_\alpha^{eq}(\mathbf{r}-\mathbf{e}_\alpha\delta_t, t-\delta_t)$ and $f_\alpha^{eq}(\mathbf{r},t)$ are obtained, $f_\alpha^{\wedge}$ can be computed by the following equation:

$$f_\alpha^{\wedge} = f_\alpha^{eq}(\mathbf{r},t) - \left(\tau_\nu - \frac{1}{2}\right)\left[f_\alpha^{eq}(\mathbf{r},t) - f_\alpha^{eq}(\mathbf{r}-\mathbf{e}_\alpha\delta_t, t-\delta_t)\right]. \quad (3.37)$$

As a result, the components of $\mathbf{R}_j$ for the flow field can be calculated via Eq. (3.21) or Eq. (3.23). The whole process of LBFS only includes algebraic summations, which ensures the simplicity of the flux reconstruction. Detailed implementation procedures of LBFS are shown below:

(a) For each cell interface, use Eq. (3.27) to compute the macroscopic flow variables ρ and $\mathbf{u}$ at $\mathbf{r}-\mathbf{e}_\alpha\delta_t$;

(b) Use Eq. (3.12) to calculate $f_\alpha^{eq}(\mathbf{r}-\mathbf{e}_\alpha\delta_t, t-\delta_t)$;

(c) Compute $\rho(\mathbf{r},t)$ and $\rho(\mathbf{r},t)\mathbf{u}(\mathbf{r},t)$ by using Eqs. (3.35) and (3.36), and further calculate $f_\alpha^{eq}(\mathbf{r},t)$ by Eq. (3.12);

(d) Calculate $f_\alpha^{\wedge}$ by using Eq. (3.37);

(e) Compute the fluxes at the cell interface by Eq. (3.21) for the 2D case and Eq. (3.23) for the 3D case.

It should be noted that the streaming distance δ_x is associated with the steaming time step δ_t through the lattice velocity $c=\delta_x/\delta_t$. For the D2Q9 and the D3Q15 models, the lattice velocity reads $c=1$. The choice of δ_x should fulfill the requirement that all streaming nodes $\mathbf{r}-\mathbf{e}_\alpha\delta_t$ must locate within the two adjacent cells sharing the interface so that extrapolation can be avoided. One simple selection criteria can be given as:

$$\delta_x = \eta \min\left\{\Delta r_L, \Delta r_R, \frac{1}{2} l_{\min}\right\}, \tag{3.38}$$

where Δr_L is the distance between the midpoint of the cell interface and the center of the left control cell; Δr_R is the distance between the midpoint of the cell interface and the center of the right control cell; and $l_{\min}$ is the shortest length of the edges of cell surfaces in the 3D case or the shortest length of the cell interfaces in the 2D case. η is a constant to restrict the interpolation distance and should be always less than 1. In practice, η=0.8 can be used.

The external forcing term $\mathbf{F}_E$ can be treated as a source term and calculated directly from the flow variables at the cell center, which is the same as that in the conventional Navier-Stokes solvers. By extruding the effects of external forces in flux reconstruction, the standard LBM without external forcing terms can be always used in LBFS.

The implementation of boundary conditions within the finite volume framework is equal to specifying the numerical fluxes at the boundaries. The same principle is applied in the present LBFS. As compared with LBM which requires tedious conversions from physical boundary conditions to the conditions for distribution functions, the implementation of boundary conditions in LBFS is relatively simpler and more straightforward.

In addition to the spatial discretization, choosing appropriate temporal schemes for Eq. (3.20) is essential for successful numerical simulations. Generally, various explicit temporal marching schemes could be the candidate. For steady flow problems, the forward Euler scheme can be applied and the local time step can be used to accelerate convergence. For unsteady flow problems, the well-known multi-stage Runge-Kutta scheme [Van Leer *et al.*, 1992] can be applied:

$$\mathbf{W}_i^{(0)} = \mathbf{W}_i^n, \tag{3.39}$$

$$\mathbf{W}_i^{(1)} = \mathbf{W}_i^{(0)} - \beta_1 \Delta t \left(-\frac{1}{\Omega_i} \sum_{j=1}^{N_f} \mathbf{R}_j S_j + \mathbf{F}_{E,i} \right)^{(0)}, \tag{3.40}$$

$$\mathbf{W}_i^{(2)} = \mathbf{W}_i^{(0)} - \beta_2 \Delta t \left(-\frac{1}{\Omega_i} \sum_{j=1}^{N_f} \mathbf{R}_j S_j + \mathbf{F}_{E,i} \right)^{(1)}, \tag{3.41}$$

$$\mathbf{W}_i^{(k+1)} = \mathbf{W}_i^{(0)} - \beta_{k+1} \Delta t \left(-\frac{1}{\Omega_i} \sum_{j=1}^{N_f} \mathbf{R}_j S_j + \mathbf{F}_{E,i} \right)^{(k)}, \tag{3.42}$$

$$\mathbf{W}_i^{(n+1)} = \mathbf{W}_i^{(k+1)}. \tag{3.43}$$

Here, β_k represent the stage coefficients. For the three-stage scheme, $\beta_1 = 0.1481$, $\beta_2 = 0.4$ and $\beta_3 = 1.0$, while for the four-stage scheme, $\beta_1 = 0.0833$, $\beta_2 = 0.2069$, $\beta_3 = 0.4265$ and $\beta_4 = 1.0$. The flowchart of LBFS for simulation of isothermal flows is shown in Figure 3.3 and the sample code for solving 2D lid-driven cavity flow is provided in **Code A**.

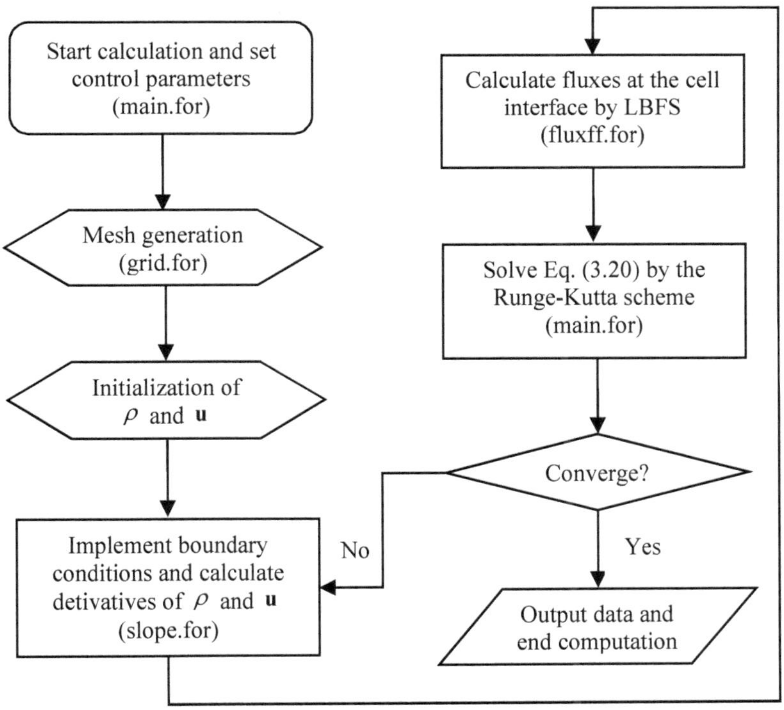

Figure 3.3. Flowchart of LBFS for simulation of isothermal flows.

3.3 Lattice Boltzmann Flux Solver for Thermal Flows

For simulation of thermal flows, the energy equation should also be incorporated into the governing equations. In this section, the thermal LBFS for solving the energy equation (3.3) is presented. Like those in the isothermal LBFS, the energy flux of equation (3.3) is calculated in a mesoscopic way by using the lattice moment of the internal energy distribution function. The macroscopic energy equation recovered from the thermal LBE reads

$$\frac{\partial \rho e}{\partial t} + \nabla \cdot \mathbf{Q} = 0, \tag{3.44}$$

where

$$\mathbf{Q} = \sum_{\alpha=0}^{M} \mathbf{e}_\alpha \hat{g}, \tag{3.45}$$

$$\hat{g}_\alpha = g_\alpha^{eq} + \left(1 - \frac{1}{2\tau_\kappa}\right) g_\alpha^{neq}, \tag{3.46}$$

$$g_\alpha^{neq} = g_\alpha - g_\alpha^{eq} = \varepsilon g_\alpha^{(1)} = -\tau_\kappa \delta_t \left(\frac{\partial}{\partial t} + \mathbf{e}_\alpha \cdot \nabla\right) g_\alpha^{eq}. \tag{3.47}$$

Here, g_α^{eq} is the equilibrium energy distribution function and g_α^{neq} is its non-equilibrium counterpart. g_α^{eq} for the 2D (D2Q9) and 3D (D3Q15) cases are respectively given by [Peng *et al.*, 2003; Peng *et al.*, 2004]

$$g_\alpha^{eq} = \begin{cases} -\dfrac{2\rho e}{3} \dfrac{|\mathbf{u}|^2}{c^2}, & \alpha = 0, \\[2ex] \dfrac{\rho e}{9}\left[\dfrac{3}{2} + \dfrac{3}{2}\dfrac{\mathbf{e}_\alpha \cdot \mathbf{u}}{c^2} + \dfrac{9}{2}\dfrac{(\mathbf{e}_\alpha \cdot \mathbf{u})^2}{c^4} - \dfrac{3}{2}\dfrac{|\mathbf{u}|^2}{c^2}\right], & \alpha = 1,2,3,4, \\[2ex] \dfrac{\rho e}{36}\left[3 + 6\dfrac{\mathbf{e}_\alpha \cdot \mathbf{u}}{c^2} + \dfrac{9}{2}\dfrac{(\mathbf{e}_\alpha \cdot \mathbf{u})^2}{c^4} - \dfrac{3}{2}\dfrac{|\mathbf{u}|^2}{c^2}\right], & \alpha = 5,6,7,8. \end{cases} \tag{3.48}$$

and

$$g_\alpha^{eq} = \begin{cases} -\dfrac{\rho e}{3}\dfrac{|\mathbf{u}|^2}{c^2}, & \alpha = 0, \\ \dfrac{\rho e}{9}\left[1 + \dfrac{\mathbf{e}_\alpha \cdot \mathbf{u}}{c^2} + \dfrac{9}{2}\dfrac{(\mathbf{e}_\alpha \cdot \mathbf{u})^2}{c^4} - \dfrac{3}{2}\dfrac{|\mathbf{u}|^2}{c^2}\right], & \alpha = 1 \sim 6, \\ \dfrac{\rho e}{72}\left[3 + 7\dfrac{\mathbf{e}_\alpha \cdot \mathbf{u}}{c^2} + \dfrac{9}{2}\dfrac{(\mathbf{e}_\alpha \cdot \mathbf{u})^2}{c^4} - \dfrac{3}{2}\dfrac{|\mathbf{u}|^2}{c^2}\right], & \alpha = 7 \sim 14. \end{cases} \tag{3.49}$$

τ_κ in Eqs. (3.46) and (3.47) is associated with the thermal diffusivity through

$$\chi = \frac{2}{3}\left(\tau_\kappa - \frac{1}{2}\right)c^2\delta_t \text{ and } \chi = \frac{5}{3}\left(\tau_\kappa - \frac{1}{2}\right)c^2\delta_t \tag{3.50}$$

for the D2Q9 model and the D3Q15 model, respectively. The derivation of Eq. (3.44) will be shown next.

3.3.1 *Governing equation for thermal LBFS*

Various models have been developed within the lattice Boltzmann framework for interpreting incompressible thermal flows, among which a simple but effective model with the BGK collision operator reads

$$g_\alpha(\mathbf{r} + \mathbf{e}_\alpha\delta_t, t + \delta_t) = g_\alpha(\mathbf{r}, t) + \frac{g_\alpha^{eq}(\mathbf{r}, t) - g_\alpha(\mathbf{r}, t)}{\tau_\kappa}, \quad \alpha = 0, 1, \ldots, M, \tag{3.51}$$

where g_α is the internal energy distribution function along the α th direction, and M is the total number of discrete particles used for g_α. The internal energy e is computed by the energy conservation law in the following form:

$$\rho e = \sum_{\alpha=0}^{M} g_\alpha . \tag{3.52}$$

By performing the Taylor-series expansion to Eq. (3.51), the resulting equation can be written as

$$\left(\frac{\partial}{\partial t}+\mathbf{e}_\alpha\cdot\nabla\right)g_\alpha+\frac{\delta_t}{2}\left(\frac{\partial}{\partial t}+\mathbf{e}_\alpha\cdot\nabla\right)^2 g_\alpha \\ +\frac{1}{\tau_\kappa\delta_t}\left(g_\alpha-g_\alpha^{eq}\right)+O\left(\delta_t^2\right)=0. \tag{3.53}$$

By introducing the multi-scale expansion to the internal energy distribution function g_α, one can obtain

$$g_\alpha=g_\alpha^{(0)}+\varepsilon g_\alpha^{(1)}+\varepsilon^2 g_\alpha^{(2)}. \tag{3.54}$$

By substituting Eqs. (2.90b)-(2.90c) and Eq. (3.54) into Eq. (3.53), the following relationships can be established at different scales:

$$O\left(\varepsilon^0\right):\ \frac{1}{\tau_\kappa\delta_t}\left(g_\alpha^{(0)}-g_\alpha^{eq}\right)=0, \tag{3.55}$$

$$O\left(\varepsilon^1\right):\ \left(\frac{\partial}{\partial t_0}+\mathbf{e}_\alpha\cdot\nabla_0\right)g_\alpha^{(0)}+\frac{1}{\tau_\kappa\delta_t}g_\alpha^{(1)}=0. \tag{3.56}$$

The energy equation (3.44) can be derived by taking the summation of Eq. (3.56) over all lattice velocity directions. Note that, in Eq. (3.44), $\mathbf{Q}$ is the thermal flux vector reproduced by the thermal LBM. By looking into the lattice moments of equilibrium internal energy distribution functions, it can be shown that $\mathbf{Q}$ is equivalent to the macroscopic thermal flux given in Eq. (3.3), i.e.,

$$\mathbf{Q}=\rho\mathbf{u}e-\chi\nabla(\rho e)=\sum_{\alpha=0}^{M}\mathbf{e}_\alpha\hat{g}_\alpha, \tag{3.57}$$

where $\hat{g}_\alpha$ is defined in Eq. (3.46). With the above relationship, the thermal LBFS has been proposed and its implementation details will be presented in the next subsection.

3.3.2 *Numerical algorithms of thermal LBFS*

Compared with the isothermal LBFS, the thermal LBFS resolves an extra energy equation (3.44), which requires reconstructing the associated thermal fluxes (3.57) on the cell interface. These fluxes will be evaluated by the local reconstruction of the thermal LBE solution. To begin with, integrating Eq. (3.44) within a control volume yields the following discretization form of the governing equation:

$$\frac{\partial(\rho e)_i}{\partial t} = -\frac{1}{\Omega_i}\sum_{j=1}^{N_f} G_j S_j \text{ , with } G_j = (\mathbf{n}\cdot\mathbf{Q})_j \,. \tag{3.58}$$

In 2D cases and with the D2Q9 model, G_j is given by

$$\begin{aligned} G_j^{2D} &= n_x\left(\hat{g}_1 - \hat{g}_3 + \hat{g}_5 - \hat{g}_6 - \hat{g}_7 + \hat{g}_8\right) \\ &\quad + n_y\left(\hat{g}_2 - \hat{g}_4 + \hat{g}_5 + \hat{g}_6 - \hat{g}_7 - \hat{g}_8\right). \end{aligned} \tag{3.59}$$

In 3D cases and with the D3Q15 model, G_j is given by

$$G_j^{3D} = \left(n_x\varphi_1 + n_y\varphi_2 + n_z\varphi_3\right)_j, \tag{3.60}$$

where φ_i are simple algebraic combinations of particle distribution functions:

$$\varphi_1 = \hat{g}_1 - \hat{g}_2 + \hat{g}_7 - \hat{g}_8 - \hat{g}_9 + \hat{g}_{10} + \hat{g}_{11} - \hat{g}_{12} - \hat{g}_{13} + \hat{g}_{14}, \tag{3.61a}$$

$$\varphi_2 = \hat{g}_3 - \hat{g}_4 + \hat{g}_7 + \hat{g}_8 + \hat{g}_9 + \hat{g}_{10} - \hat{g}_{11} - \hat{g}_{12} - \hat{g}_{13} - \hat{g}_{14}, \tag{3.61b}$$

$$\varphi_3 = \hat{g}_5 - \hat{g}_6 - \hat{g}_7 - \hat{g}_8 + \hat{g}_9 + \hat{g}_{10} - \hat{g}_{11} - \hat{g}_{12} + \hat{g}_{13} + \hat{g}_{14}. \tag{3.61c}$$

To obtain G_j, the key is to evaluate $\hat{g}$ at each cell interface. The calculation procedures of G_j are essentially the same as those of $\mathbf{R}_j$ in isothermal LBFS and will be described in detail next.

3.3.2.1 *Evaluation of* $\hat{g}_\alpha$ *at a cell interface*

The calculation procedure of $\hat{g}_\alpha$ includes the evaluations of $g_\alpha^{eq}(\mathbf{r},t)$ and $g_\alpha^{neq}(\mathbf{r},t)$, which is essentially the same as that of $\hat{f}_\alpha$ [Wang *et al.*, 2015e]. By applying the Taylor-series expansion to $g_\alpha^{eq}(\mathbf{r}-\mathbf{e}_\alpha\delta_t,t-\delta_t)$ at the space and time $(\mathbf{r},t)$ with the second-order of accuracy, we can get

$$\begin{aligned} & g_\alpha^{eq}(\mathbf{r},t)-g_\alpha^{eq}(\mathbf{r}-\mathbf{e}_\alpha\delta_t,t-\delta_t) \\ & =\delta_t\left(\frac{\partial}{\partial t}+\mathbf{e}_\alpha\cdot\nabla\right)g_\alpha^{eq}+O\left(\delta_t^2\right) \end{aligned}. \tag{3.62}$$

Substituting Eq. (3.62) into Eq. (3.47) gives

$$g_\alpha^{neq}(\mathbf{r},t)=-\tau_\kappa\left[g_\alpha^{eq}(\mathbf{r},t)-g_\alpha^{eq}(\mathbf{r}-\mathbf{e}_\alpha\delta_t,t-\delta_t)\right]. \tag{3.63}$$

Eq. (3.63) indicates that $g_\alpha^{neq}(\mathbf{r},t)$ can be evaluated by using the difference between the equilibrium internal energy distribution functions at the midpoint of the cell interface and its surrounding streaming nodes. Both $g_\alpha^{eq}(\mathbf{r},t)$ and $g_\alpha^{eq}(\mathbf{r}-\mathbf{e}_\alpha\delta_t,t-\delta_t)$ are algebraic functions of macroscopic flow variables ρ, $\mathbf{u}$, and e at different time levels and locations. As shown in the previous subsection, ρ and $\mathbf{u}$ at the streaming nodes $(\mathbf{r}-\mathbf{e}_\alpha\delta_t,t-\delta_t)$ can be obtained through interpolation using Eq. (3.27). The internal energy e at the streaming nodes can be calculated in the same way. Once ρ, $\mathbf{u}$, and e at $(\mathbf{r}-\mathbf{e}_\alpha\delta_t,t-\delta_t)$ are obtained, $g_\alpha^{eq}(\mathbf{r}-\mathbf{e}_\alpha\delta_t,t-\delta_t)$ can be computed by using the equilibrium energy distribution function given in Eq. (3.48) or Eq. (3.49).

The remaining unknown is $g_\alpha^{eq}(\mathbf{r},t)$, which is determined by $\rho(\mathbf{r},t)$, $\mathbf{u}(\mathbf{r},t)$, and $e(\mathbf{r},t)$. Here, $\rho(\mathbf{r},t)$ and $\mathbf{u}(\mathbf{r},t)$ have already been obtained in Eqs. (3.35) and (3.36) which are derived from the conservation laws and the compatibility condition. For calculation of $e(\mathbf{r},t)$, we adopt similar strategies and start from the energy conservation relationship

$$\rho(\mathbf{r},t)e(\mathbf{r},t)=\sum_{\alpha=0}^{M}g_\alpha(\mathbf{r},t)=\sum_{\alpha=0}^{M}g_\alpha^{eq}(\mathbf{r},t). \tag{3.64}$$

Here, $g_\alpha(\mathbf{r},t)$ consists of two parts: $g_\alpha^{eq}(\mathbf{r},t)$ and $g_\alpha^{neq}(\mathbf{r},t)$. It can be written as

$$\begin{aligned} g_\alpha(\mathbf{r},t) &= g_\alpha^{eq}(\mathbf{r},t) + g_\alpha^{neq}(\mathbf{r},t) \\ &= g_\alpha^{eq}(\mathbf{r},t) - \tau_\kappa\left[g_\alpha^{eq}(\mathbf{r},t) - g_\alpha^{eq}(\mathbf{r}-\mathbf{e}_\alpha\delta_t, t-\delta_t)\right]. \end{aligned} \tag{3.65}$$

Substituting Eq. (3.65) into Eq. (3.64) gives

$$\rho(\mathbf{r},t)e(\mathbf{r},t) = \sum_{\alpha=0}^{M} g_\alpha^{eq}(\mathbf{r}-\mathbf{e}_\alpha\delta_t, t-\delta_t). \tag{3.66}$$

With Eq. (3.66), the only unknown variable $e(\mathbf{r},t)$ can be calculated and then $g_\alpha^{eq}(\mathbf{r},t)$ can be obtained explicitly. By applying Eq. (3.47), $\hat{g_\alpha}$ can also be obtained:

$$\hat{g_\alpha} = g_\alpha^{eq}(\mathbf{r},t) - \left(\tau_\kappa - \frac{1}{2}\right)\left[g_\alpha^{eq}(\mathbf{r},t) - g_\alpha^{eq}(\mathbf{r}-\mathbf{e}_\alpha\delta_t, t-\delta_t)\right]. \tag{3.67}$$

It can be seen that the overall computational procedures of $\hat{g_\alpha}$ are essentially the same as those of $\hat{f_\alpha}$. They only involve the calculation of equilibrium distribution functions f_α^{eq} and g_α^{eq}. The equilibrium distribution functions at the streaming nodes around the cell interface $\mathbf{r}-\mathbf{e}_\alpha\delta_t$, which will stream to the interface center $\mathbf{r}$ in a dummy streaming time step δ_t, are used to reconstruct $\hat{f_\alpha}$ and $\hat{g_\alpha}$. This reconstruction is carried out locally at each cell interface and the streaming distance can be flexibly selected so that arbitrary meshes can be used by the LBFS for simulation. At the same time, since the reconstruction only includes equilibrium states, storages of density and energy distribution functions, which are compulsory in the conventional LBM, are avoided. This helps the LBFS save considerable computational resources, especially in the 3D computations. In addition to the spatial discretization, temporal schemes for Eq. (3.58) are selected in the same way as for the isothermal cases.

From the derivations of the isothermal LBFS and the thermal LBFS, it can be seen that, although it originates from the LBE solutions, the LBFS is completely different from the traditional LBE solvers which update the distribution functions and globally apply the streaming and the collision processes over the computational domain. The LBFS only applies the LBE solution locally at the cell interface to reconstruct the macroscopic fluxes and then updates the flow variables at cell centers within the finite volume framework. Since the FVM is applied, the LBFS can be effectively applied on arbitrary meshes for simulating flows with complex geometries, which removes the constraint of the standard LBM to uniform meshes. The direct evolution of macroscopic flow variables instead of distribution functions not only significantly reduces computational resources but also facilitates implementing physical boundary conditions. Furthermore, when the relaxation parameter τ_v is set to zero, the viscous flux vanishes and thus the LBFS can be applied to inviscid flows which is challenging for the standard LBM. Finally, the spatial and temporal discretizations are decoupled in LBFS, which enriches its flexibility in choosing more advanced time marching schemes, such as the multi-stage Runge-Kutta scheme.

3.4 Sample Applications

The LBFS has been developed with the second-order of accuracy in space. In this section, several numerical examples will be shown to demonstrate the capability and reliability of LBFS in various isothermal/thermal flow problems in both 2D and 3D conditions. It should be reaffirmed that the LBFS is applicable for both the structured and the unstructured meshes. For the structured mesh, it will be generated directly by the code. For the unstructured mesh, it will be generated by the commercial software POINTWISE with COBALT format and then transformed into a mesh input file by **Code B**.

3.4.1 *Accuracy and efficiency test*

3.4.1.1 *2D decaying vortex flow*

The 2D decaying vortex flow has been studied by using LBFS in [Shu *et al.*, 2014] to demonstrate its numerical accuracy. The solution of this flow problem can be given analytically as

$$u(x,y,t) = -u_0 \cos(\pi x/L)\sin(\pi y/L)e^{-2\pi^2 u_0 t/(ReL)}, \tag{3.68}$$

$$v(x,y,t) = +u_0 \sin(\pi x/L)\cos(\pi y/L)e^{-2\pi^2 u_0 t/(ReL)}, \tag{3.69}$$

$$\rho(x,y,t) = \rho_0 - \frac{\rho_0 u_0^2}{4c_s^2}\left[\cos(2\pi x/L) + \cos(2\pi y/L)\right]e^{-4\pi^2 u_0 t/(ReL)}, \tag{3.70}$$

where u_0 is the reference velocity, ρ_0 is the reference density, L is the reference length and Re is the Reynolds number defined as $Re=\rho_0 u_0 L / \mu$. The computational domain of this flow problem is $[-L,L]\times[-L,L]$ and six different uniform grids with the size of 21×21, 41×41, 61×61, 81×81, 101×101 and 161×161 are used to discretize the domain. The flow parameters used for simulation are $Re=10$, τ_v=0.65 and ρ_0=1 . Numerical solutions at $t = L / u_0$ are computed and the relative error of u-velocity component between the numerical result and the analytical solution is measured by the L_2 norm defined by

$$L_2(u)_{retative} = \sqrt{\frac{1}{N\times N}\sum_{i,j}\left(\frac{u_{i,j}^{numerical} - u_{i,j}^{exact}}{u_0}\right)^2}. \tag{3.71}$$

The numerical results of L_2 norms are shown in Figure 3.4, in which the mesh spacing δh is plotted in the log scale. The slope of the line is 1.971, which indicates that the accuracy of LBFS is the second order in space.

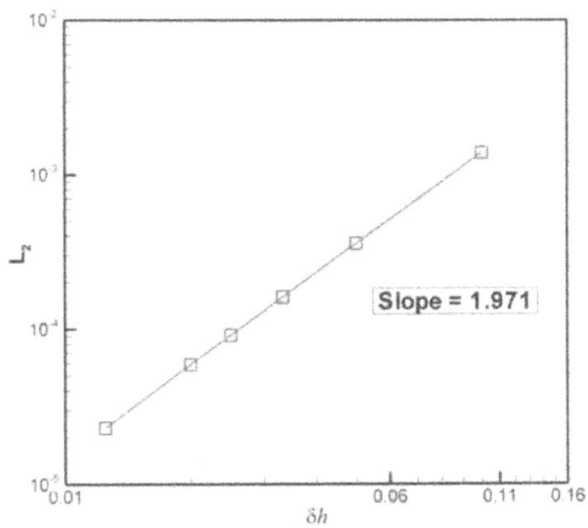

Figure 3.4. L_2 norm of the relative error of u versus δh for decaying vortex flow.

3.4.1.2 *2D lid-driven cavity flow*

The lid-driven cavity flow in a square domain is another benchmark case, which has been studied extensively in the literature [Ghia *et al.*, 1982; Huser and Biringen, 1992; Bruneau and Saad, 2006], to show the accuracy and efficiency of LBFS. Non-uniform structured mesh for the lid-driven cavity flow can be generated by the following equations:

$$x_i = \frac{1}{2}\left[1 - \cos\left(\frac{i-1}{N-1}\pi\right)\right], \quad i = 1, 2, ..., N, \tag{3.72a}$$

$$y_j = \frac{1}{2}\left[1 - \cos\left(\frac{j-1}{M-1}\pi\right)\right], \quad j = 1, 2, ..., M, \tag{3.72b}$$

where N and M are the total numbers of grid points in the x- and the y-directions, respectively. To compare the computational efficiency, the case of $Re = 1000$ is considered and three non-uniform grids of 81×81, 101×101 and 121×121 are generated by applying Eqs. (3.72a) and (3.72b). Table 3.1 compares the total iteration numbers and CPU time (seconds) on a personal Laptop (2.53GHz and 4G RAM) consumed by the LBFS and the Taylor-series expansion and least square-based lattice Boltzmann method (TLLBM) [Shu *et al.*, 2002]. Both solvers can be effectively applied on non-uniform grids. It can be seen that, when the same convergence criterion is reached, the LBFS requires fewer iteration numbers than the TLLBM and the run time per iteration by the LBFS is also less than that of TLLBM. Overall, the LBFS takes less than 40% of the computational time of TLLBM.

Since the LBFS is proposed based on the LBM, it is necessary to evaluate their performances in numerical simulations in terms of the accuracy, efficiency and stability. A comparison of the accuracy of LBFS and LBM in the 2D case has been carried out in [Wang *et al.*, 2015f] based on the lid-driven cavity flow at Re=5000, in which a non-uniform grid of 101×101 generated by Eqs. (3.72a) and (3.72b) is used by the LBFS and a uniform grid of 301×301 is used by the LBM. The finer grid used for the LBM is to ensure the numerical stability at this Reynolds number while the LBFS does not encounter such issue even on a much coarse grid. Figure 3.5 shows the pressure and the u-velocity profiles along the centerlines. Obviously, excellent agreements have been

Table 3.1. Comparison of computational time (seconds) of different schemes.

Grid size	Iteration numbers		CPU time (s) on Lenovo Laptop (2.53GHz)		
	LBFS	TLLBM	LBFS	TLLBM	LBFS/TLLBM
81×81	42924	66722	130.75	471.34	27.74%
101×101	51884	76821	252.44	764.36	33.03%
121×121	60920	87356	441.69	1145.66	38.55%

Table 3.2. Comparison of performance of LBFS and LBM on different grids for 2D lid-driven cavity flow at Re=5000.

Method	Grid size	Time (s)	Memory (Mbytes)
LBFS	301×301 (Uniform)	20838	11.05
	101×101 (Non-uniform)	909	1.23
LBM	301×301 (Uniform)	5458	23.5
	101×101 (Uniform)	Diverge	Diverge

Table 3.3. Minimum required mesh resolution of LBFS and LBM for 2D lid-driven cavity flow at different Reynolds numbers.

Re	100	1000	5000	7500
LBM	5×5	39×39	219×219	341×341
LBFS	4×4	4×4	4×4	4×4

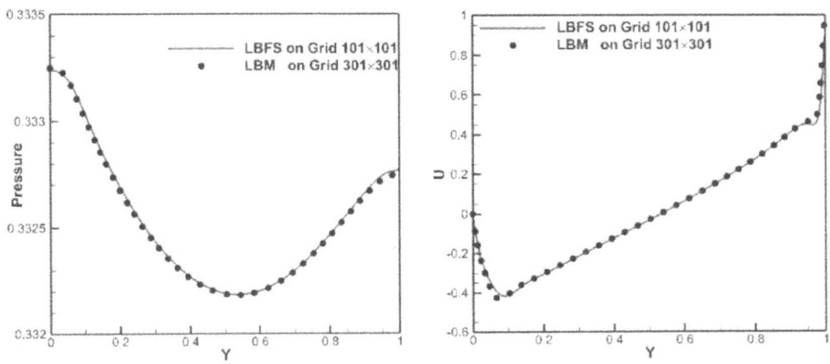

Figure 3.5. Comparison of pressure (left) and u-velocity (right) profiles on the vertical centerline for 2D lid-driven cavity flow at Re=5000.

achieved. To examine their computational efficiency, Table 3.2 shows the CPU time consumed by the LBFS and the LBM. It can be seen that the LBFS takes about 3.8 times of computational time and 47% of virtual memory used by the LBM when the same grids are applied. The expectation of reduced memory cost is validated in the numerical test. The additional computational time of LBFS on the same mesh size is consumed in the interpolation of flow variables at the cell interface and the local reconstruction of the LBE solution. When the comparable solutions are obtained, the LBFS needs a non-uniform grid of 101×101 while the LBM requires a uniform grid of 301×301. In this situation, only about 16.7% of computational time and 5.2% of virtual memory are required by the LBFS as compared with the LBM. This indicates that the LBFS needs less computational resources and can be more efficient for application on non-uniform grids. It is also noted in Table 3.2 that the LBM diverges on the uniform mesh size of 101×101 while the LBFS converges, which indicates that LBFS is more stable than LBM. This feature can be further demonstrated in Table 3.3, which examines minimum mesh sizes required by the LBFS and LBM to ensure numerical stability in simulating lid-driven cavity flows in a wide range of Reynolds numbers from 100 to 7500. The LBFS is able to maintain numerical stability for all considered Reynolds numbers on a very coarse mesh of 4×4 while the LBM needs more grid points, indicating that the LBFS can be more stable. The full code of LBFS and its instruction for this test example are provided in **Code A**.

3.4.2 *Some two-dimensional applications*

3.4.2.1 *Flow past a circular cylinder*

To demonstrate the capability of LBFS in simulating both steady and unsteady flow problems with curved boundaries, the flow past a circular cylinder, which has been studied extensively, is tested here [Shu *et al.*, 2014]. The flow characteristics of this problem are governed by the Reynolds number defined as $Re=\rho_\infty u_\infty D/\mu$, where ρ_∞, u_∞ and D are respectively the free stream density, the free stream velocity and the cylinder diameter. In order to quantify the comparison, the lift and the drag coefficients C_l and C_d are defined:

$$C_l = \frac{F_l}{\frac{1}{2}\rho_\infty u_\infty^2 D}, \tag{3.73}$$

$$C_d = \frac{F_d}{\frac{1}{2}\rho_\infty u_\infty^2 D}, \tag{3.74}$$

Table 3.4. Comparison of drag coefficient, recirculation length and separation angle for steady flow past a circular cylinder at *Re*=20 and 40.

Re	References	C_d	l_s / D	θ_s
	Dennis and Chang [1970]	2.05	0.94	43.7
	Nieuwsdadt and Keller [1973]	2.053	0.893	--
20	He and Doolen [1997]	2.152	0.921	42.96
	Shukla *et al.* [2007]	2.07	0.92	43.3
	Present	2.062	0.935	42.94
	Dennis and Chang [1970]	1.52	2.35	53.8
	Nieuwsdadt and Keller [1973]	1.54	2.18	--
40	He and Doolen [1997]	1.499	2.245	52.84
	Shukla *et al.* [2007]	1.55	2.34	52.7
	Present	1.53	2.240	52.69

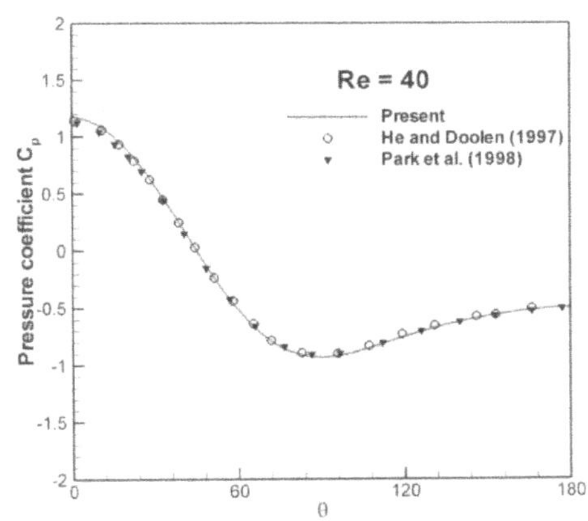

Figure 3.6. Comparison of pressure coefficient distribution on cylinder surface at *Re*=40.

$$F_l = -\int \left[\mu \frac{\partial v}{\partial x} n_x - \left(p - \mu \frac{\partial v}{\partial y} \right) n_y \right] ds, \tag{3.75}$$

$$F_d = -\int \left[\left(p - \mu \frac{\partial u}{\partial x} \right) n_x - \mu \frac{\partial u}{\partial y} n_y \right] ds, \tag{3.76}$$

where F_l and F_d are the lift and the drag forces on the cylinder, respectively; n_x and n_y are the x and the y components of the outer normal vector on the cylinder surface. The pressure coefficient C_p and the Strouhal number St for measurement of the vortex shedding frequency are respectively defined as

Table 3.5. Comparison of representative quantities for unsteady flow past a circular cylinder at *Re*=100 and 200.

Re	References	C_l	C_d	*St*
100	Braza *et al.* [1986]	±0.30	1.28 ± 0.02	0.16
	Benson *et al.* [1989]	±0.38	1.46 ± 0.01	0.17
	Ding *et al.* [2004]	±0.28	1.325 ± 0.008	0.164
	Present	±0.37	1.334 ± 0.009	0.164
200	Braza *et al.* [1986]	±0.78	1.38 ± 0.07	0.19
	Benson *et al.* [1989]	±0.65	1.45 ± 0.04	0.193
	Ding *et al.* [2004]	±0.60	1.327 ± 0.045	0.196
	Present	±0.69	1.338 ± 0.045	0.197

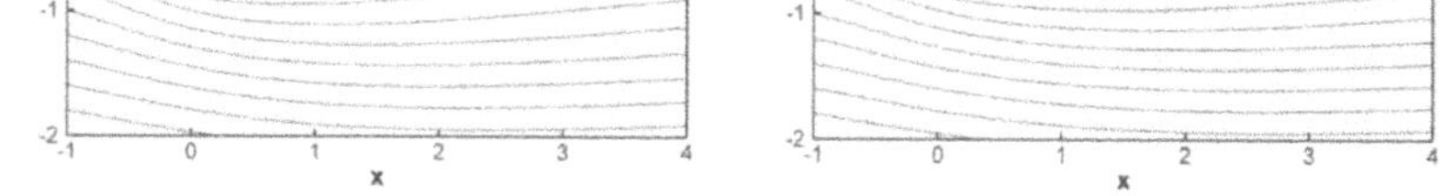

Figure 3.7. Streamlines for steady flow past a circular cylinder at *Re*=20 (left)and 40 (right).

$$C_p = \frac{p_w - p_\infty}{\frac{1}{2}\rho_\infty u_\infty^2}, \tag{3.77}$$

$$St = \frac{f_q D}{u_\infty}, \tag{3.78}$$

where p_w is the pressure on the cylinder surface and f_q is the vortex shedding frequency.

In the numerical simulation, an O-type grid is applied. The computational grids of 301×201 and 301×501 are respectively used for steady flows (*Re*=20 and 40) and unsteady flows (*Re*=100 and 200). Table 3.4 compares the drag coefficient C_d, the recirculation length l_s and the separation angle θ_s of the separation bubble behind the cylinder quantitatively. The results of previous studies carried out by Dennis and Chang [1970], Nieuwstadt and Keller [1973], He and Doolen [1997] and

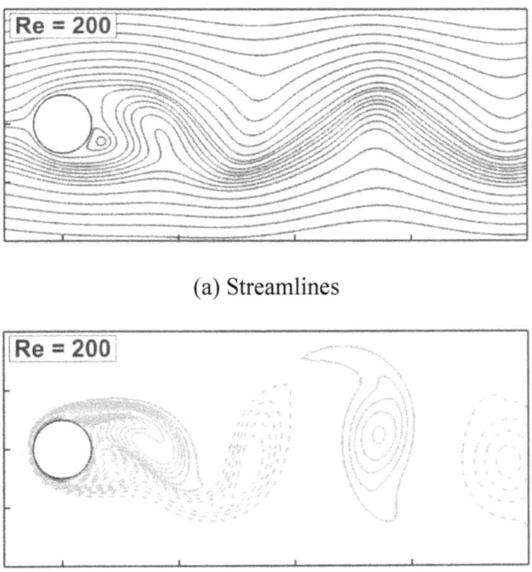

(a) Streamlines

(b) Vortices

Figure 3.8. Streamlines and vortices for unsteady flow past a circular cylinder at *Re*=200.

Shukla *et al.* [2007] are also included for comparison. Good agreements have been achieved. Figure 3.6 shows the distributions of the pressure coefficient around the cylinder surface at *Re*=40 obtained by the LBFS. The numerical solutions of He and Doolen [1997] and the experimental data of Park *et al.* [1998] are also included in the figure. Since the flow is symmetric, only the pressure coefficient on the upper surface of the cylinder is depicted. The present results also agree well with those of previous studies. Figure 3.7 depicts the streamlines around the cylinder at *Re*=20 and 40 when steady states are finally reached. With the increase of the Reynolds number, the vortex length is also increased.

When *Re* increases to 100 and 200, the flow becomes unsteady and periodic. Table 3.5 shows the quantitative comparison of the lift and the drag coefficients as well as the Strouhal number. It can be seen from this table that the present results agree well with those of Braza *et al.* [1986], Benson *et al.* [1989] and Ding *et al.* [2004]. Figure 3.8 further depicts the instantaneous streamlines and vortices behind the cylinder to demonstrate the characteristics of the flow field. Vortices are continuously shed from the rear of the cylinder and the well-known von Karman vortex street is formed. All these comparisons and observations are consistent with the results in the literature, which verifies the reliability of LBFS for simulating both steady and unsteady flows with curved boundaries.

3.4.2.2 *Natural convection in a square cavity*

The second 2D test case is used to examine the performance of TLBFS in simulating of incompressible thermal flows. In this case, the natural convection in a square cavity in a wide range of Rayleigh numbers from $Ra=10^3$ to 10^8, which has been carried out previously in the works of De Vahl Davis [1983], Shu and Xue [1998] and Peng *et al.* [2003], is simulated. Two non-dimensional parameters, i.e., the Prandtl number *Pr* and the Rayleigh number *Ra* which govern the heat transfer characteristics of this flow problem, are respectively defined as

$$Pr = \frac{\nu}{\chi}, \tag{3.79}$$

$$Ra = \frac{g\beta\Delta T L^3}{\nu\chi} = \frac{V_c^2 L^2}{\nu\chi}, \tag{3.80}$$

where L is the cavity length, ΔT is the difference of the temperature between the left and the right walls, g is the gravitational acceleration, β is the thermal expansion coefficient, and V_c is the characteristic velocity. In all cases considered here, $Pr = 0.71$ and $V_c = 0.1$. The averaged Nusselt numbers over the whole computational domain and along the line of $x = L/2$, which are usually used to examine the heat transfer rate quantitatively, are respectively defined as

$$\overline{Nu} = \frac{L}{\chi\Delta T}\frac{1}{L^2}\iint_\Omega \left(uT - \chi\frac{\partial T}{\partial x}\right)dS, \tag{3.81}$$

Table 3.6. Comparison of $\overline{Nu}$ and CPU time (seconds) obtained by the TLBFS and TLBM for natural convection in a cavity at Ra=10^3.

Grid size	$\overline{Nu}$		Iteration numbers		Run time (s)		
	TLBFS	TLBM	TLBFS	TLBM	TLBFS	TLBM	Ratio
61×61	1.112	1.114	24950	19528	142.94	261.5	54.66%
101×101	1.115	1.117	29232	29980	498.44	1175.1	42.42%
121×121	1.115	1.117	43072	34980	1128.03	2007.3	56.20%

Table 3.7. Comparison of representative quantities for natural convection in a square cavity at Ra=10^8.

	Present		Contrino *et al.* [2014]			Le Quéré [1991]
Grid Size	301^2	401^2	379^2	1019^2	1531^2	--
$\lvert\phi\rvert_{max}$	53.955	53.893	54.870	54.106	53.953	53.85
x	0.0482	0.0476	0.0469	0.0478	0.0480	0.048
y	0.5536	0.5528	0.5594	0.5545	0.5533	0.553
$Nu_{1/2}$	30.353	30.301	30.257	30.229	30.227	30.225
u_{max}	316.07	323.65	315.08	320.74	321.37	321.9
y	0.9267	0.9288	0.9239	0.9273	0.9276	0.928
v_{max}	2221.1	2222.9	2221.4	2222.1	2222.3	2222
x	0.0118	0.0119	0.0121	0.0120	0.0120	0.012

$$Nu_{1/2} = \frac{L}{\chi \Delta T} \frac{1}{L} \int_{x=L/2} \left(uT - \chi \frac{\partial T}{\partial x} \right) dl . \tag{3.82}$$

The accuracy and efficiency of TLBFS are compared with the TLBM [Peng *et al.*, 2003] by simulating this problem at Ra=10^3 on three different grids of 61×61, 101×101 and 121×121. As can be seen from Table 3.6, the relative error of the mean Nusselt number $\overline{Nu}$ obtained by the TLBFS and TLBM is within 0.2% on the same grids. However, the computational time of TLBFS is less than 57% of the time consumed by the TLBM for all cases considered, indicating that the TLBFS bears higher efficiency. More quantitative results of TLBFS, such as the maximum velocity and its position, for simulating natural convections at Ra=10^3 to 10^6 can be found in the work of Wang *et al.* [2014c]. Note that the results presented in this subsection are obtained by using the D2Q4 thermal lattice Boltzmann model and have been published in Ref. [Wang *et al.*, 2015f].

To show the capability of TLBFS for simulating natural convection problems at high Rayleigh number, the numerical results at Ra=10^8 are given in Table 3.7, including the maximum absolute value of the stream function $|\phi|_{\max}$ and its position, the mean Nusselt number $Nu_{1/2}$, the maximum velocity $u_{\max}$ along the vertical centerline and its position, and the maximum velocity $v_{\max}$ along the horizontal centerline and its

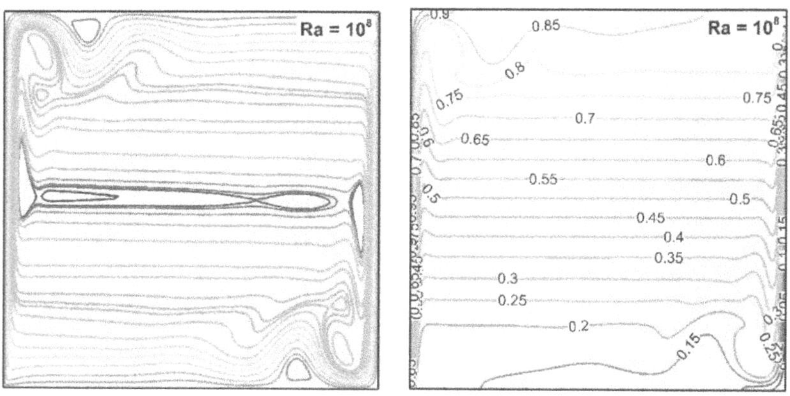

Figure 3.9. Streamlines (left) and isotherms (right) for natural convection in a square cavity at Ra=10^8.

position. The results obtained by the multi-relaxation-time TLBM [Contrino *et al.*, 2014] and the high-order pseudo-spectral Chebyshev algorithm [Le Quéré, 1991] are also included for comparison. The relative errors of $|\phi|_{\max}$ and $Nu_{1/2}$ between the TLBFS [Wang *et al.*, 2015f] on the finest mesh and the high-order pseudo-spectral Chebyshev algorithm are respectively 0.08% and 0.25%, indicating that both the velocity field and the heat transfer rate are well predicted by the TLBFS at this high Rayleigh number. Figure 3.9 shows the streamlines and isotherms at Ra=10^8. It can be seen that a large clock-wise recirculation is formed and attached to walls. The velocity and temperature boundary layers near the hot and cool walls are extremely thin. Moreover, the temperature is almost parallel to the horizontal wall in the central area of the cavity. These observations are also consistent with the results published in Refs. [Le Quéré, 1991; Contrino *et al.*, 2014; Guo *et al.*, 2002a], which further demonstrates the reliability of TLBFS for simulating natural convection flow problems at high Rayleigh numbers.

3.4.3 *Some three-dimensional applications*

3.4.3.1 *3D lid-driven cavity flow*

The reliability of LBFS for effective simulation of 3D flow problems has been demonstrated by various benchmark tests [Wang *et al.*, 2015e; Wang *et al.*, 2017b]. Here, the 3D lid-driven cavity flow at a Reynolds number of 1000 [Ku *et al.*, 1987; Ding *et al.*, 2006] is presented for illustration purposes. The accuracy and efficiency of LBFS are compared with the LBM on both uniform and non-uniform grids. As the flow is steady, the convergence criterion is set as

$$\frac{\sum_{i,j,k}\left|\left(\sqrt{u^2+v^2+w^2}\right)_{i,j,k}^{n+1}-\left(\sqrt{u^2+v^2+w^2}\right)_{i,j,k}^{n}\right|}{\sum_{i,j,k}\left(\sqrt{u^2+v^2+w^2}\right)_{i,j,k}^{n+1}}<10^{-6}. \quad (3.83)$$

The velocity profiles of u-x and v-y on the centerlines obtained by both

Table 3.8. Comparison of computational time and virtual memory of LBFS and LBM with grid-independence results for 3D lid-driven cavity flow at *Re*=1000.

Methods	Required grids	Virtual memory (Megabytes)	Maximum iteration numbers	Computational time (s)
LBFS	$61\times61\times31$	24.2	22705	4471
LBM	$101\times101\times101$	361.8	32529	33527
LBFS/LBM	11.20%	6.69%	69.8%	13.3%

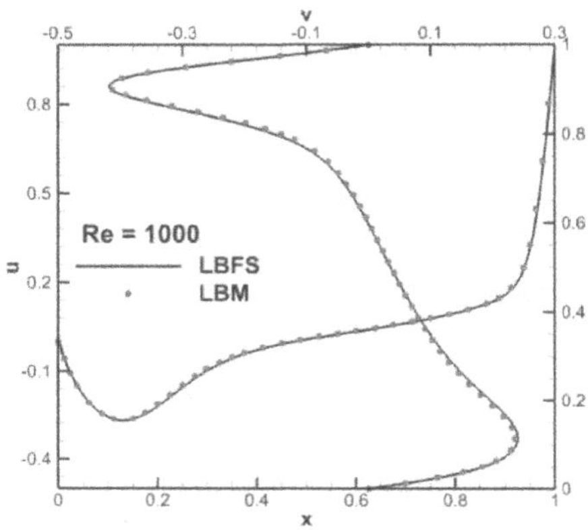

Figure 3.10. *u*-*x* and *v*-*y* on the centerlines for 3D lid-driven cavity flow at *Re*=1000 on the same uniform grids.

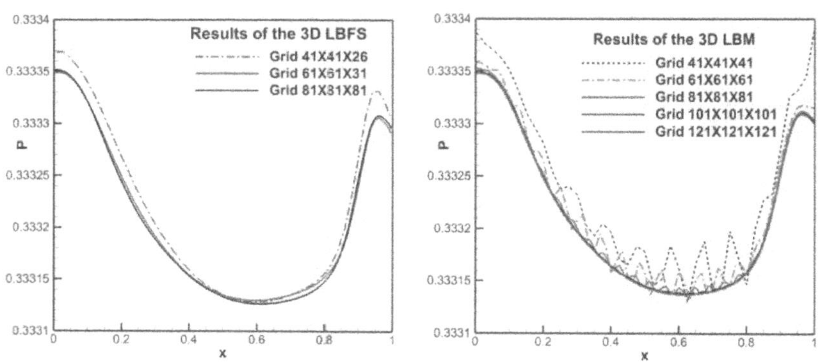

(a) LBFS (b) LBM

Figure 3.11. Pressure profiles on the horizontal centerline for 3D lid-driven cavity flow at *Re*=1000.

the LBFS [Wang *et al.*, 2015e] and the LBM on the same uniform mesh with the size of $81\times81\times81$ are depicted in Figure 3.10. Obviously, good agreements have been achieved, indicating that the LBFS and the LBM have comparable accuracy in resolving the velocity field. However, the grid-independent study reveals that, the LBFS only needs a coarse non-uniform mesh with the size of $61\times61\times31$ to achieve comparably accurate results, whilst the LBM requires a uniform mesh with the size of $101\times101\times101$, as shown in the plot of pressure profiles at Re=1000 on different meshes (see Figure 3.11). In this situation, the computational time and virtual memory of LBFS are respectively about 13.3% and 6.69% of those of LBM, as shown in Table 3.8. Moreover, unphysical oscillations of pressure exist in the solutions obtained by the LBM due to the singularity of velocity at the top corners of the cavity, which does not appear in the results of LBFS. Obviously, based on these observations, the LBFS is more efficient and robust for applications on non-uniform grids.

3.4.3.2 *3D Natural convection in a cube*

The performance of LBFS for 3D incompressible thermal flows has been tested by comparing its results with those of Fusegi *et al.* [1991], Ha and Jung [2000] and Peng *et al.* [2004] through simulating the natural convection problem in a cube, whose flow characteristics are governed by the Prandtl number (Pr) and the Rayleigh number (Ra) defined in Eqs. (3.79) and (3.80) as well. In this subsection, the results at Pr=0.071 and Ra=10^3~10^6 obtained by using the non-uniform mesh with the size of $81\times81\times81$ will be presented. The boundary conditions and initial conditions are given as: no-slip boundary conditions are applied on all walls; isothermal conditions are applied on the left and the right vertical walls while adiabatic conditions are used on four other walls. To quantitatively examine the heat transfer rate of this problem, the mean Nusselt number Nu_{mean} in the y-direction is defined as follow:

$$Nu_{mean}\left(y\right)=\int_0^1 \left.\frac{\partial T\left(y,z\right)}{\partial x}\right|_{x=1} dz \,. \tag{3.84}$$

Table 3.9 shows the representative quantities at the symmetric plane of y=0.5 for all considered cases of Ra=10^3, 10^4, 10^5 and 10^6. These properties include the maximum horizontal speed u_{max} on the vertical central line and its position z, the maximum vertical velocity w_{max} on the

Table 3.9. Comparison of representative field properties at the symmetric plane of y=0.5 for 3D natural convection.

Ra		10^3	10^4	10^5	10^6
	Present	0.132	0.200	0.142	0.082
u_{max}	Peng *et al.* [2004]	0.132	0.206	0.149	--
	Fusegi *et al.* [1991]	0.131	0.201	0.147	0.084
	Present	0.187	0.176	0.147	0.146
z	Peng *et al.* [2004]	0.188	0.163	0.136	--
	Fusegi *et al.* [1991]	0.200	0.183	0.145	0.144
	Present	0.133	0.221	0.244	0.253
w_{max}	Peng *et al.* [2004]	0.133	0.221	0.240	--
	Fusegi *et al.* [1991]	0.132	0.225	0.247	0.259
	Present	0.829	0.885	0.932	0.968
x	Peng *et al.* [2004]	0.826	0.887	0.935	--
	Fusegi *et al.* [1991]	0.833	0.883	0.935	0.967
	Present	1.092	2.289	4.622	8.921
Nu_{mean}	Peng *et al.* [2004]	1.097	2.304	4.658	--
	Fusegi *et al.* [1991]	1.105	2.302	4.646	9.012

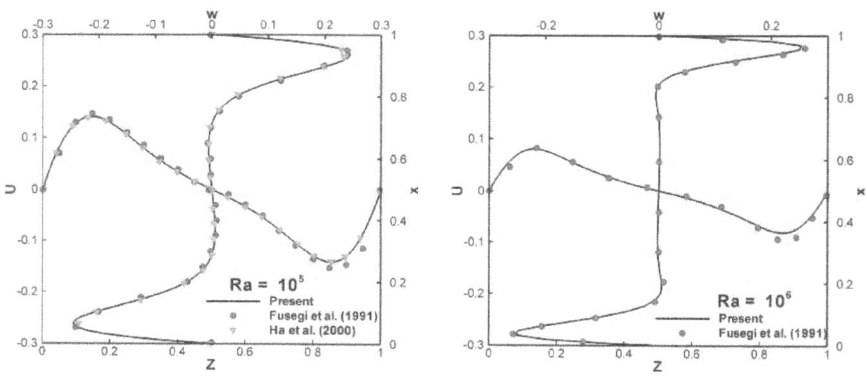

Figure 3.12. Velocity profiles at the symmetric plane of y=0.5 for 3D natural convection.

horizontal central line and its position x, and the mean Nusselt number along the line of x=1 and y=0.5. The results of Fusegi *et al.* [1991] obtained by using the third-order QUICK scheme with high resolutions are taken as the benchmark solutions, and the TLBM solutions from Peng *et al.* [2004] are also included for comparison. The relative errors between the results obtained by the LBFS and those of the third-order QUICK scheme are within 1.3%, showing that the LBFS is reliable for simulating 3D thermal flow problems. In addition, the velocity profiles on the horizontal and the vertical central lines at the symmetric plane obtained by the LBFS are depicted in Figure 3.12 and compared with the results of Fusegi *et al.* [1991] and Ha and Jung [2000]. Obviously, good agreements have been achieved. Figure 3.13 further shows the iso-surfaces of temperature at T=0.15, 0.25, 0.5, 0.75, and 0.85 to illustrate the 3D temperature field in the present natural convection problem. With the increase of Ra, the iso-surfaces are more concentrated to the left and the right walls, which is due to the enhanced heat transfer rates. This phenomenon is also in good agreement with previous reports.

3.5 Summary

This chapter presents the fundamental theory and numerical algorithms of LBFS for simulating isothermal and thermal incompressible flows on

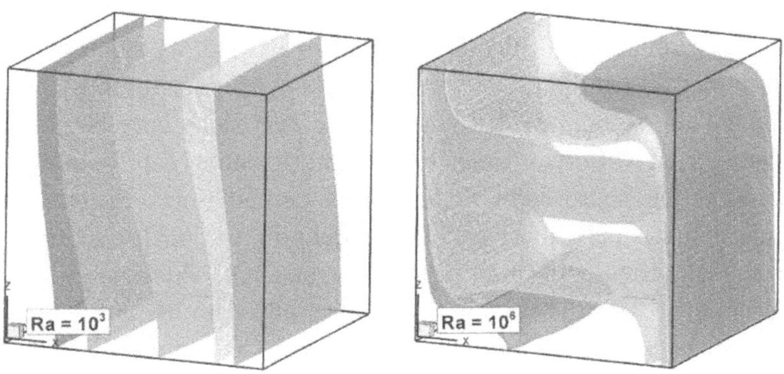

Figure 3.13. Temperature field for 3D natural convection in a cubic (Iso-surface levels: 0.15, 0.25, 0.5, 0.75 and 0.85).

arbitrary meshes. The lattice Boltzmann model and the Chapman-Enskog expansion analysis are introduced to derive equivalent macroscopic governing equations that interpret mass, momentum and energy conservation laws, and in the meantime, establish the relationships between the macroscopic fluxes and lattice moments of distribution functions. The LBFS discretizes these governing equations with the FVM to update the conservative variables at cell centers directly. The fluxes at each cell interface are evaluated by locally reconstructing the LBE solutions in a dummy streaming time step. The reconstructed solutions include the density distribution function for the density and velocity fields and the internal energy distribution function to interpret the energy variation. Both the equilibrium and the non-equilibrium parts of these two distribution functions are reconstructed to obtain the inviscid and viscous fluxes simultaneously. Through theoretical derivations, it is shown that the non-equilibrium part of the distribution function can be approximated through the difference of equilibrium distribution functions at different time levels and locations. As a consequence, the final expressions of fluxes in LBFS are the algebraic summations of equilibrium distribution functions at each cell interface and its surrounding streaming nodes. In addition to the spatial discretization, various time marching schemes, e.g. the multi-stage Runge-Kutta scheme, can be implemented to update the flow variables.

The special flux reconstruction strategy lends LBFS several distinct features as compared with the LBM and the conventional Navier-Stokes solvers for simulation of incompressible flows. First, as compared with LBM, the LBFS can be effectively applied on arbitrary meshes and can circumvent the storage of distribution functions since only equilibrium distributions at different time levels and locations are required to approximate the macroscopic fluxes at each cell interface. The streaming distance at each cell interface can be determined locally so that the constraint of lattice uniformity in LBM is removed. The time interval is also decoupled with the streaming distance, and various time marching schemes can be used as per needs. Second, as compared with the incompressible Navier-Stokes solvers, the LBFS directly solves the weakly compressible governing equations and avoids the tedious resolution of the pressure Poisson equation.

Several numerical examples, including two- and three-dimensional isothermal and thermal cases, have been presented to demonstrate the accuracy, efficiency and stability of LBFS. The obtained results are in good agreement with theoretical or numerical solutions published in the literature. It has been shown that the LBFS has the second-order of accuracy in space and can be flexibly applied on non-uniform grids. The LBFS is more stable than the LBM and consumes much less computational resources than the LBM when giving comparably accurate results.

Chapter 4

Multiphase Lattice Boltzmann Flux Solver for Two-Phase Flows

In this chapter, the idea of lattice Boltzmann flux solver (LBFS) is extended to incompressible two-phase flows, which derives the multiphase lattice Boltzmann flux solver (MLBFS) [Wang *et al.*, 2015a; Wang *et al.*, 2015b; Wang *et al.*, 2015c; Wang *et al.*, 2018a]. Multiphase flows differ from single-phase fluid flows in a way that discontinuities may exist in the physical flow fields of the density and the viscosity. Numerical interpretations of such discontinuities have always been challenging, and the difficulty amplifies in the multiphase cases with high density ratios or high Reynolds numbers. Existing multiphase models within the lattice Boltzmann framework have encountered various issues in accuracy and stability, due to which reliable solutions of lattice Boltzmann equation (LBE) for multiphase flows have been generally restricted to low-to-middle Reynolds number scenarios [Chen and Doolen, 1998; Aidun and Clausen, 2010]. Given the good performance of LBFS in single-phase flow simulations, it seems that its extension to multiphase flow simulations could possibly tackle these issues. However, due to the aforementioned physical discontinuities, such extension is not straightforward but rather difficult. Specifically, the Chapman-Enskog expansion analysis on the standard lattice Boltzmann method (LBM) with the transformed distribution function cannot recover correct macroscopic governing equations for multiphase flows: an error tensor would emerge in the viscous term. Based on different numerical strategies on this error tensor, two versions of MLBFS will be presented: one is the original MLBFS which directly compensates this error term; the other is the improved MLBFS which modifies the form of the equilibrium distribution function so that the error tensor naturally vanishes. These two versions of MLBFS both follow the idea of LBFS, which reconstructs the inviscid and the viscous fluxes at each cell interface simultaneously through the LBE solution and updates the macroscopic flow variables at cell centers. Apart from applying the MLBFS as the flow solver, the Cahn-Hilliard model will be employed to track the multiphase interfaces. The reliability and capability of MLBFS for simulating multiphase flows will be demonstrated thereafter through several numerical examples.

4.1 Macroscopic Governing Equations

The governing equations of the mass and momentum for incompressible multiphase flows, which are recovered by the multiphase lattice Boltzmann model [Shan and Chen, 1993; Briant *et al.*, 2004; Lee and Lin, 2005; Wang *et al.*, 2015a], can be given by

$$\left(\frac{\partial p}{\partial t}+\mathbf{u}\cdot\nabla p\right)+\rho c_s^2\nabla\cdot\mathbf{u}=0, \tag{4.1}$$

$$\frac{\partial \rho\mathbf{u}}{\partial t}+\nabla\cdot(\rho\mathbf{u}\mathbf{u})=-\nabla p+\nabla\cdot\left[\mu\left(\nabla\mathbf{u}+(\nabla\mathbf{u})^T\right)\right]+\mathbf{F}_s, \tag{4.2}$$

where ρ, $\mathbf{u}$, p and μ are respectively the fluid density, velocity, pressure and dynamic viscosity; $\mathbf{F}_s$ is a source term representing the body force or surface tension force. Note that the two terms in the parenthesis of Eq. (4.1) are in the order of $O(Ma^3)$ or smaller, which can be safely removed under the incompressible limit. To recover Eqs. (4.1) and (4.2), the multiphase LBM [Shan and Chen, 1994; Swift *et al.*, 1996; Briant and Yeomans, 2004; Yuan and Schaefer, 2006; Zheng *et al.*, 2006; Yu and Fan, 2009] usually includes complex external forcing terms with additional gradients of density and/or velocity. In contrast, the MLBFS based on the standard LBE can alleviate this tediousness. The Chapman-Enskog expansion analysis to the standard multiphase LBE will be presented in the next section to derive the final expressions of governing equations adopted by the MLBFS for numerical simulations.

4.2 MLBFS for Flow Field

In this section, two versions of MLBFS will be presented. The first version is the original MLBFS which solves the macroscopic governing equations recovered by the standard LBE with compensation of the error term. The second version is an improved MLBFS, which introduces a modified equilibrium distribution function to stabilize the solver and facilitate implementation.

4.2.1 *Multiphase lattice Boltzmann model and Chapman-Enskog expansion analysis*

The development of MLBFS [Wang *et al.*, 2015a] starts from the LBE without considering any external forcing term, which is much simpler than the multiphase lattice Boltzmann models used in the LBM. For simulation of immiscible two-phase fluid flows, the standard LBE without the external forcing term is given by

$$f_\alpha\left(\mathbf{r}+\mathbf{e}_\alpha\delta_t, t+\delta_t\right)=f_\alpha\left(\mathbf{r},t\right)+\frac{f_\alpha^{eq}\left(\mathbf{r},t\right)-f_\alpha\left(\mathbf{r},t\right)}{\tau},\quad \alpha=0,1,\ldots,N, \tag{4.3}$$

where $\mathbf{r}=(x,y,z)$ represents a physical location, f_α is the particle distribution function and f_α^{eq} is its equilibrium state, τ is the single relaxation parameter, δ_t is the streaming time step, $\mathbf{e}_\alpha$ is the particle velocity in the αth direction and N is the number of discrete particle velocities. The equilibrium state f_α^{eq} is given as follows:

$$\begin{aligned} f_\alpha^{eq} &= w_\alpha\left[p+\rho\Gamma_\alpha\right] \\ &= w_\alpha\left[p+\rho c_s^2\left(\frac{\mathbf{e}_\alpha\cdot\mathbf{u}}{c_s^2}+\frac{\left(\mathbf{e}_\alpha\cdot\mathbf{u}\right)^2-\left(c_s\left|\mathbf{u}\right|\right)^2}{2c_s^4}\right)\right]. \end{aligned} \tag{4.4}$$

Here, both w_α and c_s depend on the lattice velocity model. Note that f_α^{eq} used here for multiphase flows is different from the one for single-phase flows. The D2Q9 and D3Q15 lattice velocity models are commonly applied for the 2D and 3D simulations, respectively, which have been introduced in **Chapter 3**. For the D2Q9 model, $w_0 = 4/9$, $w_{1\sim4} = 1/9$, $w_{5\sim8} = 1/36$ and $c_s = c/\sqrt{3}$; and for the D3Q15 model, $w_0 = 2/9$, $w_{1\sim6} = 1/9$, $w_{7\sim14} = 1/72$ and $c_s = c/\sqrt{3}$.

To guarantee the mass and momentum conservation laws, the compatibility condition for f_α and f_α^{eq} should be satisfied at any physical location

$$p = \sum_{\alpha=0}^{N} f_\alpha^{eq} = \sum_{\alpha=0}^{N} f_\alpha, \tag{4.5}$$

$$\rho c_s^2 \mathbf{u} = \sum_{\alpha=0}^{N} \mathbf{e}_\alpha f_\alpha^{eq} = \sum_{\alpha=0}^{N} \mathbf{e}_\alpha f_\alpha. \tag{4.6}$$

Two important notions should be made here. First, the density distribution function and its associated equilibria have been modified through the incompressible transformation, due to which the zeroth moment recovers the pressure instead of the density. And secondly, the macroscopic equations recovered from this LBE are slightly different from the governing equations for multiphase flows. In the following, we will show such difference through the Chapman-Enskog expansion analysis and rebuild the consistency by imposing proper amending strategies.

The multi-scale expansions to the particle distribution function, temporal derivative and spatial derivative in consecutive order of ε give the following relationships:

$$f_\alpha = f_\alpha^{(0)} + \varepsilon f_\alpha^{(1)} + \varepsilon^2 f_\alpha^{(2)}, \tag{4.7}$$

$$\frac{\partial}{\partial t} = \varepsilon \frac{\partial}{\partial t_0} + \varepsilon^2 \frac{\partial}{\partial t_1}, \tag{4.8}$$

$$\nabla = \varepsilon \nabla_0. \tag{4.9}$$

Applying the second-order Taylor series expansion in both space and time to Eq. (4.3), one can obtain the differential equation for the distribution function

$$\begin{aligned}&\left(\frac{\partial}{\partial t}+\mathbf{e}_\alpha\cdot\nabla\right)f_\alpha+\frac{\delta_t}{2}\left(\frac{\partial}{\partial t}+\mathbf{e}_\alpha\cdot\nabla\right)^2 f_\alpha\\&\quad+\frac{1}{\tau\delta_t}\left(f_\alpha-f_\alpha^{eq}\right)+O\left(\delta_t^2\right)=0.\end{aligned}\tag{4.10}$$

Substituting the multi-scale relationships of Eqs. (4.7)-(4.9) into Eq. (4.10) and reorganizing the resultant equations in the orders of ε^0, ε^1 and ε^2, one can obtain

$$O\left(\varepsilon^0\right):\ \frac{1}{\tau\delta_t}\left(f_\alpha^{(0)}-f_\alpha^{eq}\right)=0,\tag{4.11}$$

$$O\left(\varepsilon^1\right):\ \frac{D}{Dt_0}f_\alpha^{(0)}+\frac{1}{\tau\delta_t}f_\alpha^{(1)}=0,\tag{4.12}$$

$$O\left(\varepsilon^2\right):\ \frac{\partial f_\alpha^{(0)}}{\partial t_1}+\left(1-\frac{1}{2\tau}\right)\frac{D}{Dt_0}f_\alpha^{(1)}+\frac{1}{\tau\delta_t}f_\alpha^{(2)}=0,\tag{4.13}$$

where

$$\frac{D}{Dt_0}=\frac{\partial}{\partial t_0}+\mathbf{e}_\alpha\cdot\nabla_0.\tag{4.14}$$

By taking summations of the zeroth moments of Eqs. (4.12) and (4.13) over α respectively, the governing equations on the time scales of t_0 and t_1 can be derived as follows:

$$\frac{\partial p}{\partial t_0}+\nabla_0\cdot\left(\sum_{\alpha=0}^{N}\mathbf{e}_\alpha f_\alpha^{eq}\right)=0,\tag{4.15}$$

$$\frac{\partial p}{\partial t_1}=0.\tag{4.16}$$

By multiplying Eq. (4.15) with ε and Eq. (4.16) with ε^2 and further consolidating the two resultant equations, the pressure evolution equation can be derived

$$\frac{\partial p}{\partial t}+\nabla\cdot\left(\rho c_s^2\mathbf{u}\right)=0. \tag{4.17}$$

Similarly, by taking the first moment summation of Eqs. (4.12) and (4.13) in the lattice velocity space $\mathbf{e}_\alpha$ over α respectively, the resulting equations can be given as follows:

$$\frac{\partial \rho c_s^2\mathbf{u}}{\partial t_0}+\nabla_0\cdot\mathbf{\Pi}^{(0)}=\mathbf{0}, \tag{4.18}$$

$$\frac{\partial \rho c_s^2\mathbf{u}}{\partial t_1}+\left(1-\frac{1}{2\tau}\right)\nabla_0\cdot\mathbf{\Pi}^{(1)}=\mathbf{0}, \tag{4.19}$$

where $\mathbf{\Pi}^{(0)}$ and $\mathbf{\Pi}^{(1)}$ are the second-order tensors

$$\Pi_{jk}^{(0)}=\sum_{\alpha=0}^{N}\left(\mathbf{e}_\alpha\right)_j\left(\mathbf{e}_\alpha\right)_k f_\alpha^{eq}, \tag{4.20}$$

$$\Pi_{jk}^{(1)}=-\tau\delta_t\sum_{\alpha=0}^{N}\left(\mathbf{e}_\alpha\right)_j\left(\mathbf{e}_\alpha\right)_k\frac{D}{Dt_0}f_\alpha^{eq}. \tag{4.21}$$

Again, by multiplying Eq. (4.18) with ε and Eq. (4.19) with ε^2 and further summarizing the two resultant equations, the momentum equation can be derived

$$\frac{\partial \rho c_s^2\mathbf{u}}{\partial t}+\nabla\cdot\mathbf{\Pi}=\mathbf{0}, \tag{4.22}$$

where $\mathbf{\Pi}$ is the momentum flux tensor defined by

$$\begin{aligned}\Pi_{jk}&=\Pi_{jk}^{(0)}+\varepsilon\left(1-\frac{1}{2\tau}\right)\Pi_{jk}^{(1)}\\&=\sum_{\alpha=0}^{N}\left(\mathbf{e}_\alpha\right)_j\left(\mathbf{e}_\alpha\right)_k\left[f_\alpha^{eq}+\left(1-\frac{1}{2\tau}\right)f_\alpha^{neq}\right],\end{aligned} \tag{4.23}$$

$$f_\alpha^{neq}=-\varepsilon\tau\delta_t\left(\frac{\partial}{\partial t_0}+\mathbf{e}_\alpha\cdot\nabla_0\right)f_\alpha^{eq}. \tag{4.24}$$

The macroscopic expressions of momentum fluxes in Eq. (4.23) should be derived analytically by using the lattice velocity model and the equilibrium distribution function. The second moment of the equilibrium distribution function can be derived as

$$\Pi_{jk}^{(0)} = \sum_{\alpha=0}^{N} (\mathbf{e}_\alpha)_j (\mathbf{e}_\alpha)_k f_\alpha^{eq} = c_s^2 \left(p\delta_{jk} + \rho u_j u_k \right). \tag{4.25}$$

The next task is to reveal the macroscopic form of the tensor $\Pi_{jk}^{(1)}$ in Eq. (4.23), which starts from

$$\begin{aligned} \Pi_{jk}^{(1)} &= -\tau\delta_t \sum_{\alpha=0}^{N} (\mathbf{e}_\alpha)_j (\mathbf{e}_\alpha)_k \left(\frac{\partial}{\partial t_0} + \mathbf{e}_\alpha \cdot \nabla_0 \right) f_\alpha^{eq} \\ &= -\tau\delta_t \left[\frac{\partial}{\partial t_0} \Pi_{jk}^{(0)} + \frac{\partial}{\partial x_{0l}} \sum_{\alpha=0}^{N} (\mathbf{e}_\alpha)_j (\mathbf{e}_\alpha)_k (\mathbf{e}_\alpha)_l f_\alpha^{eq} \right]. \end{aligned} \tag{4.26}$$

The first term $\partial \Pi_{jk}^{(0)} / \partial t_0$ on the right-hand side of Eq. (4.26) can be given explicitly as

$$\begin{aligned} \frac{\partial}{\partial t_0} \Pi_{jk}^{(0)} &= \frac{\partial}{\partial t_0} \left(pc_s^2 \delta_{jk} + \rho c_s^2 u_j u_k \right) \\ &= c_s^2 \delta_{jk} \frac{\partial p}{\partial t_0} + u_j \frac{\partial}{\partial t_0} \left(\rho c_s^2 u_k \right) + u_k \frac{\partial}{\partial t_0} \left(\rho c_s^2 u_j \right) - u_j u_k \frac{\partial}{\partial t_0} \left(\rho c_s^2 \right) \\ &= c_s^2 \delta_{jk} \frac{\partial p}{\partial t_0} + u_j \left[-\frac{\partial}{\partial x_{0k}} \left(pc_s^2 \right) - \frac{\partial}{\partial x_{0m}} \left(\rho c_s^2 u_k u_m \right) \right] \\ &\quad + u_k \left[-\frac{\partial}{\partial x_{0j}} \left(pc_s^2 \right) - \frac{\partial}{\partial x_{0m}} \left(\rho c_s^2 u_j u_m \right) \right] - u_j u_k \frac{\partial}{\partial t_0} \left(\rho c_s^2 \right) \\ &= c_s^2 \delta_{jk} \frac{\partial p}{\partial t_0} - c_s^2 \left[u_j \frac{\partial p}{\partial x_{0k}} + u_k \frac{\partial p}{\partial x_{0j}} \right] - \frac{\partial}{\partial x_{0m}} \left(\rho c_s^2 u_j u_k u_m \right) \\ &\quad - u_j u_k \frac{\partial}{\partial x_{0m}} \left(\rho c_s^2 u_m \right) - u_j u_k \frac{\partial}{\partial t_0} \left(\rho c_s^2 \right) \end{aligned} \tag{4.27}$$

Brought down

Continued

$$
\begin{aligned}
&= c_s^2 \delta_{jk} \left[-\rho c_s^2 \nabla_0 \cdot \mathbf{u} - c_s^2 u_m \frac{\partial \rho}{\partial x_{0m}} \right] \\
&\quad - c_s^2 \left[u_j \frac{\partial p}{\partial x_{0k}} + u_k \frac{\partial p}{\partial x_{0j}} \right] - \frac{\partial}{\partial x_{0m}} \left(\rho c_s^2 u_j u_k u_m \right) \\
&= -\rho c_s^4 \delta_{jk} \nabla_0 \cdot \mathbf{u} - \delta_{jk} c_s^4 u_m \frac{\partial \rho}{\partial x_{0m}} \\
&\quad - c_s^2 \left[u_j \frac{\partial p}{\partial x_{0k}} + u_k \frac{\partial p}{\partial x_{0j}} \right] - \frac{\partial}{\partial x_{0m}} \left(\rho c_s^2 u_j u_k u_m \right).
\end{aligned}
\tag{4.27}
$$

Since the incompressible multiphase flows are considered here, the low Mach number (Ma) assumption is valid so that minor terms in the order of $O(Ma^3)$, such as $\partial\left(\rho c_s^2 u_j u_k u_m\right)/\partial x_{0m}$ in the above equations, can be neglected.

The second term on the right-hand side of Eq. (4.26) can also be given explicitly as

$$
\begin{aligned}
&\frac{\partial}{\partial x_{0l}} \sum_{\alpha=0}^{N} (\mathbf{e}_\alpha)_j (\mathbf{e}_\alpha)_k (\mathbf{e}_\alpha)_l f_\alpha^{eq} = c_s^4 \frac{\partial}{\partial x_{0l}} \left(\rho \Delta_{jklm} u_m \right) \\
&= c_s^4 \frac{\partial}{\partial x_{0l}} \left(\rho \delta_{jk} \delta_{lm} + \rho \delta_{jl} \delta_{km} + \rho \delta_{jm} \delta_{kl} \right) u_m \\
&= \rho c_s^4 \left(\delta_{jk} \nabla_0 \cdot \mathbf{u} + \frac{\partial u_j}{\partial x_{0k}} + \frac{\partial u_k}{\partial x_{0j}} \right) + c_s^4 \left(\delta_{jk} u_l + \delta_{jl} u_k + \delta_{kl} u_j \right) \frac{\partial \rho}{\partial x_{0l}}.
\end{aligned}
\tag{4.28}
$$

The final expression of $\Pi_{jk}^{(1)}$ can be obtained by consolidating Eqs. (4.27) and (4.28)

$$
\begin{aligned}
\Pi_{jk}^{(1)} = -\tau \delta_t \Bigg[& \rho c_s^4 \left(\frac{\partial u_j}{\partial x_{0k}} + \frac{\partial u_k}{\partial x_{0j}} \right) \\
& - c_s^2 \left(u_j \frac{\partial}{\partial x_{0k}} + u_k \frac{\partial}{\partial x_{0j}} \right) \left(p - \rho c_s^2 \right) \Bigg].
\end{aligned}
\tag{4.29}
$$

Therefore, the momentum equation can be recovered by substituting Eqs. (4.25) and (4.29) into Eqs. (4.22) and (4.23)

$$\frac{\partial \rho \mathbf{u}}{\partial t}+\nabla\cdot(\rho\mathbf{u}\mathbf{u})=-\nabla p+\nabla\cdot\left[\mu\left(\nabla\mathbf{u}+(\nabla\mathbf{u})^T-\mathbf{\Pi}^e/c_s^2\right)\right], \quad (4.30)$$

where the dynamic viscosity is associated with the relaxation time through

$$\mu=\rho c_s^2\left(\tau-\frac{1}{2}\right)\delta_t, \quad (4.31)$$

and the error tensor is given by

$$\Pi_{jk}^e=\frac{1}{\rho}\left(u_j\frac{\partial}{\partial x_k}+u_k\frac{\partial}{\partial x_j}\right)\left(p-\rho c_s^2\right). \quad (4.32)$$

Note that the term related to the pressure is actually in the third-order of Mach number, i.e.,

$$\frac{1}{\rho}\left(u_j\frac{\partial}{\partial x_k}+u_k\frac{\partial}{\partial x_j}\right)p\sim O\left(Ma^3\right). \quad (4.33)$$

Hence it can be removed under the incompressible limit. Comparing the recovered macroscopic governing equations (Eqs. (4.17) and (4.30)) and the true macroscopic governing equations for multiphase flows (Eqs. (4.1) and (4.2)), the inconsistencies lie in an error term in the continuity equation and the error tensor in the momentum equation as shown in Eq. (4.32). To accurately recover multiphase flow behaviors, the development of MLBFS should consider offsetting these extra terms.

4.2.2 *Original MLBFS*

By compensating error terms shown in Eqs. (4.17) and (4.30), the original MLBFS proposed in the work of Wang *et al.* [2015a] directly solves the following governing equations:

$$\frac{\partial p}{\partial t}+\nabla\cdot\left(\sum_{\alpha=0}^{N}\mathbf{e}_\alpha f_\alpha^{eq}\right)+\mathbf{u}\cdot\nabla\left(p-\rho c_s^2\right)=0, \tag{4.34}$$

$$\frac{\partial \rho\mathbf{u}}{\partial t}+\frac{1}{c_s^2}\nabla\cdot\left(\mathbf{\Pi}-\mu\mathbf{\Pi}^e\right)-\mathbf{F}_s=\mathbf{0}, \tag{4.35}$$

where

$$\mathbf{\Pi}=\sum_{\alpha=0}^{N}\mathbf{e}_\alpha\mathbf{e}_\alpha\left[f_\alpha^{eq}+\left(1-\frac{1}{2\tau}\right)f_\alpha^{neq}\right], \tag{4.36}$$

$$f_\alpha^{neq}=-\tau\delta_t\left(\frac{\partial}{\partial t}+\mathbf{e}_\alpha\cdot\nabla\right)f_\alpha^{eq}. \tag{4.37}$$

Note that Eqs. (4.34) and (4.35) are the alternative forms of Eqs. (4.1) and (4.2). The last term in Eq. (4.34) and the error tensor $\mathbf{\Pi}^e$ in Eq. (4.35) are introduced to compensate errors since the governing equations recovered by the standard LBE differ from Eqs. (4.1) and (4.2).

The MLBFS applies the cell-centered finite volume method (FVM) to discretize Eqs. (4.34) and (4.35) and updates the flow variables $\mathbf{W}=\left(p,\rho\mathbf{u}\right)^T$ at each cell center with appropriate time-marching schemes. By applying the Green-Gauss theorem, the integration of Eqs. (4.34) and (4.35) over a control volume i gives the following formula in the 2D case:

$$\frac{\partial \mathbf{W}_i}{\partial t}=-\frac{1}{\Omega_i}\sum_{j=1}^{N_f}\mathbf{R}_j S_j+\mathbf{F}_{E,i}, \tag{4.38}$$

$$\mathbf{F}_{E,i}=\left(-\mathbf{u}\cdot\nabla\left(p-\rho c_s^2\right),F_{sx},F_{sy}\right)_i^T, \tag{4.39}$$

where Ω_i is the volume of the ith control cell, N_f represents the number of cell surfaces, and S_j denotes the area of the jth control surface enclosed the cell i. The expression of the flux vector $\mathbf{R}_j$ at each cell interface depends on the lattice velocity model used. For simulation of

2D multiphase flows, the D2Q9 model is commonly applied and the flux vector $\mathbf{R}_j$ reads

$$\mathbf{R}_j = \frac{1}{c_s^2}\begin{bmatrix} n_x\psi_1 + n_y\psi_2 \\ n_x\left(\phi_{11} - \mu\Pi_{11}^e\right) + n_y\left(\phi_{12} - \mu\Pi_{12}^e\right) \\ n_x\left(\phi_{21} - \mu\Pi_{21}^e\right) + n_y\left(\phi_{22} - \mu\Pi_{22}^e\right) \end{bmatrix}_j . \tag{4.40}$$

Here

$$\psi_1 = f_1^{eq} - f_3^{eq} + f_5^{eq} - f_6^{eq} - f_7^{eq} + f_8^{eq},$$

$$\psi_2 = f_2^{eq} - f_4^{eq} + f_5^{eq} + f_6^{eq} - f_7^{eq} - f_8^{eq},$$

$$\phi_{11} = f_1^{\wedge} + f_3^{\wedge} + f_5^{\wedge} + f_6^{\wedge} + f_7^{\wedge} + f_8^{\wedge},$$

$$\phi_{12} = \phi_{21} = f_5^{\wedge} - f_6^{\wedge} + f_7^{\wedge} - f_8^{\wedge},$$

$$\phi_{22} = f_2^{\wedge} + f_4^{\wedge} + f_5^{\wedge} + f_6^{\wedge} + f_7^{\wedge} + f_8^{\wedge},$$

n_x and n_y are respectively the x and y components of the unit outer normal vector, and the term $f_\alpha^{\wedge}$ is defined as

$$f_\alpha^{\wedge} = f_\alpha^{eq} + \left(1 - \frac{1}{2\tau}\right) f_\alpha^{neq}. \tag{4.41}$$

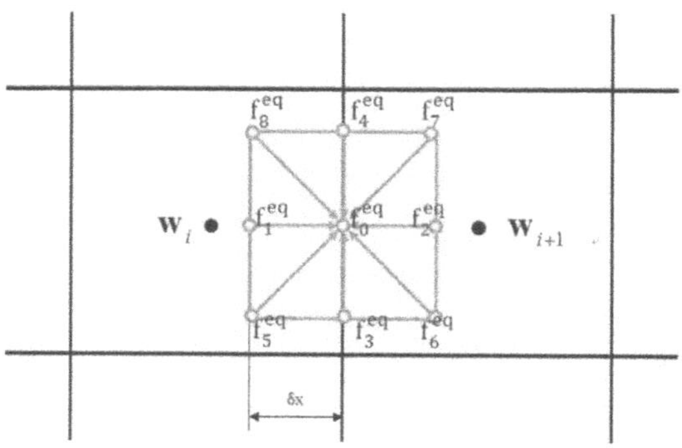

Figure 4.1. Local reconstruction of the LBE solution at a cell interface.

For simulations of 3D multiphase flows, the D3Q15 model can be used and the associated flux vector $\mathbf{R}_j$ can be obtained accordingly.

The key in the solution procedure of MLBFS is to reconstruct the flux vector $\mathbf{R}_j$ at each cell interface. As shown in Eq. (4.40), there are three unknowns: f_α^{eq}, $f_\alpha^{\wedge}$ and Π_{jk}^{e}. Since Π_{jk}^{e} includes the velocity and derivatives of the density and pressure, which can be easily obtained through interpolations, the numerical value of Π_{jk}^{e} can be directly computed at each cell interface. Therefore, the key issue for the calculation of the flux vector $\mathbf{R}_j$ is to accurately evaluate f_α^{eq} and $f_\alpha^{\wedge}$. And the evaluation procedure is essentially the same as that in LBFS for simulation of single-phase fluid flows described in **Chapter 3**. In the following, we briefly recall this procedure.

As shown in Figure 4.1, f_α^{eq} and $f_\alpha^{\wedge}$ are evaluated at the midpoint of the interface between two adjacent cells. $f_\alpha^{\wedge}$ defined in Eq. (4.41) includes both the equilibrium and the non-equilibrium parts of the distribution function, i.e., f_α^{eq} and f_α^{neq}. The non-equilibrium part f_α^{neq} at the interface center is calculated by applying the second-order Taylor-series expansion to Eq. (4.37), which gives

$$f_\alpha^{neq}(\mathbf{r},t) = -\tau\left[f_\alpha^{eq}(\mathbf{r},t) - f_\alpha^{eq}(\mathbf{r}-\mathbf{e}_\alpha\delta_t, t-\delta_t)\right], \tag{4.42}$$

where $\mathbf{r}$ represents the location of the interface center and $\mathbf{r}-\mathbf{e}_\alpha\delta_t$ denotes the location of the streaming nodes around the interface center defined by the lattice velocity model. The equilibrium distribution function $f_\alpha^{eq}(\mathbf{r}-\mathbf{e}_\alpha\delta_t, t-\delta_t)$ can be evaluated from Eq. (4.4) using the physical properties ρ, $\mathbf{u}$ and p at the corresponding locations. And these physical properties can be obtained by interpolation from flow variables defined at cell centers. $f_\alpha^{eq}(\mathbf{r},t)$ can also be obtained in a similar way by using Eq. (4.4) with the macroscopic flow variables ρ, $\mathbf{u}$ and p at $(\mathbf{r},t)$. These properties are reconstructed by the following formulations using the LBE solution [Wang *et al.*, 2015a]:

$$p(\mathbf{r},t) = \sum_{\alpha=0}^{N} f_\alpha^{eq}(\mathbf{r}-\mathbf{e}_\alpha\delta_t, t-\delta_t), \tag{4.43}$$

$$\rho(\mathbf{r},t)\mathbf{u}(\mathbf{r},t)c_s^2=\sum_{\alpha=0}^{N}\mathbf{e}_\alpha f_\alpha^{eq}(\mathbf{r}-\mathbf{e}_\alpha\delta_t,t-\delta_t), \tag{4.44}$$

$$\rho(\mathbf{r},t)=\sum_{\alpha=0}^{N}w_\alpha\rho(\mathbf{r}-\mathbf{e}_\alpha\delta_t,t-\delta_t). \tag{4.45}$$

With f_α^{eq} and f_α^{neq}, all terms in the flux vector $\mathbf{R}_j$ related to the distribution functions can be obtained. Since the additional tensor Π_{jk}^{e} is approximated at each cell interface through the finite difference method (FDM), all elements in the flux vector $\mathbf{R}_j$ have been resolved, which completes the flux reconstruction in MLBFS.

It is noteworthy that the compensation tensor Π_{jk}^{e} involves space gradients of various flow variables. The presence of this term not only complicates the flux reconstruction process, but also poses negative effects on numerical stability. Alternative strategies will be introduced in the following subsections to resolve this issue.

4.2.3 *Analysis of flux reconstruction by the original MLBFS*

Recall the fluxes reconstructed by the standard LBE solution

$$\begin{aligned}\Pi_{jk}&=\sum_{\alpha=0}^{N}(\mathbf{e}_\alpha)_j(\mathbf{e}_\alpha)_k\left(f_\alpha^{eq}+\left(1-\frac{1}{2\tau}\right)f_\alpha^{neq}\right)\\&=c_s^2\left[\rho\mathbf{uu}+p\mathbf{I}-\mu\left(\nabla\mathbf{u}+(\nabla\mathbf{u})^T\right)\right]+\mu\mathbf{\Pi}^e.\end{aligned} \tag{4.46}$$

The error tensor $\mathbf{\Pi}^e$ originates from the fact that the equation of state for single-phase flows no longer holds in multiphase scenarios, i.e. $p\neq\rho c_s^2$.

With the expression of the equilibrium distribution function, one can easily check that, in Eq. (4.46), the second moment of the equilibrium distribution function can accurately reproduce the inviscid flux

$$\sum_{\alpha=0}^{N}(\mathbf{e}_\alpha)_j(\mathbf{e}_\alpha)_k f_\alpha^{eq}=c_s^2(\rho\mathbf{uu}+p\mathbf{I}). \tag{4.47}$$

Subtracting Eq. (4.47) from Eq. (4.46), one can obtain

$$\sum_{\alpha=0}^{N}(\mathbf{e}_\alpha)_j(\mathbf{e}_\alpha)_k\left(1-\frac{1}{2\tau}\right)f_\alpha^{neq}=-\mu c_s^2\left[\nabla\mathbf{u}+(\nabla\mathbf{u})^T\right]+\mu\mathbf{\Pi}^e. \quad (4.48)$$

Substituting Eqs. (4.31) and (4.42) into Eq. (4.48) gives

$$\begin{aligned}&\frac{\mu}{\rho c_s^2\delta_t}\sum_{\alpha=0}^{N}(\mathbf{e}_\alpha)_j(\mathbf{e}_\alpha)_k\left[f_\alpha^{eq}(\mathbf{r},t)-f_\alpha^{eq}(\mathbf{r}-\mathbf{e}_\alpha\delta_t,t-\delta_t)\right]\\&=\mu c_s^2\left[\nabla\mathbf{u}+(\nabla\mathbf{u})^T\right]-\mu\mathbf{\Pi}^e.\end{aligned} \quad (4.49)$$

Obviously, when the density ratio is one and the equation of state for the pressure and density is valid, the error tensor varnishes. In the following, we will show that the error tensor can be eliminated by introducing a new equilibrium distribution function in Eq. (4.49).

By applying the second-order Taylor-series expansion to the density at each cell interface, one can obtain

$$\rho(\mathbf{r}-\mathbf{e}_\alpha\delta_t,t-\delta_t)=\rho(\mathbf{r},t-\delta_t)-\mathbf{e}_\alpha\delta_t\cdot\nabla\rho+\mathrm{O}(\delta_t^2). \quad (4.50)$$

Substituting Eq. (4.50) into $f_\alpha^{eq}(\mathbf{r}-\mathbf{e}_\alpha\delta_t,t-\delta_t)$ gives

$$\begin{aligned}f_\alpha^{eq}(\mathbf{r}-\mathbf{e}_\alpha\delta_t,t-\delta_t)&=\omega_\alpha\left[p+\left(\rho(\mathbf{r},t-\delta_t)-\mathbf{e}_\alpha\delta_t\cdot\nabla\rho\right)\Gamma_\alpha\right]\\&=\omega_\alpha\left[p+\rho(\mathbf{r},t-\delta_t)\Gamma_\alpha\right]-\omega_\alpha\Gamma_\alpha\mathbf{e}_\alpha\delta_t\cdot\nabla\rho\\&=f_\alpha^{eq\_\mathrm{m}}(\mathbf{r}-\mathbf{e}_\alpha\delta_t,t-\delta_t)-\omega_\alpha\Gamma_\alpha\mathbf{e}_\alpha\delta_t\cdot\nabla\rho,\end{aligned} \quad (4.51)$$

where $f_\alpha^{eq\_\mathrm{m}}(\mathbf{r}-\mathbf{e}_\alpha\delta_t,t-\delta_t)$ is the so-called new equilibrium distribution function

$$f_\alpha^{eq\_\mathrm{m}}(\mathbf{r}-\mathbf{e}_\alpha\delta_t,t-\delta_t)=\omega_\alpha\left[p+\rho(\mathbf{r},t-\delta_t)\Gamma_\alpha\right]. \quad (4.52)$$

It can be seen that $f_\alpha^{eq\_\mathrm{m}}(\mathbf{r}-\mathbf{e}_\alpha\delta_t,t-\delta_t)$ is evaluated by the density at the midpoint of the cell interface instead of the density at the streaming nodes around the cell interface. The second term on the right-hand side of Eq. (4.51) can be removed when it is substituted into Eq. (4.49). Details are illustrated as follows:

$$
\begin{aligned}
&\frac{\mu}{\rho c_s^2 \delta_t} \sum_{\alpha=0}^{N} (\mathbf{e}_\alpha)_j (\mathbf{e}_\alpha)_k \left[f_\alpha^{eq}(\mathbf{r},t) - f_\alpha^{eq}(\mathbf{r}-\mathbf{e}_\alpha\delta_t, t-\delta_t) \right] \\
&= \frac{\mu}{\rho c_s^2 \delta_t} \sum_{\alpha=0}^{N} (\mathbf{e}_\alpha)_j (\mathbf{e}_\alpha)_k \left[f_\alpha^{eq}(\mathbf{r},t) - f_\alpha^{eq\_m}(\mathbf{r}-\mathbf{e}_\alpha\delta_t, t-\delta_t) \right] \\
&+ \frac{\mu}{\rho c_s^2} \sum_{\alpha=0}^{N} (\mathbf{e}_\alpha)_j (\mathbf{e}_\alpha)_k \omega_\alpha \Gamma_\alpha (\mathbf{e}_\alpha \cdot \nabla\rho).
\end{aligned}
\tag{4.53}
$$

Here, the last term in Eq. (4.53) can be further simplified by using the expression of Γ_α

$$
\begin{aligned}
&\frac{\mu}{\rho c_s^2} \sum_{\alpha=0}^{N} (\mathbf{e}_\alpha)_j (\mathbf{e}_\alpha)_k \omega_\alpha \Gamma_\alpha (\mathbf{e}_\alpha \cdot \nabla\rho) \\
&= \frac{\mu}{\rho c_s^2} \sum_{\alpha=0}^{N} \omega_\alpha (\mathbf{e}_\alpha)_j (\mathbf{e}_\alpha)_k (\mathbf{e}_\alpha)_l (\mathbf{e}_\alpha)_m u_m \frac{\partial\rho}{\partial x_l} \\
&= \frac{\mu}{\rho c_s^2} c_s^4 \left(\delta_{jk}\delta_{lm} + \delta_{jl}\delta_{km} + \delta_{jm}\delta_{kl} \right) u_m \frac{\partial\rho}{\partial x_l} \\
&= \frac{\mu}{\rho} c_s^2 \left(\delta_{jk} \mathbf{u}\cdot\nabla\rho + u_j \frac{\partial\rho}{\partial x_k} + u_k \frac{\partial\rho}{\partial x_j} \right).
\end{aligned}
\tag{4.54}
$$

As is shown in **Section 4.2.1**, the Chapman-Enskog expansion analysis of the standard LBE for multiphase flows with density contrast gives the following expression for $\mathbf{u}\cdot\nabla\rho$:

$$
\frac{\partial p}{\partial t} + \mathbf{u}\cdot\nabla\rho c_s^2 + \rho c_s^2 \nabla\cdot\mathbf{u} = 0. \tag{4.55}
$$

Within the incompressible limit $\rho c_s^2 \nabla\cdot\mathbf{u} = 0$, Eq. (4.55) can be rewritten as

$$
\mathbf{u}\cdot\nabla\rho = -\frac{1}{c_s^2}\frac{\partial p}{\partial t}. \tag{4.56}
$$

Substituting Eq. (4.56) into Eq. (4.54) gives the approximation as follows:

$$\begin{aligned}&\frac{\mu}{\rho c_s^2}\sum_{\alpha=0}^{N}(\mathbf{e}_\alpha)_j(\mathbf{e}_\alpha)_k\omega_\alpha\Gamma_\alpha(\mathbf{e}_\alpha\cdot\nabla\rho)\\&=\frac{\mu}{\rho}c_s^2\left(-\frac{\delta_{jk}}{c_s^2}\frac{\partial p}{\partial t}+u_j\frac{\partial\rho}{\partial x_k}+u_k\frac{\partial\rho}{\partial x_j}\right).\end{aligned}\tag{4.57}$$

It should be noted that Eq. (4.57) is applied locally and implicitly at each cell interface to approximate the viscous flux, due to which the time derivative term $\partial p/\partial t$ can be removed. Another reason for ignoring $\partial p/\partial t$ is that the pressure itself is a relatively small term scaled by the square of Mach number, i.e., $p\propto Ma^2$. By removing the time derivative term, the remaining terms on the right-hand side of Eq. (4.57) are equivalent to the error tensor $\mu\mathbf{\Pi}^e$ under the incompressible limit. Therefore, Eq. (4.57) can be finally written as

$$\begin{aligned}&\frac{\mu}{\rho c_s^2}\sum_{\alpha=0}^{N}(\mathbf{e}_\alpha)_j(\mathbf{e}_\alpha)_k\omega_\alpha\Gamma_\alpha(\mathbf{e}_\alpha\cdot\nabla\rho)\\&=\frac{\mu}{\rho}c_s^2\left(u_j\frac{\partial\rho}{\partial x_k}+u_k\frac{\partial\rho}{\partial x_j}\right)=-\mu\mathbf{\Pi}^e.\end{aligned}\tag{4.58}$$

With Eq. (4.58), Eq. (4.53) can be simplified as

$$\begin{aligned}&\frac{\mu}{\rho c_s^2\delta_t}\sum_{\alpha=0}^{N}(\mathbf{e}_\alpha)_j(\mathbf{e}_\alpha)_k\left[f_\alpha^{eq}(\mathbf{r},t)-f_\alpha^{eq}(\mathbf{r}-\mathbf{e}_\alpha\delta_t,t-\delta_t)\right]\\&=\frac{\mu}{\rho c_s^2\delta_t}\sum_{\alpha=0}^{N}(\mathbf{e}_\alpha)_j(\mathbf{e}_\alpha)_k\left[f_\alpha^{eq}(\mathbf{r},t)-f_\alpha^{eq\_m}(\mathbf{r}-\mathbf{r}_\alpha\delta_t,t-\delta_t)\right]-\mu\Pi_{jk}^e.\end{aligned}\tag{4.59}$$

By comparing Eq. (4.49) and Eq. (4.59), the viscous fluxes can be reconstructed by the following formulation without explicitly approximating the error tensor:

$$\begin{aligned}&\frac{\mu}{\rho c_s^2\delta_t}\sum_{\alpha=0}^{N}(\mathbf{e}_\alpha)_j(\mathbf{e}_\alpha)_k\left[f_\alpha^{eq}(\mathbf{r},t)-f_\alpha^{eq\_m}(\mathbf{r}-\mathbf{e}_\alpha\delta_t,t-\delta_t)\right]\\&=\mu c_s^2\left[\nabla\mathbf{u}+(\nabla\mathbf{u})^T\right].\end{aligned}\tag{4.60}$$

With this improvement, the fluxes of MLBFS can be calculated in the same way as those of LBFS for simulating single-phase flows and the final expression of the flux reconstruction reads

$$\mathbf{\Pi}^{cc} = \rho\mathbf{u}\mathbf{u} + p\mathbf{I} - \mu\left(\nabla\mathbf{u} + (\nabla\mathbf{u})^T\right) = \sum_{\alpha=0}^{N} \mathbf{e}_\alpha \mathbf{e}_\alpha f_\alpha^{\wedge}, \tag{4.61}$$

where

$$f_\alpha^{\wedge} = f_\alpha^{eq}(\mathbf{r},t) - \left(\tau - \frac{1}{2}\right)\left(f_\alpha^{eq}(\mathbf{r},t) - f_\alpha^{eq\_m}(\mathbf{r} - \mathbf{e}_\alpha\delta_t, t - \delta_t)\right). \tag{4.62}$$

The above analysis reveals a novel strategy to remove the error tensor by modifying the equilibrium distribution function. Based on this strategy, the improved MLBFS has been developed in [Wang *et al.*, 2015b; Wang *et al.*, 2015c]. The improved MLBFS greatly simplifies the flux reconstruction procedures and shows its advantages in simulating multiphase flows with large density ratios.

4.2.4 *Improved MLBFS*

In this subsection, details of the improved MLBFS are described. The improvement is characterized by the modification made on the equilibrium distribution function. The governing equations of the improved MLBFS read

$$\frac{\partial p}{\partial t} + \nabla\cdot\left(\sum_{\alpha=0}^{N} \mathbf{e}_\alpha f_\alpha^{eq}\right) + \mathbf{u}\cdot\nabla\left(p - \rho c_s^2\right) = 0, \tag{4.63}$$

$$\frac{\partial \rho\mathbf{u}}{\partial t} + \frac{1}{c_s^2}\nabla\cdot\left(\mathbf{\Pi}^{cc}\right) - \mathbf{F}_s = \mathbf{0}, \tag{4.64}$$

where $\mathbf{\Pi}^{cc}$ is defined by Eq. (4.61). A modified equilibrium distribution function is introduced to calculate the stress tensor as shown in Eqs. (4.61) and (4.62). The major change of the modified equilibrium distribution function is to use the density at the midpoint of the cell center instead of the local density at the streaming nodes in the evaluation. This modification does not pose any difficulty in program implementation, but does offset the error tensor in the recovered macroscopic equation and thus simplifies the computation.

In the improved MLBFS, Eqs. (4.63) and (4.64) are discretized by applying the cell-centered FVM. The flow variables $\mathbf{W} = (p, \rho\mathbf{u})^T$ are defined at each cell center and updated with time by

$$\frac{\partial \mathbf{W}_i}{\partial t} = -\frac{1}{\Omega_i}\sum_{j=1}^{N_f}\mathbf{R}_j S_j + \mathbf{F}_{E,i}, \tag{4.65}$$

$$\mathbf{F}_{E,i} = \left(-\mathbf{u}\cdot\nabla\left(p - \rho c_s^2\right), \mathbf{F}_s\right)_i^T, \tag{4.66}$$

where the flux vector $\mathbf{R}_j$ at each cell interface is given by

$$\mathbf{R}_j = \begin{bmatrix} \mathbf{n}\cdot\sum_{\alpha=0}^{N}\mathbf{e}_\alpha f_\alpha^{eq} \\ \mathbf{n}\cdot\sum_{\alpha=0}^{N}\mathbf{e}_\alpha\mathbf{e}_\alpha \hat{f}_\alpha \end{bmatrix}_j. \tag{4.67}$$

Detailed expressions of $\mathbf{R}_j$ depend on the lattice velocity model. For both 2D multiphase simulations with the D2Q9 model and 3D modellings with the D3Q15 model, the expressions of $\mathbf{R}_j$ are very similar to those in the LBFS for single-phase fluid flows, which have been given in **Chapter 3**. The main differences are that the equilibrium distribution function f_α^{eq} should be Eq. (4.4) and the distribution function at the cell interface $\hat{f}_\alpha$ is approximated by Eq. (4.62).

The evolution of the flux vector $\mathbf{R}_j$ in the improved MLBFS follows the same procedure as the original MLBFS. In Eq. (4.67), $\hat{f}_\alpha$ is already in a discrete form and thus the unknowns in $\mathbf{R}_j$ are $f_\alpha^{eq}(\mathbf{r},t)$ and $f_\alpha^{eq\_m}(\mathbf{r}-\mathbf{e}_\alpha\delta_t, t-\delta_t)$. $f_\alpha^{eq\_m}(\mathbf{r}-\mathbf{e}_\alpha\delta_t, t-\delta_t)$ can be calculated by using Eq. (4.52) and $f_\alpha^{eq}(\mathbf{r},t)$ is determined by Eq. (4.4) and Eqs. (4.43)-(4.45). With $f_\alpha^{eq}(\mathbf{r},t)$ and $f_\alpha^{eq\_m}(\mathbf{r}-\mathbf{e}_\alpha\delta_t, t-\delta_t)$, the flux vector $\mathbf{R}_j$ can be obtained easily through simple algebraic summations. Once the flux vector $\mathbf{R}_j$ is obtained, the macroscopic flow variables $\mathbf{W}$ can be updated by solving the ordinary differential equation (4.65) with a suitable time marching scheme, such as the well-known multi-stage Runge-Kutta scheme.

4.3 Solution of Cahn-Hilliard Equation

For simulation of multiphase flows using fixed grid points, extra interfacial tracking algorithms should be introduced to evolve the fluid interface between different phases. One category of interfacial tracking strategy is the diffuse interface method in which the following Cahn-Hilliard model should be resolved:

$$\frac{\partial C}{\partial t}+\nabla\cdot(\mathbf{u}C)=M\nabla^2\mu_C, \tag{4.68}$$

where C is the volume fraction of the heavier fluid defined in the range of [0, 1]; M is the mobility and μ_C is the chemical potential. The local density of the fluid ρ is obtained by a linear combination of the heavier and lighter fluids of density ρ_H and ρ_L using the volume fraction C

$$\rho=C\rho_H+(1-C)\rho_L. \tag{4.69}$$

The local relaxation parameter τ can also be defined by using the relaxation parameters of τ_H and τ_L for two fluids and the volume fraction C

$$\frac{1}{\tau}=\frac{C}{\tau_H}+\frac{(1-C)}{\tau_L}. \tag{4.70}$$

The chemical potential μ_C is determined by minimizing the total free energy function $F(C,\nabla C)$ in the system, which is constituted by the energies on the fluid-fluid and the fluid-wall interfaces

$$F(C,\nabla C)=\int_\Omega\left(E_0(C)+\frac{\kappa}{2}|\nabla C|^2\right)d\Omega+\int_S\varphi(C_S)dS. \tag{4.71}$$

In Eq. (4.71), the bulk free energy $E_0(C)$ is a function of the volume fraction C

$$E_0 = \beta C^2 (C-1)^2 . \tag{4.72}$$

The free energy approach assumes that the fluid interface is in an equilibrium state with minimum energy loss and thus the interface profile can be described by the variation of the total energy function. With this assumption, the chemical potential is obtained through the canonical variation of Eq. (4.71)

$$\mu_C = \frac{\partial E_0}{\partial C} - \kappa \nabla^2 C = 2\beta C(C-1)(2C-1) - \kappa \nabla^2 C , \tag{4.73}$$

where the two constants β and κ can be determined through the relationships of the interface thickness ξ and the interfacial tension σ as follows:

$$\sigma = \sqrt{2\kappa\beta}/6 , \tag{4.74}$$

$$\kappa = \beta \xi^2 /8 . \tag{4.75}$$

With the definition of the chemical potential, the surface tension force applied in the momentum equation can be written in a vector form

$$\mathbf{F}_s = -C \nabla \mu_C . \tag{4.76}$$

To obtain the solution of the Cahn-Hilliard equation numerically, two physical boundary conditions should be imposed for this fourth-order partial differential equation. The first boundary condition is to enforce the mass conservation law at the wall. Mathematically, this is achieved by setting the gradient of the chemical potential to be zero, i.e.,

$$\mathbf{n} \cdot \nabla \mu_C \big|_{wall} = 0 , \tag{4.77}$$

where $\mathbf{n}$ is the outer normal vector of the wall. Eq. (4.77) ensures that there is no mass flux across the wall. The second physical boundary condition is to minimize the total free energy contributed by the specified wall free energy so that the wettability of walls can be varied. Mathematically, this is achieved by imposing restrictions to the normal gradient of the volume fraction C with the following relationship:

$$\mathbf{n}\cdot\nabla C\big|_{wall}=\omega\left(C_{wall}-C_{wall}^{2}\right)\sqrt{2\beta/\kappa}\,, \tag{4.78}$$

where ω is defined as the non-dimensional wetting potential and determined by the Young's Law for surface wettability. Suppose that θ^{eq} is the equilibrium contact angle at the three-phase contact line, the wetting potential can be calculated by the following equation [Lee and Liu, 2010]:

$$\cos\theta^{eq}=-\omega\,. \tag{4.79}$$

The equilibrium contact angle θ^{eq} represents the hydrophilic or hydrophobic properties of the solid wall, which is a given physical parameter.

Numerical discretization of Eq. (4.68) can be carried out by either the FDM or the FVM. For applications on uniform grids and with the second-order of accuracy, these two schemes are equivalent. However, it is usually preferred to apply high-order schemes to discretize the Cahn-Hilliard equation so that delicate flow structures of the fluid interface can be recovered on relatively coarse meshes. In this regard, the high-order FDM is preferred on the structured meshes. In the present application, the fifth-order weighted non-essentially oscillation (WENO) scheme [Shu and Osher, 1988; Liu *et al.*, 1994] is applied for the discretization of the convection term $\nabla\cdot(\mathbf{u}C)$ and the central difference scheme is used to discretize the diffusion term $M\nabla^{2}\mu_{C}$. For general structured meshes, the numerical discretization requires a coordinate transformation from the physical space to the computational space. As a result, the convection term $\nabla\cdot(\mathbf{u}C)$ can be rewritten as follows:

$$\nabla\cdot(\mathbf{u}C)=\frac{\partial}{\partial\xi}\hat{F}+\frac{\partial}{\partial\eta}\hat{G}+\frac{\partial}{\partial\zeta}\hat{H}\,. \tag{4.80}$$

With J representing the Jacobian of the transformation, we have

$$\hat{F}=\frac{1}{J}\left[\frac{\partial\xi}{\partial x}(uC)+\frac{\partial\xi}{\partial y}(vC)+\frac{\partial\xi}{\partial z}(wC)\right], \tag{4.81}$$

$$\hat{G}=\frac{1}{J}\left[\frac{\partial\eta}{\partial x}(uC)+\frac{\partial\eta}{\partial y}(vC)+\frac{\partial\eta}{\partial z}(wC)\right], \tag{4.82}$$

$$\hat{H}=\frac{1}{J}\left[\frac{\partial\zeta}{\partial x}(uC)+\frac{\partial\zeta}{\partial y}(vC)+\frac{\partial\zeta}{\partial z}(wC)\right]. \tag{4.83}$$

After the coordinate transformation, the convection terms can be approximated by the WENO scheme. A similar procedure of coordinate transformation can be used for the diffusion term, after which the central difference scheme is applied for space discretization. In addition to the space discretization, the temporal term can be discretized by applying the multi-stage Runge-Kutta scheme so that the volume fraction C can be updated.

Up to now, both the original and the improved MLBFS as the flow solver and the Cahn-Hilliard model for the phase field have been presented. The solution procedure of the improved MLBFS can be summarized as follows:

(1) Identify the streaming distance δ_x at each cell interface. All the points surrounding the cell interface should fall into the adjacent control volumes and the streaming distance can be different from interface to interface;
(2) Calculate the macroscopic flow properties ρ, $\mathbf{u}$ and p at the locations $(\mathbf{r}-\mathbf{e}_\alpha\delta_t)$ around the considered cell interface;
(3) Compute the equilibrium function $f_\alpha^{eq\_m}(\mathbf{r}-\mathbf{e}_\alpha\delta_t, t-\delta_t)$ according to Eq. (4.52);
(4) Calculate ρ, $\mathbf{u}$ and p at the interface center $(\mathbf{r},t)$ by Eqs. (4.43)-(4.45) and compute $f_\alpha^{eq}(\mathbf{r},t)$ by Eq. (4.4);
(5) Compute the relaxation parameter τ by Eq. (4.31);
(6) Calculate $f_\alpha^{\wedge}$ by Eq. (4.62) and further compute the flux vector $\mathbf{R}_j$ at each cell interface by Eq. (4.67);
(7) Compute the convective and diffusion terms in Eq. (4.68) by the WENO scheme and the central difference scheme, respectively;
(8) Update the flow variables and the volume fraction by using the multi-stage Runge-Kutta scheme to solve Eqs. (4.65) and (4.68);
(9) Repeat steps (2)-(8) until final solutions are obtained.

4.4 Sample Applications

Both the original and the improved MLBFS have been systematically validated by simulating various multiphase flow problems [Wang *et al.*, 2015a; Wang *et al.*, 2015b; Wang *et al.*, 2015c; Wang *et al.*, 2018a], such as 2D two-layered flows in a channel and concentric cylinders; Rayleigh-Taylor instability; droplet splashing on a thin film; binary collision of identical droplets and droplets impacting on a dry surface with different wettability. The MLBFS has also been extended and improved continuously to enrich its applicability to more challenging multiphase flows and supercritical fluid flows with variable densities [Cao, 2016; Cao and Zhang, 2017]. These progresses of MLBFS have been made within the same framework, and their applications to various problems can be found in the literature [Wang *et al.*, 2016a; Yuan *et al.*,

Table 4.1. Comparison of CPU time on different grids for two-phase co-current flows in a channel at density ratio of 10.

Grid		u_{max} Present	u_{max} Analytical	Relative error	CPU Time (s)
Non-uniform	41×11	0.001452	0.0015	3.2%	128.1
	81×11	0.001480		1.33%	547.9
	121×11	0.001487		0.87%	907.1
Uniform	121×11	0.001469		2.07%	1465.2

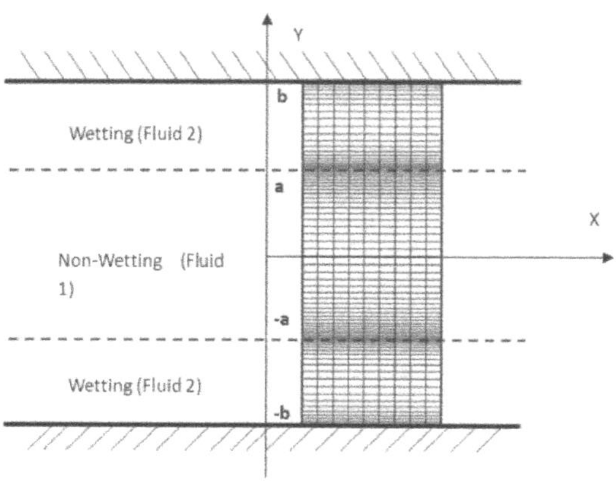

Figure 4.2. Schematic diagram for two-phase co-current flows in a channel.

2017; Wang *et al.*, 2018b]. In this section, we will present several sample applications of MLBFS to demonstrate its reliability and capability in simulating multiphase flows with large density ratios and high Reynolds numbers. The numerical accuracy of MLBFS will be examined quantitatively by comparing the numerical results with theoretical, experimental and numerical data. Note that most of the results shown below have been published in the literature.

4.4.1 *Some two-dimensional applications*

4.4.1.1 *Two-phase co-current flows in a channel*

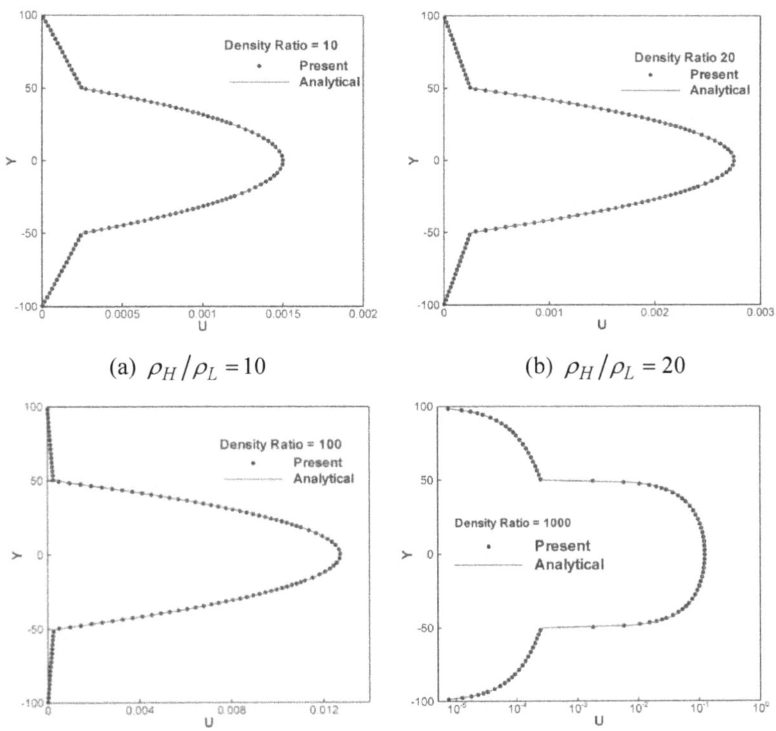

(a) $\rho_H/\rho_L = 10$ (b) $\rho_H/\rho_L = 20$

(c) $\rho_H/\rho_L = 100$ (d) $\rho_H/\rho_L = 1000$

Figure 4.3. Velocity profiles for two-phase co-current flows in a channel with ρ_H/ρ_L =10, 20, 100 and 1000 (d: the velocity is shown in the log scale).

Numerical simulations of the two-phase co-current fluid flow in a straight channel have been carried out extensively by both the MLBFS [Wang *et al.*, 2015a] and other multiphase lattice Boltzmann models [Huang and Lu, 2009; Huang *et al.*, 2013] due to the availability of the analytical solution for comparison. As illustrated in Figure 4.2, the channel is filled with two-layered fluids: the lighter fluid of density ρ_L is placed symmetrically in the inner non-wetting region of the channel and the heavier fluid of density ρ_H is located near the upper and lower wetting walls. The present simulations using the MLBFS have refined grids near the fluid interfaces and the solid walls so that the computational accuracy can be improved. Four different density ratios of 10, 20, 100 and 1000 are used and the external forces are exerted on the non-wetting fluid to drive the flow. Table 4.1 shows the maximum horizontal velocity $u_{\max}$ obtained on both the uniform and non-uniform grids and the CPU time on different grids. It can be seen that the numerical result on the non-uniform grid of 81×11 is more accurate than that on the uniform grid of 121×11 as compared with the analytical solution. And less computational time is consumed on the coarse mesh. The velocity profiles obtained for the four considered density ratios are shown in Figure 4.3, which show good agreement with the analytical solutions. This comparison verifies the accuracy of MLBFS for simulation of multiphase flows.

4.4.1.2 *2D Rayleigh-Taylor instability*

The performance of MLBFS is further examined through its applications to the 2D Rayleigh-Taylor instability of binary fluids driven by the gravity. This problem has been investigated theoretically, numerically and experimentally by many researchers [Daly, 1967; Sharp, 1984; Holdych *et al.*, 1998; Guermond and Quartapelle, 2000; Nourgaliev *et al.*, 2004; Ding *et al.*, 2007b]. The computational domain is taken as a rectangular domain of the size $[-d/2,d/2]\times[-2d,2d]$, which is filled by two layers of fluids: the heavy fluid of density ρ_H on top of the light one of density ρ_L filling the lower part of the domain. Here d is the width of the channel. Initial disturbance of $y(x)=0.1d\cos(2\pi x/d)$ has been imposed on the interface to trigger the instability. No-slip boundary conditions are applied on the top and the bottom walls and periodic conditions are enforced on side boundaries. The reference length and time are respectively taken as d and $T=\sqrt{d/g}$. The Reynolds number

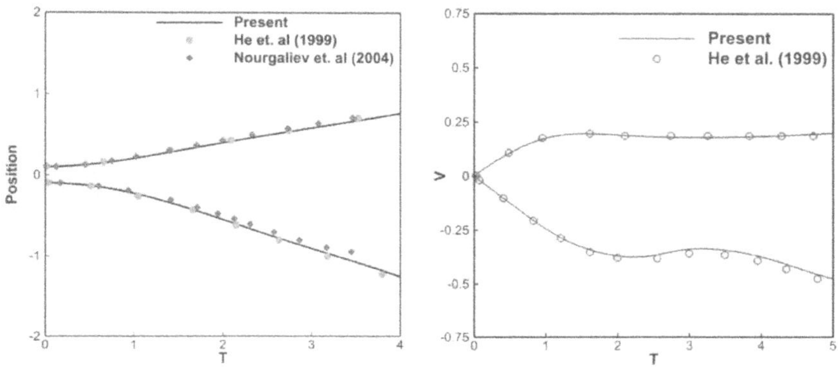

Figure 4.4. Positions (left) and velocities (right) of the rising bubble front and the falling spike tip for Rayleigh-Taylor instability at Re=256.

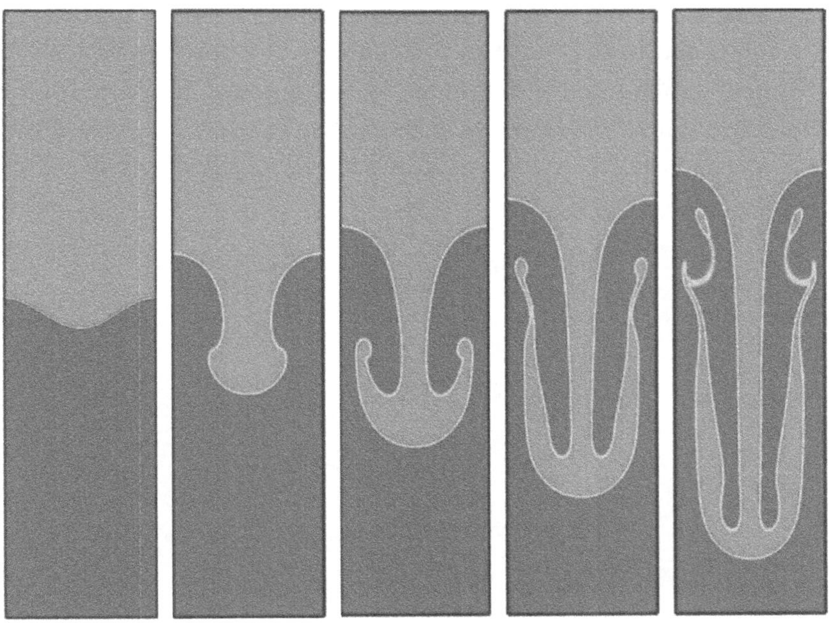

(a) T=0 (b) T=2 (c) T=3 (d) T=4 (e) T=5

Figure 4.5. Fluid interfaces at different times for Rayleigh-Taylor instability at Re=256.

is defined as $Re = \sqrt{gd}d/v$ and the Atwood number is $At = (\rho_H - \rho_L)/(\rho_H + \rho_L)$. Here, the Reynolds number is taken as *Re*=256 and the Atwood number is set as $At = 0.5$.

Figure 4.4 compares the interface positions and velocities of the rising bubble front and the falling spike tip respectively. The numerical results obtained by the mesoscopic LBM [He *et al.*, 1999] and the macroscopic Navier-Stokes solver [Nourgaliev *et al.*, 2004] are also included for comparison. Obviously, good agreements have been achieved. The evolution of the fluid interface is further shown in Figure 4.5. Typical features such as the falling spike, the rising bubble, the rolling-up of the heavy fluid and the secondary vortices due to the Kelvin-Helmholtz instability can be clearly observed and they agree well with previous reports in the literature [Tryggvason, 1988; Guermond and Quartapelle, 2000; Nourgaliev *et al.*, 2004; Ding *et al.*, 2007b].

4.4.1.3 *2D Droplet splashing on a thin film*

Numerical simulation of a single droplet splashing on a thin film is challenging due to the interfacial complexity. In this subsection, the MLBFS is applied in this problem to demonstrate its capability in simulating multiphase flows with large density ratio. Several lattice Boltzmann models [Li *et al.*, 2012; Shao *et al.*, 2014; Ba *et al.*, 2016]

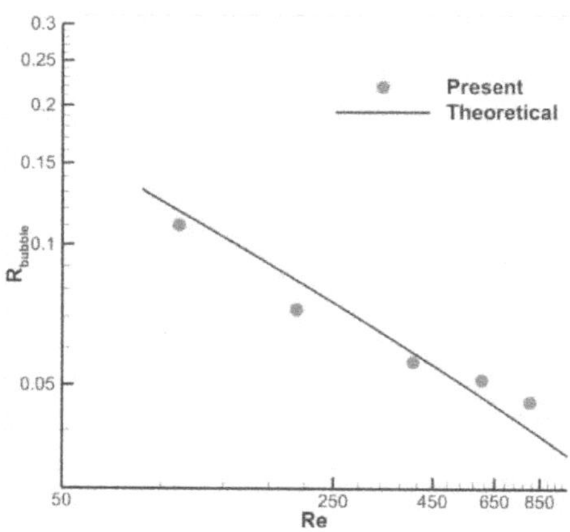

Figure 4.6. Comparison of radius of the entrapped bubble with the theoretical prediction for droplet splashing on a thin film at $100 \le Re \le 800$.

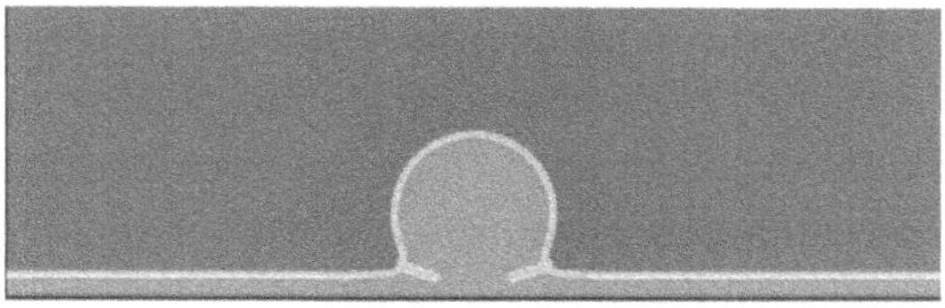

(a) T=0.1

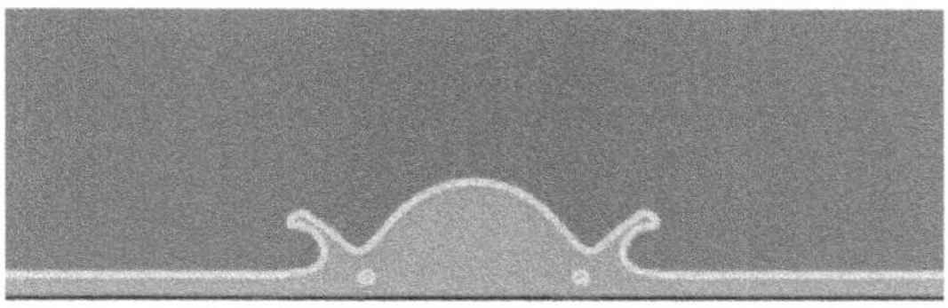

(b) T=0.5

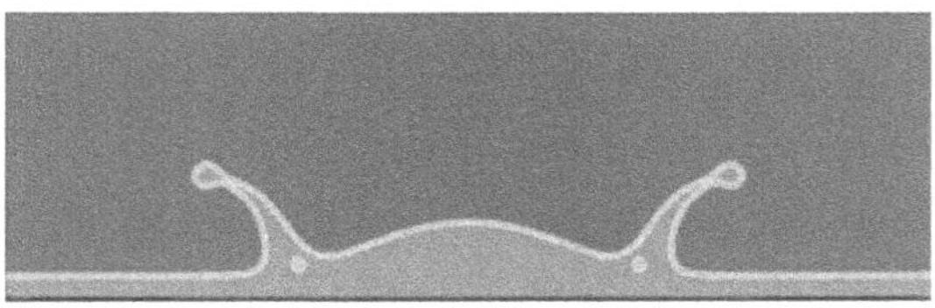

(c) T=1.0

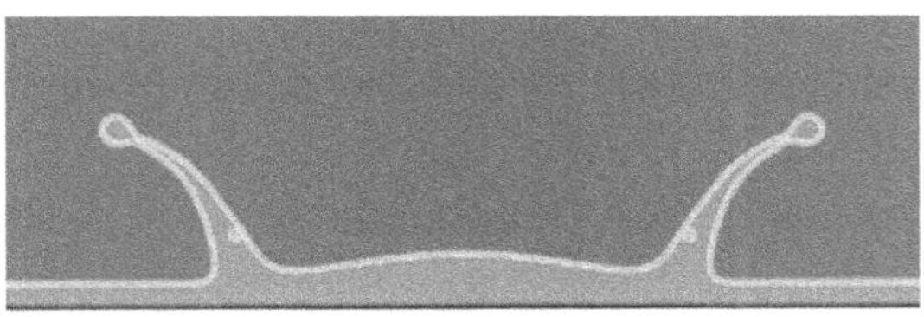

(d) T=1.6

(e) T=2.1

Figure 4.7. Evolution of the fluid interface for droplet splashing on a thin film at Re=800 and We=150.

have been applied to study this flow problem in the 2D case. Due to the sharp change of the density across the phase interface, numerical methods usually introduce considerable diffusion to maintain stability. Therefore, it is usually difficult to capture the phenomena of the tiny entrapped bubble during splashing which has been confirmed by recent experimental observations [Bouwhuis *et al.*, 2012; Tran *et al.*, 2013].

Two non-dimensional parameters, namely, the Reynolds number and the Weber number, are formulated to govern the flow characteristics, which are respectively defined as

$$Re = \frac{2\rho UR}{\mu}, \tag{4.84}$$

and

$$We = \frac{2\rho_H U^2 R}{\sigma}, \tag{4.85}$$

where R is the radius of the droplet. ρ_H is the density of the droplet and the thin film, which is 1000 times the density of the lighter phase. U is the reference velocity and σ is the surface tension. Different Reynolds numbers from Re=100 to 800 are considered. More details about the computational parameters used for simulations can be found in the work of Wang *et al.* [2018a].

From experimental studies [Bouwhuis *et al.*, 2012; Tran *et al.*, 2013], the size of the entrapped bubble varies with the Reynolds number and approximately satisfies the following equation:

$$R_{bubble} \sim a\left(\frac{\mu_L}{\mu_H}\frac{1}{Re}\right)^{4/9} R_{drop}, \tag{4.86}$$

where a is a constant. Numerical results for the cases of Re=100~800 obtained by the MLBFS are shown in Figure 4.6. As can be seen, good agreements with the experimental prediction have been achieved. Figure 4.7 shows the evolution of the fluid interface at Re=800 and We=150. It can be clearly seen that the entrapped air bubble is generated upon the impact of the droplet on the thin film and gradually moves out of the droplet through the upper surface of the liquid crown. More details about the numerical results can be found in the work of Wang *et al.* [2018a], which further shows that the MLBFS is an effective tool for simulating 2D multiphase flows with large density ratios and complex interfacial changes.

4.4.2 *Some three-dimensional applications*

4.4.2.1 *3D Droplet spreading on a flat plate with different wettability*

The performances of the improved MLBFS will be demonstrated by considering a 3D droplet spreading on a flat plate with different wettability. This case is a good numerical example to examine the wetting boundary condition for the Cahn-Hilliard model. Different wettability can be achieved by setting the contact angle of the equilibrium droplet with the flat wall. Numerical simulations have been carried out by using a grid size of $101 \times 101 \times 61$ in the work of Wang *et al.* [2015b]. The density ratio between the droplet and the ambient air is 10 and the thickness of the fluid interface is 3. Initially, the droplet placed on the plate is in the shape of semi-sphere. Figure 4.8 shows the numerical results and the analytical solutions for the droplet in the final equilibrium state. As can be seen, the numerical results obtained by the improved MLBFS agree well with the analytical solution. The

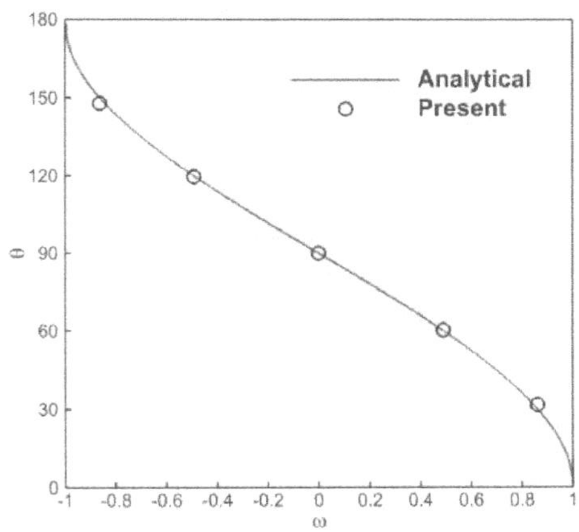

Figure 4.8. Comparison of equilibrium contact angles with the analytical data.

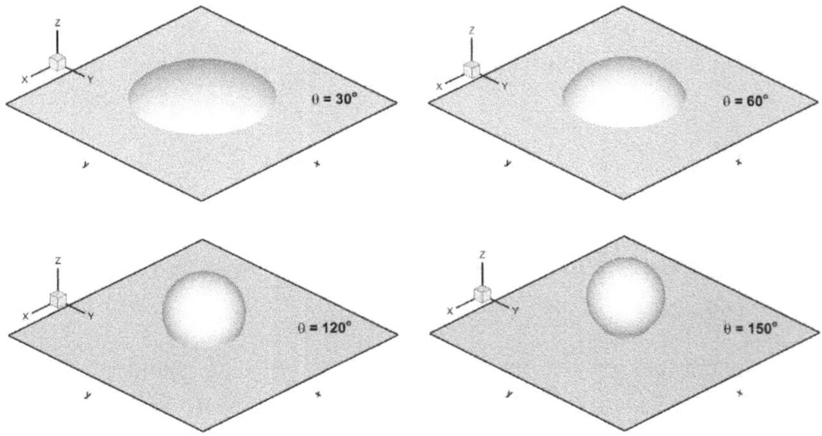

Figure 4.9. Equilibrium states of the droplet obtained by the improved MLBFS.

morphologies of the droplet at four different equilibrium contact angles ranging from hydrophobic to hydrophilic states are shown in Figure 4.9. This numerical example demonstrates the effectiveness of the improved MLBFS for simulation of multiphase flows interacting with solid walls.

4.4.2.2 *A micro-droplet impacting on a dry surface*

The potential of the improved MLBFS for more practical engineering problems will be demonstrated through the example of a micro-droplet impacting on a dry surface, which is involved in the ink jet printing and fuel spray in an engine. The size of the droplet is usually in the order of 10 μm and the typical Reynolds number is 10^3. The non-dimensional parameters defined in Eqs. (4.84) and (4.85), i.e. the Reynolds number *Re* and the Weber number *We*, also characterize this problem. Numerical simulations by the improved MLBFS [Wang *et al.*, 2015b] mimic the settings in previous experiments [Cossali *et al.*, 1997]: the diameter of the droplet is 50.5 μm, the initial speed before impacting on the wall is 12.2 *m/s* and the equilibrium angle between the droplet and wall is 107°, the density ratio between the droplet and air is 841.8 and the viscosity ratio is 51. With these parameters, the Reynolds number and Weber number are given as Re=685 and We=103, respectively. To reduce computational efforts, one-fourth of the droplet is considered and a grid

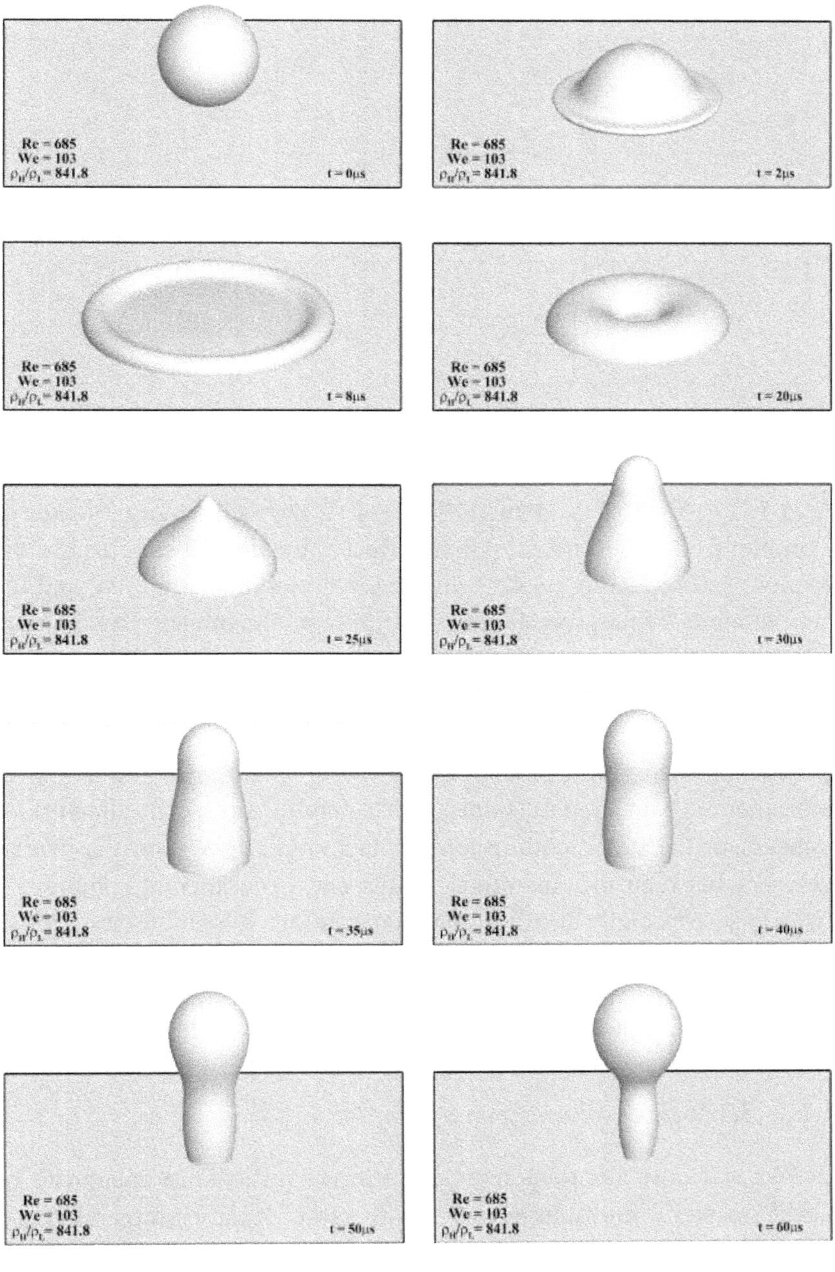

Figure 4.10. Evolution of interface morphologies for 3D micro-droplet impacting on a dry surface with *Re*=685 and *We*=103.

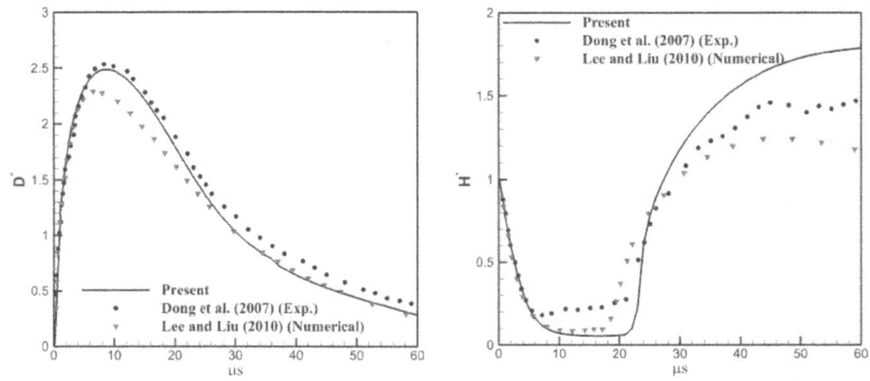

Figure 4.11. Comparison of impact diameter (left) and height (right) for 3D micro-droplet impacting on a dry surface with *Re*=685 and *We*=103.

size of 151×151×151 is adopted. Figure 4.10 shows the morphologies of the droplet from t=0 to 60 μs. As can be seen, the droplet experiences the spreading and retracting process due to the initial impact inertial and the hydrophobicity of the wall. Figure 4.11 presents the impact diameter and the height of the droplet with both experimental and numerical solutions [Cossali *et al.*, 1997; Lee and Liu, 2010]. As can be seen from Figure 4.11, the present results of the impacting diameter agree well with the experimental measurement for t<20 μs. Discrepancies for the height of the droplet arise in the late stage, which may be caused by the surface roughness or the stated equilibrium contact angle. However, the overall agreement between the numerical results and experimental solutions is satisfactory, especially in the initial stage before retraction. With these observations and quantitative comparisons, it can be verified that the improved MLBFS is suitable for practical applications in unsteady multiphase flow problems with large density ratios.

4.4.2.3 *3D Droplet splashing on a thin film*

The last test case for demonstrating the reliability and capability of MLBFS to study multiphase flows with large density ratios and high Reynolds numbers is a real 3D droplet splashing on a thin film [Yarin, 2006], in which the density ratio of the liquid to the ambient air is 1000.

In this case, a spherical droplet of diameter D normally hits a thin film with an impingement velocity U. Complex interfacial interactions and satellite droplets are generated near the rim of the liquid crown due to the Rayleigh-Taylor instability. Due to strong 3D non-linear effects, any small disturbance will grow and induce asymmetric flow features. To recover such asymmetricity, the computational domain should cover the whole physical flow domain. In simulations [Wang *et al.*, 2015b], the

Table 4.2. Flow parameters for 3D droplet splashing on a thin film.

Cases	Micro droplet		Re	We	ρ_H/ρ_L	μ_H/μ_L
	Speed U (m/s)	Size D (μm)				
Case 1	8.930	50.00	500	55.40	1000.0	100.0
Case 2	17.86	50.00	1000	221.5		
Case 3	32.25	55.37	2000	800.0		
Case 4	18.14	147.7	3000	675.0		

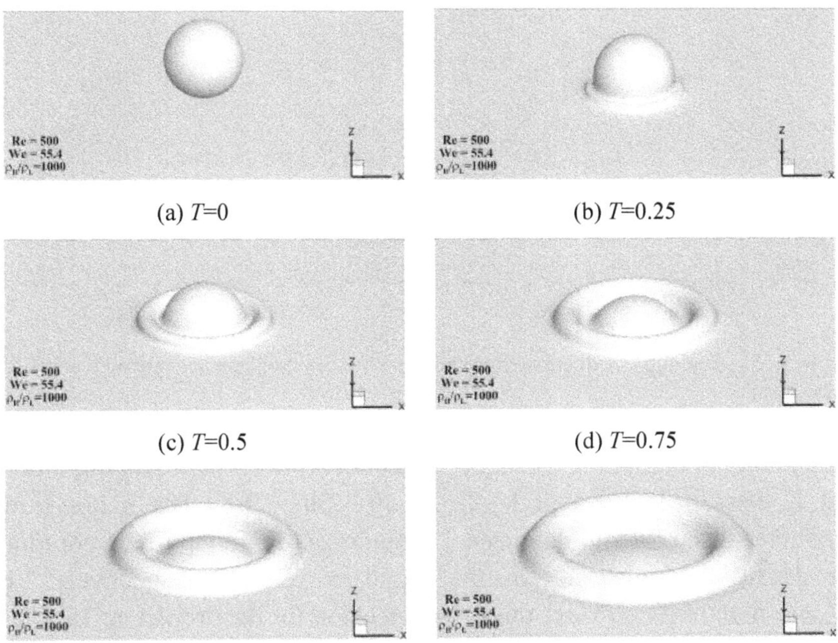

Figure 4.12. Evolution of interface morphologies for 3D droplet splashing on a thin film at Re=500.

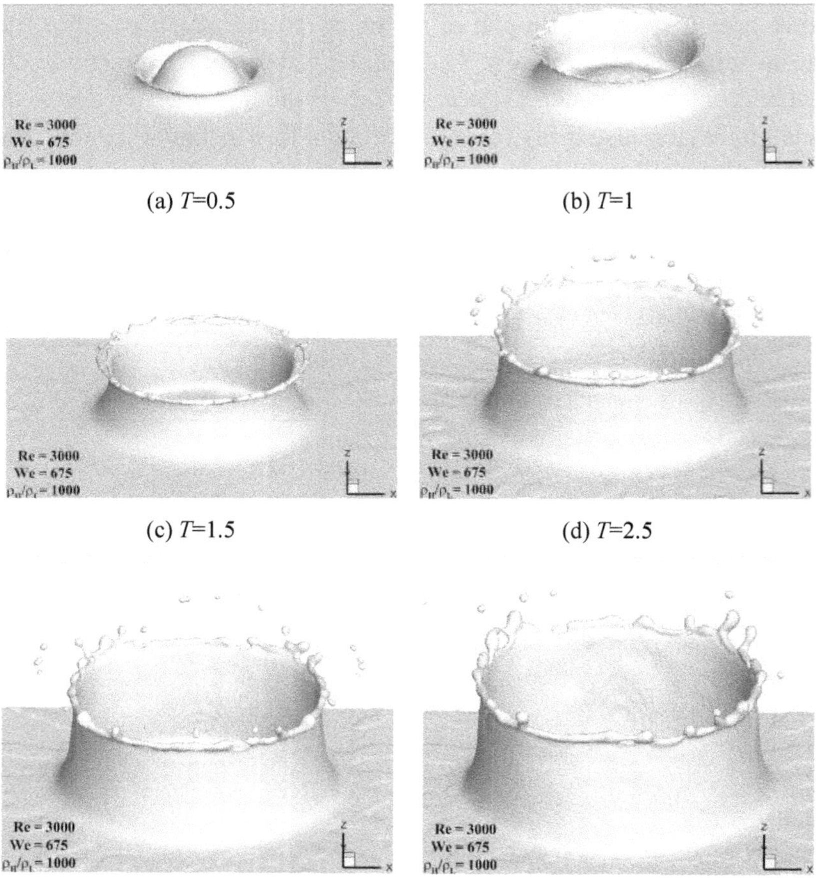

(a) T=0.5 (b) T=1

(c) T=1.5 (d) T=2.5

(e) T=3 (f) T=4

Figure 4.13. Evolution of interface morphologies for 3D droplet splashing on a thin film at Re=3000.

computational domain is taken as [-2.5D, 2.5D]×[-2.5D, 2.5D]×[0, 2.6D] and is discretized by a grid size of $501\times501\times261$. The major flow parameters used for simulations are summarized in Table 4.2. In addition, the density and viscosity of the liquid droplet are respectively 1.0 g/cm^3 and 0.893 $mPa\cdot s$, the surface tension for the droplet at $25^\circ C$ is 0.072 N/m. Following the suggestions of Rieber and Frohn [1999], a random disturbance with Gaussian distribution is introduced into the initial flow field for the droplet and the film to mimic the disturbance in real physics.

The evolutions of fluid interfaces at *Re*=500 and 3000 are shown in Figures. 4.12 and 4.13, respectively. At low Reynolds number (e.g., *Re*=500), the evolution reflects a deposition process, in which the surface wave formed by the interactions between the droplet and the thin film moves outwards from the impact center. At higher Reynolds number (e.g., *Re*=3000), a high-speed splashing liquid sheet ejects from the impact neck and grows with time. Due to the surface tension and the viscous forces, this liquid sheet could be unstable and finally breaks up into minor droplets due to the Rayleigh-Taylor instability. Figure 4.14 further shows the radial growth of the splashing lamella in cases with *Re*=1000 to 3000. The theoretical prediction by Yarin [2006] for evolution of the radius states that:

$$\frac{r}{D} = (6H)^{0.25}\sqrt{T}, \tag{4.87}$$

where H is the height of the thin film normalized by the diameter of the droplet. Figure 4.14 shows that the present numerical results agree well with the theoretical prediction. All these evidences demonstrate that the improved MLBFS is an effective tool for simulation of multiphase flows with large density ratios and complex interfacial interactions.

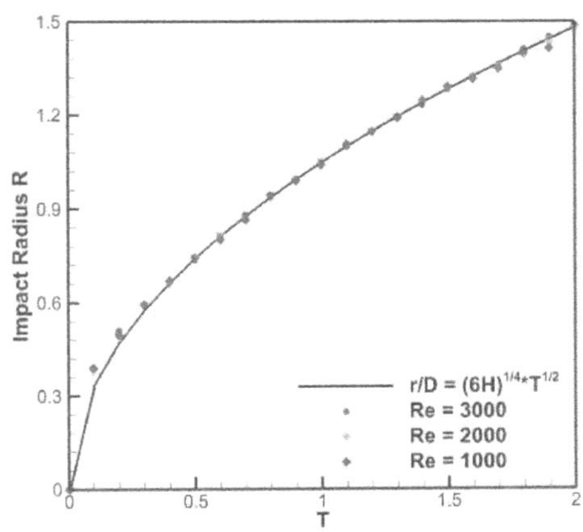

Figure 4.14. Evolution of impact radius for 3D droplet splashing on a thin film at different Reynolds numbers.

4.5 Summary

In this chapter, two versions of MLBFS, i.e., the original MLBFS based on the standard LBM and the improved MLBFS with a modified equilibrium distribution function, are presented for simulation of multiphase flows with large density ratios. Both solvers are proposed within the framework of FVM and update the macroscopic flow variables at cell centers. The numerical fluxes at the cell interface are evaluated through the local reconstruction of the LBE solution. Due to the inconsistency of fluxes reproduced by the standard LBM without including external forcing terms, the original MLBFS incorporates additional terms for compensation purposes. These terms involve space gradients of the density and the velocity, which could deteriorate the numerical stability and certainly complicates the implementation. Efforts are then made in the improved version of MLBFS to fix these flaws. Specifically, the improved MLBFS utilizes a modified equilibrium distribution function to evaluate the viscous fluxes. The modification is reflected by using the density at the midpoint of the cell interface to evaluate the equilibrium distribution function at the streaming nodes around the interface. With this improvement, the error tensor vanishes under the incompressible limit, and the overall implementation procedure of the improved MLBFS is essentially the same as that of LBFS presented in **Chapter 3** for simulating single-phase flow problems.

Various numerical examples, including both 2D and 3D multiphase flows with high density ratios and complex interfacial interactions, are presented to demonstrate the reliability of the original and the improved MLBFS. Good agreements between the results of MLBFS and the reference data in the literature are achieved, which validates the accuracy and stability of the proposed solvers. It has been shown that the MLBFS has great potential for practical applications in multiphase flow problems.

Chapter 5

Maxwellian Function-Based Gas Kinetic Flux Solver

As indicated in **Chapter 2**, there are two important issues when recovering the Euler/Navier-Stokes equations from the Boltzmann equation. One is the truncation of the distribution function in terms of the collision time scale. The other is the correct recovery of the fluid density, momentum, energy and their fluxes from the moment relationships of the equilibrium state. In solvers that are based on the discrete velocity Boltzmann equation (DVBE) or the lattice Boltzmann equation (LBE) [Mei and Shyy, 1998; Dixit and Babu, 2006; Gan *et al.*, 2008], the Maxwellian distribution function is replaced by a lattice Boltzmann model for the purpose of simplification, while the Boltzmann equation is still used in evolving the flow variables. The direct resolution of the Boltzmann equation brings a huge number of discrete governing equations, whose size is usually larger than that of the Euler/Navier-Stokes equations. To alleviate this tediousness, truncation can be made to the distribution function in terms of the collision time scale. The LBFS shown in the previous two chapters is one exploration in this direction. However, the previous LBFS is established under the incompressible limit and is thus restricted to incompressible flows. Extra efforts should be made to extend such a strategy to compressible flows.

In this chapter, the Maxwellian function-based gas kinetic scheme (M-GKS) [Xu, 2002; Tang *et al.*, 2004; Jiang and Qian, 2012; Li *et al.*, 2015b; Li *et al.*, 2017; Pan *et al.*, 2018] and Maxwellian function-based gas kinetic flux solver (M-GKFS) [Sun *et al.*, 2015; Sun *et al.*, 2016; Sun *et al.*, 2017] will be introduced. The principle of these two methods is similar to that of LBFS. The difference is that the numerical fluxes on the cell interface are reconstructed by the local solution to the Boltzmann equation instead of the LBE. In the flux reconstruction process, the Maxwellian distribution function is taken as the equilibrium state. Since the Maxwellian distribution function is applicable to both incompressible and compressible flows, these two solvers expect applicability in a wide span of flow speed.

5.1 BGK-Boltzmann Equation, Navier-Stokes Equations and Their Connection

In order to apply the local solution to the Boltzmann equation to reconstruct the numerical fluxes of the Navier-Stokes equations, the connection between these two systems of governing equations should be established. The Boltzmann equation with the Bhatnagar-Gross-Krook (BGK) collision model can be written as [Bhatnagar *et al.*, 1954]

$$\frac{\partial f}{\partial t} + \boldsymbol{\xi} \cdot \nabla f = \frac{g - f}{\tau}, \tag{5.1}$$

where f is the gas distribution function that varies with the physical space $\mathbf{x} = (x_1, x_2, ..., x_D)$, particle velocity space $(\boldsymbol{\xi}, \boldsymbol{\zeta})$, and time t. $\boldsymbol{\xi} = (\xi_1, \xi_2, ..., \xi_D)$ is the translational particle velocity in D-dimensional space and $\boldsymbol{\zeta} = (\zeta_1, \zeta_2, ..., \zeta_K)$ represents the rest components of the translational particle velocity (ξ_1, ξ_2, ξ_3) in 3-dimensional space and the rotational particle velocity, i.e., $K = 3 - D + N$, where N denotes the number of rotational degrees of freedom. g is the equilibrium state approached by f through particle collisions within a collision time scale τ. For the BGK collision model, the equilibrium state can be expressed by the Maxwellian distribution function

$$g = \rho \left(\frac{\lambda}{\pi} \right)^{\frac{D+K}{2}} e^{-\lambda \left[\sum_{i=1}^{D} (\xi_i - u_i)^2 + \sum_{j=1}^{K} \zeta_j^2 \right]}. \tag{5.2}$$

Here, ρ is the density of the mean flow; u_i is the component of the macroscopic flow velocity in the ith direction of a D-dimensional space; $\lambda = m/(2kT) = 1/(2RT)$, where m, k, R and T are the molecular mass, the Boltzmann constant, the gas constant and the temperature, respectively. The specific heat ratio γ can be computed by

$$\gamma = \frac{b+2}{b} = \frac{K+D+2}{K+D}, \tag{5.3}$$

where b is the total number of degrees of freedom of molecules. For monatomic gas and diatomic gas, b takes the value of 3 and 5, respectively.

At the macroscopic level, the fluid system conserves the mass, momentum and energy. Adopting these conservation laws on the control volume gives the following continuity, momentum and energy equations in differential forms:

$$\frac{\partial \rho}{\partial t} + \nabla \cdot (\rho \mathbf{u}) = 0, \tag{5.4}$$

$$\frac{\partial \rho \mathbf{u}}{\partial t} + \nabla \cdot (\rho \mathbf{u}\mathbf{u} + p\mathbf{I}) = \nabla \cdot \mathbf{\Pi}, \tag{5.5}$$

$$\frac{\partial \rho E}{\partial t} + \nabla \cdot \left[(\rho E + p)\mathbf{u} \right] = \nabla \cdot (\kappa \nabla T) + \nabla \cdot (\mathbf{u} \cdot \mathbf{\Pi}), \tag{5.6}$$

where $\mathbf{\Pi}$ denotes the stress tensor given by

$$\mathbf{\Pi} = \mu \left[\left(\nabla \mathbf{u} + (\nabla \mathbf{u})^T \right) - \frac{2}{3} (\nabla \cdot \mathbf{u}) \mathbf{I} \right] + \mu_b (\nabla \cdot \mathbf{u}) \mathbf{I}. \tag{5.7}$$

Here, $\mathbf{u}$, p and E are respectively the velocity vector, the pressure and the total energy of the mean flow. And the total energy of the ideal gas can be expressed as $E = p/[(\gamma-1)\rho] + \mathbf{u} \cdot \mathbf{u}/2$. μ, μ_b and κ are the dynamic viscosity, the bulk viscosity and the thermal conductivity, respectively. $\mathbf{I}$ is the unit tensor. $\mathbf{u} = (u, v)$ is for the two-dimensional (2D) case and $\mathbf{u} = (u, v, w)$ is for the three-dimensional (3D) case, where u, v and w are the velocity components in the x-, y- and z- directions, respectively. For simplicity, Eqs. (5.4)-(5.6) can be rewritten into the following unified form:

$$\frac{\partial \mathbf{W}}{\partial t} + \nabla \cdot \mathbf{F} = 0, \tag{5.8}$$

with

$$\mathbf{W} = \begin{bmatrix} \rho \\ \rho \mathbf{u} \\ \rho E \end{bmatrix}, \tag{5.9}$$

$$\mathbf{F} = \begin{bmatrix} \rho \mathbf{u} \\ \rho \mathbf{uu} + p\mathbf{I} - \mathbf{\Pi} \\ (\rho E + p)\mathbf{u} - \kappa \nabla T - \mathbf{u} \cdot \mathbf{\Pi} \end{bmatrix}. \tag{5.10}$$

According to the Chapman-Enskog expansion analysis shown in **Chapter 2**, the following relationships between the Boltzmann equation and the Navier-Stokes equations can be established:

$$\mathbf{W} = \langle \boldsymbol{\varphi}_a f \rangle = \langle \boldsymbol{\varphi}_a g \rangle, \tag{5.11}$$

$$\mathbf{F} = \langle \boldsymbol{\xi} \boldsymbol{\varphi}_a f \rangle, \tag{5.12}$$

where $\boldsymbol{\varphi}_a$ stands for the moment vector as follows:

$$\boldsymbol{\varphi}_\alpha = \left(1, \boldsymbol{\xi}, \frac{1}{2} \left(\sum_{i=1}^{D} \xi_i^2 + \sum_{j=1}^{K} \zeta_j^2 \right) \right)^T. \tag{5.13}$$

The symbol $\langle \cdot \rangle$ denotes the integral over the particle velocity space, i.e.,

$$\langle \phi \rangle = \int_{-\infty}^{+\infty} \phi d\Xi. \tag{5.14}$$

Here, ϕ is an arbitrary function and $d\Xi$ is the volume element in the particle velocity space. Note that f in Eqs. (5.11) and (5.12) is the distribution function truncated to the Navier-Stokes level, namely the first order of collision time scale. In addition, the dynamic viscosity and thermal conductivity are associated with the collision time scale through

$$\mu = \tau p, \tag{5.15}$$

$$\kappa = \tau p\left(c_V + R\right) = \mu c_p, \tag{5.16}$$

where $c_V = R/(\gamma - 1)$ and $c_p = \gamma R/(\gamma - 1)$ are specific heat capacities at constant volume and at constant pressure, respectively.

With the established correlation between the Boltzmann equation and the Navier-Stokes equations, the local solution to the Boltzmann equation can be used to reconstruct the numerical fluxes of the Navier-Stokes equations. With the finite volume method (FVM), the discrete form of Eq. (5.8) can be written as

$$\frac{\partial \mathbf{W}_i}{\partial t} = -\frac{1}{\Omega_i}\sum_{j=1}^{N_f} \mathbf{F}_{nj} S_j, \tag{5.17}$$

where i is the index of the control volume, Ω_i and N_f represent the volume and the number of surfaces of the control volume i, respectively. S_j denotes the area of the jth interface of the control volume. In the 2D case, the vector of conservative variables $\mathbf{W}$ and the vector of fluxes $\mathbf{F}_n$ are given by

$$\mathbf{W} = \langle \boldsymbol{\varphi}_a f \rangle = \left(\rho, \rho u, \rho v, \rho E\right)^T, \tag{5.18}$$

$$\mathbf{F}_n = \langle \mathbf{n} \cdot \boldsymbol{\xi} \boldsymbol{\varphi}_a f \rangle = \left(F_\rho, F_{\rho u}, F_{\rho v}, F_{\rho E}\right)^T, \tag{5.19}$$

where the moment vector $\boldsymbol{\varphi}_a$ in the 2D scenario in the global Cartesian coordinate system can be expressed as

$$\boldsymbol{\varphi}_\alpha = \left(1, \xi_x, \xi_y, \frac{1}{2}\left(\sum_{i=1}^{D} \xi_i^2 + \sum_{j=1}^{K} \zeta_j^2\right)\right)^T. \tag{5.20}$$

Here, ξ_x and ξ_y are the components of $\boldsymbol{\xi}$ in the x- and the y- directions, respectively. $\mathbf{n} = \left(n_x, n_y\right)$ denotes the unit normal vector of the cell interface in the global Cartesian coordinate system.

By resolving Eq. (5.17), the conservative variable vector $\mathbf{W}$ defined at the cell center can be updated. The key issue in this resolution process is to evaluate the flux vector $\mathbf{F}_n$ on the cell interface from the conservative variables at cell centers. For the convenience of derivation and application, the local coordinate system defined at the midpoint of the cell interface is introduced. In the 2D case, x_1-axis points to the normal direction and x_2-axis is aligned with the tangential direction of the cell interface. Under the local coordinate system, the conservative variable vector and the flux vector can be respectively expressed as

$$\overline{\mathbf{W}} = \langle \overline{\boldsymbol{\varphi}}_a f \rangle = (\rho, \rho u_1, \rho u_2, \rho E)^T, \tag{5.21}$$

$$\overline{\mathbf{F}} = \langle \mathbf{n} \cdot \boldsymbol{\xi} \overline{\boldsymbol{\varphi}}_a f \rangle = \left(F_\rho, F_{\rho u_1}, F_{\rho u_2}, F_{\rho E}\right)^T, \tag{5.22}$$

where $\overline{\boldsymbol{\varphi}}_a$ is the moment vector in the local coordinate system, i.e.,

$$\overline{\boldsymbol{\varphi}}_a = \left(1, \xi_1, \xi_2, \frac{1}{2}\left(\sum_{i=1}^{D} \xi_i^2 + \sum_{j=1}^{K} \varsigma_j^2\right)\right)^T. \tag{5.23}$$

Here, u_1 and u_2 are the velocity components of the mean flow expressed in the local coordinate system; ξ_1 and ξ_2 are the components of the particle velocity vector expressed in the local coordinate system. A coordinate transformation can be performed to correlate the local coordinate system with the global Cartesian coordinate system, which gives

$$\begin{cases} u = n_x u_1 - n_y u_2, \\ v = n_x u_2 + n_y u_1, \end{cases} \tag{5.24}$$

$$\begin{cases} \xi_x = n_x \xi_1 - n_y \xi_2, \\ \xi_y = n_x \xi_2 + n_y \xi_1, \end{cases} \tag{5.25}$$

$$\begin{cases} F_{\rho u} = n_x F_{\rho u_1} - n_y F_{\rho u_2}, \\ F_{\rho v} = n_x F_{\rho u_2} + n_y F_{\rho u_1}. \end{cases} \tag{5.26}$$

Moreover, the normal components of the mean flow velocity and the particle velocity in the local coordinate can be calculated by

$u_1 = n_x u + n_y v$ and $\xi_1 = n_x \xi_x + n_y \xi_y$, respectively. Eq. (5.26) correlates the flux vector $\mathbf{F}_n$ in the global Cartesian coordinate system and the flux vector $\overline{\mathbf{F}}$ in the local coordinate system, which will be utilized in constructing different flux solvers in the following sections.

5.2 Maxwellian Function-Based GKS

The method, which uses the local solution to the Boltzmann equation with Maxwellian distribution function to reconstruct the numerical fluxes of the Euler/Navier-Stokes equations, is usually called GKS in most of the literature [Xu *et al.*, 2005; Li *et al.*, 2011; Pan *et al.*, 2016a; Li *et al.*, 2019b]. To differ this method from other GKS with different equilibrium distribution functions, we named it as M-GKS in this book. There are two common types of M-GKS: the kinetic flux vector scheme (KFVS) and the gas kinetic BGK scheme. The KFVS method, which is also referred as equilibrium-flux method (EFM) Representative KFVS schemes have been proposed by Pullin [1980], Mandal and Deshpande [1994] and Chou and Baganoff [1997]. These schemes generally involve two resolving steps: free transport and collision. In the free transport step, the collisionless Boltzmann equation is solved to evaluate the fluxes of macroscopic governing equations. In the collision step, the collision process is modelled through a numerical time interval. It has been demonstrated that the KFVS scheme holds good positivity property in simulating flows with strong shock waves and expansion waves. Nevertheless, it has been noted that the KFVS scheme is usually less accurate than the Riemann solvers [Roe, 1981; Toro *et al.*, 1994]. As reported by Xu [1998], the main reason may be that the numerical dissipation of KFVS scheme is proportional to the mesh spacing. Consequently, the KFVS scheme could only give accurate solutions in cases where the physical viscosity is sufficiently larger than the numerical viscosity.

To overcome the drawbacks of KFVS, the gas kinetic BGK scheme, which utilizes the BGK collision model, was proposed by Prendergast and Xu [1993] and then nurtured by other researchers [Xu *et al.*, 1995; Chae *et al.*, 2000; May *et al.*, 2007; Pan *et al.*, 2016b; Li *et al.*, 2019a].

In the gas kinetic BGK scheme, the particle collision is scaled by the real particle collision time which is a function of the dynamic viscosity and pressure. In other words, instead of using a numerical time step to measure the particle collision process in the KFVS scheme, the gas kinetic BGK scheme aims to recover the real physical collision. It was demonstrated by Xu [2001] that the entropy condition is always satisfied by the gas kinetic BGK scheme since the gas evolution process is always towards the equilibrium state. In the smooth regions, the gas kinetic BGK scheme gives accurate solutions to the Navier-Stokes equations. In the discontinuous regions, the delicate dissipative mechanism in the gas kinetic BGK scheme generates stable and crisp shock transition. Moreover, it has been demonstrated that the unphysical oscillations around the shock wave such as the "carbuncle phenomenon" vanish in the gas kinetic BGK scheme for simulation of hypersonic flows [Xu *et al.*, 2005; Xu *et al.*, 2010]. Owing to these distinctive features, the gas kinetic BGK scheme has been successfully applied to various flow problems, including low-speed flows [Su *et al.*, 1999; Xu and He, 2003; Guo *et al.*, 2008], complicated 3D flows [Ruan and Jameson, 2002; Tian *et al.*, 2007; Li and Fu, 2011, Pan and Xu, 2015] and multi-material flows [Jiang and Ni, 2004; Li *et al.*, 2005a; Jiang and Ni, 2007; Ni *et al.*, 2012; Pan *et al.*, 2012]. In this section, the gas kinetic BGK scheme will be introduced in detail.

5.2.1 *Local integral solution to Boltzmann equation on the cell interface*

In the gas kinetic BGK scheme, the local integral solution to the Boltzmann equation is used to reconstruct the macroscopic numerical fluxes on the cell interface. Suppose that the cell interface is located at $\mathbf{x}=\mathbf{0}$, the integral solution to the Boltzmann equation with the BGK collision model (5.1) yields

$$
\begin{aligned}
f(\mathbf{0},\xi,t) &= \frac{1}{\tau}\int_0^t g\left(-\xi(t-t'),\xi,t'\right)e^{-(t-t')/\tau}dt' \\
&\quad + e^{-t/\tau} f_0\left(-\xi t,\xi,0\right),
\end{aligned}
\tag{5.27}
$$

where $f_0(-\xi t,\xi,0)$ is the initial distribution function at the streaming nodes around the cell interface. In the above equation, free transport and collision are coupled in space and time. In order to obtain the solution f, the functions g and f_0 have to be calculated.

According to the Chapman-Enskog expansion analysis shown in **Chapter 2**, when maintaining consistency with the Navier-Stokes equations, the initial distribution function around the cell interface can be expanded to the following form:

$$f_0(\mathbf{x},\xi,0)=g(\mathbf{0},\xi,0)+\mathbf{x}\cdot\nabla g(\mathbf{0},\xi,0)-\tau Dg(\mathbf{0},\xi,0). \tag{5.28}$$

In general cases, the flow variables and their derivatives could be discontinuous across the cell interface. Thus, the left and the right states are separately considered when specifying the initial distribution function [Xu *et al.*, 2005]

$$f_0(\mathbf{x},\xi,0)=\begin{cases} g^L\left[1+a^L x_1+b^L x_2-\tau\left(a^L\xi_1+b^L\xi_2+A^L\right)\right], & x_1\le 0,\\ g^R\left[1+a^R x_1+b^R x_2-\tau\left(a^R\xi_1+b^R\xi_2+A^R\right)\right], & x_1>0. \end{cases} \tag{5.29}$$

where x_1 and x_2 are the coordinate components expressed in the local coordinate system. g^L and g^R are the equilibrium state on the left and the right sides of the cell interface, respectively. They can be calculated straightforwardly from the conservative variables at the corresponding physical locations by Eq. (5.2). $g^L a^L$ and $g^L b^L$ are related to the spatial derivatives of g^L on the center of the left cell and expressed in the normal direction and the tangential direction of the cell interface, respectively; and $g^L A^L$ is connected to the temporal derivative of g^L. Similarly, $g^R a^R$, $g^R b^R$ and $g^R A^R$ are related to either the spatial or the temporal derivatives of g^R. In Eq. (5.29), the terms with τ represent the non-equilibrium parts of the distribution function, which have no contribution to the calculation of conservative flow variables, i.e.,

$$\left\langle\left(a^L\xi_1+b^L\xi_2+A^L\right)\bar{\boldsymbol{\varphi}}_a g^L\right\rangle=\mathbf{0}, \tag{5.30}$$

$$\left\langle\left(a^R\xi_1+b^R\xi_2+A^R\right)\bar{\boldsymbol{\varphi}}_a g^R\right\rangle=\mathbf{0}. \tag{5.31}$$

Eqs. (5.30) and (5.31) suggest that A^L and A^R can be accurately evaluated when a^L, b^L, g^L, a^R, b^R and g^R are determined.

In addition, the equilibrium distribution function at time t around the cell interface can be obtained from the expansion of a reference equilibrium state

$$g(\mathbf{x},\xi,t) = g_0\left[1+(1-H[x_1])\overline{a}^L x_1 + H[x_1]\overline{a}^R x_1 + \overline{b}x_2 + \overline{A}t\right], \quad (5.32)$$

where $H[x]$ is the Heaviside function defined as

$$H[x] = \begin{cases} 0, & x \le 0, \\ 1, & x > 0. \end{cases} \quad (5.33)$$

Here $g_0 = g(\mathbf{0},\xi,0)$ is a local Maxwellian distribution function located at the cell interface, which can be determined by the corresponding conservative variables, $\overline{\mathbf{W}}^{face}$. $g_0\overline{a}^L$ and $g_0\overline{a}^R$ are related to the spatial derivatives of g_0 in the normal direction on the left and the right sides of the cell interface, respectively. $g_0\overline{b}$ is the spatial derivative of g_0 in the tangential direction of the cell interface and $g_0\overline{A}$ is the temporal derivative of g_0.

The dependence of a^L, b^L, ..., $\overline{A}$ on the particle velocities can be obtained from the Taylor series expansion of a Maxwellian distribution function, which gives

$$\begin{aligned}
a^L &= a_0^L + a_1^L\xi_1 + a_2^L\xi_2 + a_3^L\frac{1}{2}\left(\sum_{i=1}^{D}\xi_i^2 + \sum_{j=1}^{K}\varsigma_j^2\right) = a_\alpha^L\overline{\boldsymbol{\varphi}}_a, \\
b^L &= b_0^L + b_1^L\xi_1 + b_2^L\xi_2 + b_3^L\frac{1}{2}\left(\sum_{i=1}^{D}\xi_i^2 + \sum_{j=1}^{K}\varsigma_j^2\right) = b_\alpha^L\overline{\boldsymbol{\varphi}}_a, \\
&\cdots \\
\overline{A} &= \overline{A}_0 + \overline{A}_1\xi_1 + \overline{A}_2\xi_2 + \overline{A}_3\frac{1}{2}\left(\sum_{i=1}^{D}\xi_i^2 + \sum_{j=1}^{K}\varsigma_j^2\right) = \overline{A}_\alpha\overline{\boldsymbol{\varphi}}_a,
\end{aligned} \quad (5.34)$$

where a_0^L, a_1^L, ..., $\overline{A}_3$ are local constants to be determined by the conservative flow variables and their first-order derivatives at cell centers. With the initial reconstruction, the macroscopic status on the left and the right sides of the interface can be obtained by the following interpolations:

$$\begin{aligned}\mathbf{W}^L &= \mathbf{W}_c^L + \Psi_c^L\left(\mathbf{x}_b - \mathbf{x}_c^L\right)\cdot\nabla\mathbf{W}_c^L,\\ \mathbf{W}^R &= \mathbf{W}_c^R + \Psi_c^R\left(\mathbf{x}_b - \mathbf{x}_c^R\right)\cdot\nabla\mathbf{W}_c^R.\end{aligned} \tag{5.35}$$

Here, $\mathbf{W}_c^L$ and $\mathbf{W}_c^R$ are the conservative flow variables at centers of the left and the right cells, respectively; $\nabla\mathbf{W}_c^L$ and $\nabla\mathbf{W}_c^R$ are their corresponding first-order derivatives. $\mathbf{x}_c^L$, $\mathbf{x}_c^R$ and $\mathbf{x}_b$ are the coordinates of the left cell center, the right cell center and the midpoint of the cell interface, respectively. Ψ_c^L and Ψ_c^R are the limiter functions utilized in the left and the right cells, respectively. By using the relation between the gas distribution function and the macroscopic variables, we can get

$$\left\langle\overline{\varphi}_a g^L\right\rangle = \overline{\mathbf{W}}^L,\ \left\langle\overline{\varphi}_a g^R\right\rangle = \overline{\mathbf{W}}^R, \tag{5.36}$$

$$\left\langle\overline{\varphi}_a g^L a^L\right\rangle = \frac{\partial\overline{\mathbf{W}}_c^L}{\partial x_1},\ \left\langle\overline{\varphi}_a g^L b^L\right\rangle = \frac{\partial\overline{\mathbf{W}}_c^L}{\partial x_2}, \tag{5.37a}$$

$$\left\langle\overline{\varphi}_a g^R a^R\right\rangle = \frac{\partial\overline{\mathbf{W}}_c^R}{\partial x_1},\ \left\langle\overline{\varphi}_a g^R b^R\right\rangle = \frac{\partial\overline{\mathbf{W}}_c^R}{\partial x_2}. \tag{5.37b}$$

Note that, the conservative flow variables and their derivatives in Eqs. (5.36) and (5.37a)-(5.37b) are expressed in the local coordinate system, which can be obtained by the coordinate transformation using the values in the global coordinate system. With the recurrence relationships shown in **Appendix A**, a^L, b^L, a^R and b^R can be uniquely determined from Eqs. (5.37a) and (5.37b). Detailed expressions of these coefficients can be found in **Appendix B**. With these coefficients, Eqs. (5.30) and (5.31) can be reformulated into the following form to calculate A^L and A^R:

$$\left\langle\overline{\varphi}_a g^L A^L\right\rangle = -\left\langle\left(a^L\xi_1 + b^L\xi_2\right)\overline{\varphi}_a g^L\right\rangle, \tag{5.38}$$

$$\left\langle \bar{\varphi}_a g^R A^R \right\rangle = -\left\langle \left(a^R \xi_1 + b^R \xi_2 \right) \bar{\varphi}_a g^R \right\rangle . \tag{5.39}$$

Since a^L, b^L, a^R and b^R have been determined, the terms on the right-hand side of Eqs. (5.38) and (5.39) can be obtained easily by the recurrence relationships shown in **Appendix A**. After that, A^L and A^R can be calculated in the same way as a^L shown in **Appendix B**.

To calculate g_0, the coefficients $\bar{a}^L$, $\bar{a}^R$ and $\bar{b}$, as well as the conservative variables on the cell interface have to be determined in advance. Employing the limit $t \to 0$ to Eq. (5.27) and substituting its solution into Eq. (5.11) and Eq. (5.21), we have

$$\begin{aligned} \overline{\mathbf{W}}^{face} &= \langle \bar{\varphi}_a g_0 \rangle \\ &= \langle \bar{\varphi}_a f(\mathbf{0}, \xi, 0) \rangle \\ &= \left\langle \bar{\varphi}_a g^L \right\rangle_{>0} + \left\langle \bar{\varphi}_a g^R \right\rangle_{<0}, \end{aligned} \tag{5.40}$$

where the symbols $\langle \cdot \rangle_{>0}$ and $\langle \cdot \rangle_{<0}$ respectively denote the integrals over the positive and the negative half domains of the ξ_1 particle velocity space and over the whole velocity spaces of remaining particle velocities, which are given by

$$\langle \phi \rangle_{>0} = \int_{\xi_1 > 0} \int_{-\infty}^{+\infty} \phi d\Xi , \tag{5.41}$$

$$\langle \phi \rangle_{<0} = \int_{\xi_1 < 0} \int_{-\infty}^{+\infty} \phi d\Xi . \tag{5.42}$$

Since g^L and g^R have been determined earlier, the moments on the right-hand side of Eq. (5.40) can be evaluated explicitly by the recurrence relationships shown in **Appendix A**. Therefore, the conservative variables on the cell interface $\overline{\mathbf{W}}^{face}$ can be obtained, with which g_0 is uniquely determined from Eq. (5.2). Then, $\bar{a}^L$ and $\bar{a}^R$ can be calculated through the following relations:

$$\left\langle \bar{\varphi}_a g_0 \bar{a}^L \right\rangle = \frac{\overline{\mathbf{W}}^{face} - \overline{\mathbf{W}}_c^L}{\left(\mathbf{x}_b - \mathbf{x}_c^L \right) \cdot \mathbf{n}}, \tag{5.43}$$

$$\left\langle \overline{\boldsymbol{\varphi}}_a g_0 \overline{a}^R \right\rangle = \frac{\overline{\mathbf{W}}^{face} - \overline{\mathbf{W}}_c^R}{\left(\mathbf{x}_b - \mathbf{x}_c^R\right)\cdot \mathbf{n}}. \tag{5.44}$$

Furthermore, the coefficient $\overline{b}$ can be approximated by

$$\left\langle \overline{\boldsymbol{\varphi}}_a g_0 \overline{b} \right\rangle = \left\langle \overline{\boldsymbol{\varphi}}_a g^L b^L \right\rangle_{>0} + \left\langle \overline{\boldsymbol{\varphi}}_a g^R b^R \right\rangle_{<0}. \tag{5.45}$$

Since the terms on the right-hand side of Eqs. (5.43)-(5.45) are known, the detailed expressions of $\overline{a}^L$, $\overline{a}^R$ and $\overline{b}$ can be calculated in the same way as a^L shown in **Appendix B**.

By now, the only remaining unknown parameter in the initial gas distribution function $f_0\left(\mathbf{x},\xi,0\right)$ and the equilibrium state $g\left(\mathbf{x},\xi,t\right)$ is $\overline{A}$. To evaluate this coefficient, we substitute Eqs. (5.29) and (5.32) into Eq. (5.27) and derive the following expression of the gas distribution function $f\left(\mathbf{0},\xi,t\right)$ on the cell interface:

$$\begin{aligned}
f\left(\mathbf{0},\xi,t\right) &= \left(1-e^{-t/\tau}\right)g_0 \\
&+\left(\tau\left(-1+e^{-t/\tau}\right)+te^{-t/\tau}\right) \\
&\times\left(\left(\overline{a}^L H[\xi_1]+\overline{a}^R\left(1-H[\xi_1]\right)\right)\xi_1+\overline{b}\xi_2\right)g_0 \\
&+\tau\left(t/\tau-1+e^{-t/\tau}\right)\overline{A}g_0 \\
&+e^{-t/\tau}\Big[\left(1-(t+\tau)\left(a^L\xi_1+b^L\xi_2\right)\right)H[\xi_1]g^L \\
&+\left(1-(t+\tau)\left(a^R\xi_1+b^R\xi_2\right)\right)\left(1-H[\xi_1]\right)g^R \\
&+\left(-\tau A^L H[\xi_1]g^L-\tau A^R\left(1-H[\xi_1]\right)g^R\right)\Big].
\end{aligned} \tag{5.46}$$

The only unknown in the above expression is $\overline{A}$. Since both $f\left(\mathbf{0},\xi,t\right)$ (Eq. (5.46)) and $g\left(\mathbf{x},\xi,t\right)$ (Eq. (5.32)) contain $\overline{A}$, the integration of the collision term on the cell interface over a time step Δt gives

$$\int_0^{\Delta t}\left\langle\left(g\left(\mathbf{0},\xi,t\right)-f\left(\mathbf{0},\xi,t\right)\right)\overline{\boldsymbol{\varphi}}_a\right\rangle dt=\mathbf{0}, \tag{5.47}$$

where Δt is the time interval for solving the Navier-Stokes equations. After some manipulations, Eq. (5.47) can be rewritten as

$$\begin{aligned}\left\langle \overline{\varphi}_a g_0 \overline{A} \right\rangle &= \gamma_1 \left\langle \overline{\varphi}_\alpha g_0 \right\rangle \\ &+\gamma_2 \left(\left\langle \left(\overline{a}^L H[\xi_1] + \overline{a}^R (1 - H[\xi_1]) \right) \xi_1 \overline{\varphi}_\alpha g_0 \right\rangle + \left\langle \overline{b} \xi_2 \overline{\varphi}_\alpha g_0 \right\rangle \right) \\ &+\gamma_3 \left\langle \left(g^L H[\xi_1] + g^R (1 - H[\xi_1]) \right) \overline{\varphi}_\alpha \right\rangle \\ &+\gamma_4 \left\langle \left(a^L g^L H[\xi_1] + a^R g^R (1 - H[\xi_1]) \right) \xi_1 \overline{\varphi}_\alpha \right\rangle \\ &+\gamma_4 \left\langle \left(b^L g^L H[\xi_1] + b^R g^R (1 - H[\xi_1]) \right) \xi_2 \overline{\varphi}_\alpha \right\rangle \\ &+\gamma_5 \left\langle \left(A^L g^L H[\xi_1] + A^R g^R (1 - H[\xi_1]) \right) \overline{\varphi}_\alpha \right\rangle,\end{aligned} \tag{5.48}$$

where

$$\gamma_0 = \Delta t - \tau\left(1 - e^{-\Delta t/\tau}\right),$$

$$\gamma_1 = -\left(1 - e^{-\Delta t/\tau}\right)/\gamma_0,$$

$$\gamma_2 = \left(-\Delta t + 2\tau\left(1 - e^{-\Delta t/\tau}\right) - \Delta t e^{-\Delta t/\tau}\right)/\gamma_0,$$

$$\gamma_3 = \left(1 - e^{-\Delta t/\tau}\right)/\gamma_0,$$

$$\gamma_4 = \left(\Delta t e^{-\Delta t/\tau} - 2\tau\left(1 - e^{-\Delta t/\tau}\right)\right)/\gamma_0,$$

$$\gamma_5 = -\tau\left(1 - e^{-\Delta t/\tau}\right)/\gamma_0.$$

Since all coefficients on the right-hand side of Eq. (5.48) have been determined previously, the coefficient $\overline{A}$ can be calculated in the same way as a^L shown in **Appendix B**.

5.2.2 *Evaluation of fluxes on the cell interface by M-GKS*

After determining the local integral solution to the Boltzmann equation, the numerical fluxes of the Navier-Stokes equations on the cell interface can be computed by substituting Eq. (5.46) into Eq. (5.22), which yields

$$\begin{aligned}
\overline{\mathbf{F}} &= \frac{1}{\Delta t}\int_0^{\Delta t}\left\langle \xi_1 \overline{\boldsymbol{\varphi}}_\alpha f(\mathbf{0},\boldsymbol{\xi},t)\right\rangle dt \\
&= \frac{1}{\Delta t}\Bigg[\left\langle\left(\tau\gamma_0\gamma_1 - \tau\gamma_0\overline{A} + \Delta t + \frac{\overline{A}}{2}\Delta t^2\right)\xi_1\overline{\boldsymbol{\varphi}}_\alpha g_0\right\rangle \\
&\quad + \tau\gamma_0\gamma_2\left(\left\langle\left(\overline{a}^L H[\xi_1] + \overline{a}^R\left(1 - H[\xi_1]\right)\right)\xi_1^2\overline{\boldsymbol{\varphi}}_\alpha g_0\right\rangle\right) \\
&\quad + \tau\gamma_0\gamma_2\left\langle \overline{b}\,\xi_1\xi_2\overline{\boldsymbol{\varphi}}_\alpha g_0\right\rangle \\
&\quad + \tau\gamma_0\gamma_3\left\langle\left(g^L H[\xi_1] + g^R\left(1 - H[\xi_1]\right)\right)\xi_1\overline{\boldsymbol{\varphi}}_\alpha\right\rangle \\
&\quad + \tau\gamma_0\gamma_4\left\langle\left(a^L g^L H[\xi_1] + a^R g^R\left(1 - H[\xi_1]\right)\right)\xi_1^2\overline{\boldsymbol{\varphi}}_\alpha\right\rangle \\
&\quad + \tau\gamma_0\gamma_4\left\langle\left(b^L g^L H[\xi_1] + b^R g^R\left(1 - H[\xi_1]\right)\right)\xi_1\xi_2\overline{\boldsymbol{\varphi}}_\alpha\right\rangle \\
&\quad + \tau\gamma_0\gamma_5\left\langle\left(A^L g^L H[\xi_1] + A^R g^R\left(1 - H[\xi_1]\right)\right)\xi_1\overline{\boldsymbol{\varphi}}_\alpha\right\rangle\Bigg].
\end{aligned} \tag{5.49}$$

Since $f(\mathbf{0},\boldsymbol{\xi},t)$ is a time-dependent function, the time integration has been carried out in the above equation. The detailed expression of Eq. (5.49) can be obtained by using the recurrence relationships shown in **Appendix A**.

As shown in **Chapter 2**, the Prandtl number of the Boltzmann equation with the BGK collision model is fixed at 1. Thus, the gas kinetic BGK scheme corresponds to the unit Prandtl number. Some corrections should be made to reflect the true Prandtl number. Since the explicit expression of the time-dependent gas distribution function $f(\mathbf{0},\boldsymbol{\xi},t)$ has been obtained, the time-dependent heat flux can be evaluated as follows:

$$q = \frac{1}{\Delta t}\int_0^{\Delta t}\left\langle \frac{1}{2}\left(\xi_1 - u_1^{face}\right)\left(\sum_{i=1}^{D}\left(\xi_i - u_i^{face}\right)^2 + \sum_{j=1}^{K}\zeta_j^2\right) f(\mathbf{0},\boldsymbol{\xi},t)\right\rangle dt, \tag{5.50}$$

where u_i^{face} is the velocity component of the mean flow on the cell interface, which can be obtained from $\overline{\mathbf{W}}^{face}$. Then, the Prandtl number for the gas kinetic BGK scheme can be equivalently adjusted by modifying the energy flux

$$F_{\rho E}^{correct} = F_{\rho E} + \left(\frac{1}{Pr} - 1 \right) q, \tag{5.51}$$

where Pr is the practical Prandtl number and $F_{\rho E}$ is the energy flux given by Eq. (5.49). This Prandtl number correction technique is more like a physical model in the discretized level [Xu, 2001], and is applied at every time step when evaluating the numerical fluxes of the Navier-Stokes equations on the cell interface.

Theoretically, the dissipative flow structure, such as the thickness of the shock wave, is solely determined by the physical viscosity, and thus should be independent of the cell size and time step size used in a numerical scheme. It implies that the collision time scale τ in Eq. (5.49) should be calculated uniquely by Eq. (5.15). However, even if the Navier-Stokes equations are solved with sufficient accuracy, the physical structure can still be smeared by the numerical resolution when the cell size is not fine enough to capture the wave structure. For example, the thickness of the physical shock wave may be replaced by the numerical cell size. In such a situation, we cannot solve the Navier-Stokes equations with the original physical viscosity. The effective viscosity should be a combination of both the physical and the numerical viscosities. To achieve this goal, in the gas kinetic BGK scheme [Xu, 2001], the collision time scale τ is modified as

$$\tau = \frac{\mu^{face}}{p^{face}} + \frac{\left| p^L - p^R \right|}{\left| p^L + p^R \right|} \Delta t, \tag{5.52}$$

where μ^{face} and p^{face} are the dynamic viscosity and the pressure on the cell interface, which are functions of $\overline{\mathbf{W}}^{face}$. p^L and p^R are respectively the pressure on the left and the right sides of the cell interface, which can be determined by $\overline{\mathbf{W}}^L$ and $\overline{\mathbf{W}}^R$.

The gas kinetic BGK scheme is characterized by some distinct features. First, the gas kinetic BGK scheme bears the intrinsic kinetic nature due to its origination from the kinetic theory, which makes it particularly suitable for modeling the non-linear physics in complex fluid flows. Second, the use of a real collision time in the BGK model enriches its consistency with the true physics, which allows the recovery of a stable and crisp shock transition in the discontinuous region while maintaining accurate Navier-Stokes solutions in the smooth region. However, limitations exist in the gas kinetic BGK scheme. Firstly, it is usually more complicated and less efficient than the conventional Navier-Stokes solvers. This is due to its tedious flux reconstruction process which involves many terms and coefficients associated with the conservative variables and their derivatives. It was reported by Guo *et al.* [2008] that, for 2D incompressible flow problems, the gas kinetic BGK scheme is about 10 times and 3 times slower than the LBM for steady and unsteady flow calculations, respectively. Secondly, the explicit formulations for the numerical fluxes via the gas kinetic BGK scheme are extremely complicated. Tang [2012] pointed out that the explicit expressions for the numerical fluxes in terms of the macroscopic flow variables and their derivatives in the gas kinetic BGK scheme took more than eight pages after weeks of derivation. The work of Tang [2012] was limited to the 2D case and it can be imagined that the full formulations in the 3D scenario would be even more complex.

In the literature, several works have been done to simplify the original gas kinetic BGK scheme. Chae *et al.* [2000] neglected a time evolution term in the integral solution of the BGK model and claimed improving computational efficiency and convergence rate. May *et al.* [2007] proposed a new formulation to calculate the initial non-equilibrium terms, and introduced a new time derivative to reduce the CPU time. In the work of Tang [2012], the spatial derivatives of the equilibrium distribution function across the cell interface were assumed to be continuous rather than piecewise linear in the conventional gas kinetic BGK scheme; and all time-related terms were not considered in the calculation of numerical fluxes. There are some other efforts to improve the accuracy of the gas kinetic BGK scheme, such as incorporating the discontinuous Galerkin (DG) method [Liu and Xu, 2007; Ni *et al.*, 2008], the Runge-Kutta scheme [Li *et al.*, 2010] and other novel numerical strategies into the scheme.

5.3 Maxwellian Function-Based GKFS

To further simplify the original gas kinetic BGK scheme while keeping its intrinsic advantages, Shu and his co-workers [Sun *et al.*, 2015; Sun *et al.*, 2016; Sun *et al.*, 2017] proposed the M-GKFS. One novel strategy in the M-GKFS is its simple but effective approximation of the non-equilibrium distribution function. The distribution function on the cell interface can be expressed as $f = f^{eq} + f^{neq}$, where $f^{eq} = g$ is the equilibrium part and f^{neq} is the non-equilibrium part. The equilibrium distribution function f^{eq} can be simply calculated from the Maxwellian distribution function using the conservative variables. Meanwhile, the compatibility condition indicates that the non-equilibrium distribution function f^{neq} has no contribution to the calculation of conservative variables but will influence the numerical fluxes. The Chapman-Enskog expansion analysis in **Chapter 2** establishes an important relationship between the equilibrium and the non-equilibrium distribution functions, i.e., $f^{neq} = -\tau D f^{eq} = -\tau\left(\partial f^{eq}/\partial t + \xi \cdot \nabla f^{eq}\right)$, which can be numerically approximated by the difference of equilibrium distribution functions on the cell interface and its surrounding streaming nodes. This treatment provides an effective way to simplify the flux reconstruction process in M-GKFS. In this section, the details of M-GKFS will be presented.

5.3.1 *Local asymptotic solution to Boltzmann equation on the cell interface*

In M-GKFS, the local asymptotic solution to the Boltzmann equation is used to calculate the distribution function f and then reconstruct the numerical fluxes on the cell interface. Suppose that the cell interface is located at $\mathbf{x} = \mathbf{0}$, the asymptotic solution to the Boltzmann equation with the BGK collision model (5.1) yields

$$f(\mathbf{0},\xi,t) = g(\mathbf{0},\xi,t) - \tau D g(\mathbf{0},\xi,t). \tag{5.53}$$

By applying the Taylor series expansion in time and physical space to approximate $Dg(\mathbf{0},\xi,t)$, Eq. (5.53) can be rewritten as

$$f(\mathbf{0},\xi,t)=g(\mathbf{0},\xi,t)+\tau_0\left[g(-\xi\delta_t,\xi,t-\delta_t)-g(\mathbf{0},\xi,t)\right], \quad (5.54)$$

where $\tau_0=\tau/\delta_t$ is the dimensionless collision time and δ_t is the streaming time step. $g(\mathbf{0},\xi,t)$ is the equilibrium distribution function on the cell interface at time t, and $g(-\xi\delta_t,\xi,t-\delta_t)$ is the equilibrium distribution function at the surrounding point of the cell interface at time $t-\delta_t$. Here, $t-\delta_t=t^n$ can be treated as the initial time of each iteration. As a result, $g(-\xi\delta_t,\xi,t-\delta_t)$ can be calculated directly from the initial reconstruction of flow variables, i.e.,

$$g(-\xi\delta_t,\xi,t-\delta_t)=\begin{cases} g^L\left[1-a^L\xi_1\delta_t-b^L\xi_2\delta_t\right], & \xi_1\geq 0, \\ g^R\left[1-a^R\xi_1\delta_t-b^R\xi_2\delta_t\right], & \xi_1<0. \end{cases} \quad (5.55)$$

where g^L and g^R are the equilibrium states on the left and the right sides of the cell interface, which satisfy the following relation:

$$g(\mathbf{0},\xi,t-\delta_t)=\begin{cases} g^L, & \xi_1\geq 0, \\ g^R, & \xi_1<0. \end{cases} \quad (5.56)$$

The coefficients a^L, b^L, a^R and b^R can be determined in the same way as those in Eqs. (5.36) and (5.37a)-(5.37b). Substituting Eqs. (5.55) and (5.56) into Eq. (5.54), we have

$$\begin{aligned} f(\mathbf{0},\xi,t)&=g(\mathbf{0},\xi,t)+\tau_0\left[g(\mathbf{0},\xi,t-\delta_t)-g(\mathbf{0},\xi,t)\right] \\ &\quad-\tau\left[g^L\left(a^L\xi_1+b^L\xi_2\right)H[\xi_1]\right. \\ &\quad\left.+g^R\left(a^R\xi_1+b^R\xi_2\right)\left(1-H[\xi_1]\right)\right]. \end{aligned} \quad (5.57)$$

Note that, in Eq. (5.57), the equilibrium distribution function on the cell interface $g(\mathbf{0},\xi,t)$ is still unknown. Substituting Eqs. (5.56) and (5.57) into Eq. (5.21) and adopting the compatibility condition, we have

$$\begin{aligned} \overline{\mathbf{W}}^{face}&=\left\langle\overline{\boldsymbol{\varphi}}_a g(\mathbf{0},\xi,t)\right\rangle \\ &=\left\langle\overline{\boldsymbol{\varphi}}_a g^L\left[1-a^L\xi_1\delta_t-b^L\xi_2\delta_t\right]\right\rangle_{>0}+\left\langle\overline{\boldsymbol{\varphi}}_a g^R\left[1-a^R\xi_1\delta_t-b^R\xi_2\delta_t\right]\right\rangle_{<0}. \end{aligned} \quad (5.58)$$

Since g^L, a^L, b^L, g^R, a^R and b^R have been determined earlier, the above moments on the right-hand side of Eq. (5.58) can be evaluated explicitly. Therefore, the conservative variables on the cell interface $\overline{\mathbf{W}}^{face}$ can be obtained, from which $g(\mathbf{0},\xi,t)$ is uniquely determined. Eq. (5.57) shows that once the equilibrium distribution functions on the cell interface and its surrounding points are known, the distribution function on the cell interface can be evaluated.

As shown in Eqs. (5.57) and (5.58), the calculation of the distribution function on the cell interface involves the streaming time step δ_t. The principle of determining δ_t is to avoid extrapolation, which requires that the streaming nodes around the cell interface must locate within the two cells that share the interface. However, for the Maxwellian distribution function, the particle velocity ranges from -∞ to +∞. Therefore, the streaming time step δ_t cannot be simply determined. Sun *et al.* [2015] investigated different resolution strategies and suggested to assume $\delta_t \to 0$ and replace the term $g(\mathbf{0},\xi,t)-g(\mathbf{0},\xi,t-\delta_t)$ in Eq. (5.57) by the temporal derivative of $g(\mathbf{0},\xi,t)$.

By taking the limit $\delta_t \to 0$ in Eq. (5.58), we have

$$\overline{\mathbf{W}}^{face} = \langle \overline{\boldsymbol{\varphi}}_a g(\mathbf{0},\xi,t) \rangle = \langle \overline{\boldsymbol{\varphi}}_a g^L \rangle_{>0} + \langle \overline{\boldsymbol{\varphi}}_a g^R \rangle_{<0}. \tag{5.59}$$

In fact, Eq. (5.59) is the same as Eq. (5.40), which is used to calculate the equilibrium distribution function on the cell interface for the gas kinetic BGK scheme. The detailed expressions of Eq. (5.59) can be obtained by using the recurrence relationships shown in **Appendix A**, which yields

$$\rho^{face} = \rho^L \beta_0^L + \rho^R \beta_0^R, \tag{5.60}$$

$$(\rho u_1)^{face} = \rho^L \beta_1^L + \rho^R \beta_1^R, \tag{5.61}$$

$$(\rho u_2)^{face} = \rho^L \beta_0^L \chi_1^L + \rho^R \beta_0^R \chi_1^R, \tag{5.62}$$

$$\begin{aligned}(\rho E)^{face} &= \rho^L \left(\beta_2^L + \beta_0^L \chi_2^L + \beta_0^L \eta_2^L\right)/2 \\ &+ \rho^R \left(\beta_2^R + \beta_0^R \chi_2^R + \beta_0^R \eta_2^R\right)/2,\end{aligned} \tag{5.63}$$

where

$$\beta_0^L = \left[\frac{1}{2}erfc\left(-\sqrt{\lambda}u_1\right)\right]^L,\ \beta_1^L = u_1^L\beta_0^L + \left[\frac{1}{2}\frac{e^{-\lambda u_1^2}}{\sqrt{\pi\lambda}}\right]^L,$$

$$\beta_2^L = u_1^L\beta_1^L + \frac{1}{2\lambda^L}\beta_0^L,\ \beta_3^L = u_1^L\beta_2^L + \frac{2}{2\lambda^L}\beta_1^L,$$

$$\beta_4^L = u_1^L\beta_3^L + \frac{3}{2\lambda^L}\beta_2^L,\ \beta_5^L = u_1^L\beta_4^L + \frac{4}{2\lambda^L}\beta_3^L,$$

$$\beta_6^L = u_1^L\beta_5^L + \frac{5}{2\lambda^L}\beta_4^L.$$

$$\beta_0^R = \left[\frac{1}{2}erfc\left(\sqrt{\lambda}u_1\right)\right]^R,\ \beta_1^R = u_1^R\beta_0^R - \left[\frac{1}{2}\frac{e^{-\lambda u_1^2}}{\sqrt{\pi\lambda}}\right]^R,$$

$$\beta_2^R = u_1^R\beta_1^R + \frac{1}{2\lambda^R}\beta_0^R,\ \beta_3^R = u_1^R\beta_2^R + \frac{2}{2\lambda^R}\beta_1^R,$$

$$\beta_4^R = u_1^R\beta_3^R + \frac{3}{2\lambda^R}\beta_2^R,\ \beta_5^R = u_1^R\beta_4^R + \frac{4}{2\lambda^R}\beta_3^R,$$

$$\beta_6^R = u_1^R\beta_5^R + \frac{5}{2\lambda^R}\beta_4^R.$$

$$\gamma_1 = u_1,\ \gamma_2 = u_1^2 + \frac{1}{2\lambda},\ \gamma_3 = u_1\gamma_2 + \frac{2}{2\lambda}\gamma_1,$$

$$\gamma_4 = u_1\gamma_3 + \frac{3}{2\lambda}\gamma_2,\ \gamma_5 = u_1\gamma_4 + \frac{4}{2\lambda}\gamma_3.$$

$$\chi_1 = u_2,\ \chi_2 = u_2^2 + \frac{1}{2\lambda},\ \chi_3 = u_2\chi_2 + \frac{2}{2\lambda}\chi_1,$$

$$\chi_4 = u_2\chi_3 + \frac{3}{2\lambda}\chi_2,\ \chi_5 = u_2\chi_4 + \frac{4}{2\lambda}\chi_3.$$

$$\eta_2=\frac{K}{2\lambda},\ \eta_4=\frac{3K}{4\lambda^2}+\frac{K(K-1)}{4\lambda^2}.$$

The above coefficients will also be applied to calculate the numerical fluxes on the cell interface. Once the conservative variables on the cell interface $\overline{\mathbf{W}}^{face}$ are obtained, $g(\mathbf{0},\boldsymbol{\xi},t)$ can be determined from Eq. (5.2).

In addition, noting that $[g(\mathbf{0},\boldsymbol{\xi},t)-g(\mathbf{0},\boldsymbol{\xi},t-\delta_t)]/\delta_t$ approximates the temporal derivative of $g(\mathbf{0},\boldsymbol{\xi},t)$, Eq. (5.57) can be rewritten as

$$\begin{aligned} f(\mathbf{0},\boldsymbol{\xi},t) &= g(\mathbf{0},\boldsymbol{\xi},t)(1-\tau B) \\ &\quad -\tau\Big[g^L\left(a^L\xi_1+b^L\xi_2\right)H[\xi_1] \\ &\quad +g^R\left(a^R\xi_1+b^R\xi_2\right)\left(1-H[\xi_1]\right)\Big], \end{aligned} \tag{5.64}$$

where $Bg(\mathbf{0},\boldsymbol{\xi},t)=\partial g(\mathbf{0},\boldsymbol{\xi},t)/\partial t$. The coefficient B depends on the particle velocities. Applying the Taylor series expansion to the Maxwellian distribution function, we have

$$B=B_0+B_1\xi_1+B_2\xi_2+B_3\frac{1}{2}\left(\sum_{i=1}^{D}\xi_i^2+\sum_{j=1}^{K}\varsigma_j^2\right)=B_\alpha\overline{\boldsymbol{\varphi}}_a, \tag{5.65}$$

where B_0 , B_1 , B_2 and B_3 are local constants to be determined. Substituting Eq. (5.64) into Eq. (5.21) and adopting the compatibility condition, we have

$$\begin{aligned} \left\langle\overline{\boldsymbol{\varphi}}_a Bg(\mathbf{0},\boldsymbol{\xi},t)\right\rangle &= -\Big[\left\langle\overline{\boldsymbol{\varphi}}_a g^L\left(a^L\xi_1+b^L\xi_2\right)\right\rangle_{>0} \\ &\quad +\left\langle\overline{\boldsymbol{\varphi}}_a g^R\left(a^R\xi_1+b^R\xi_2\right)\right\rangle_{<0}\Big]. \end{aligned} \tag{5.66}$$

Since g^L , a^L , b^L , g^R , a^R , b^R and $g(\mathbf{0},\boldsymbol{\xi},t)$ have been determined earlier, the coefficient B can be evaluated from Eq. (5.66) by using the relationships shown in **Appendix B**. For the convenience of application, we denote the terms on the right-hand side of Eq. (5.66) as $\hat{\mathbf{H}}=\left(\hat{h}_0,\hat{h}_1,\hat{h}_2,\hat{h}_3\right)^T$, which can be given explicitly as follows:

$$\hat{h}_0 = -\left(\rho^L \hbar_0^L + \rho^R \hbar_0^R\right), \tag{5.67}$$

$$\hat{h}_1 = -\left(\rho^L \hbar_1^L + \rho^R \hbar_1^R\right), \tag{5.68}$$

$$\hat{h}_2 = -\left(\rho^L \hbar_2^L + \rho^R \hbar_2^R\right), \tag{5.69}$$

$$\hat{h}_3 = -\left(\rho^L \hbar_3^L + \rho^R \hbar_3^R\right), \tag{5.70}$$

where

$$\begin{aligned}\hbar_0 =& a_0\beta_1 + a_1\beta_2 + a_2\beta_1\chi_1 + \frac{1}{2}a_3\left(\beta_3 + \beta_1\chi_2 + \beta_1\eta_2\right)\\ &+ b_0\beta_0\chi_1 + b_1\beta_1\chi_1 + b_2\beta_0\chi_2 + \frac{1}{2}b_3\left(\beta_2\chi_1 + \beta_0\chi_3 + \beta_0\chi_1\eta_2\right),\end{aligned}$$

$$\begin{aligned}\hbar_1 =& a_0\beta_2 + a_1\beta_3 + a_2\beta_2\chi_1 + \frac{1}{2}a_3\left(\beta_4 + \beta_2\chi_2 + \beta_2\eta_2\right)\\ &+ b_0\beta_1\chi_1 + b_1\beta_2\chi_1 + b_2\beta_1\chi_2 + \frac{1}{2}b_3\left(\beta_3\chi_1 + \beta_1\chi_3 + \beta_1\chi_1\eta_2\right),\end{aligned}$$

$$\begin{aligned}\hbar_2 =& a_0\beta_1\chi_1 + a_1\beta_2\chi_1 + a_2\beta_1\chi_2 + \frac{1}{2}a_3\left(\beta_3\chi_1 + \beta_1\chi_3 + \beta_1\chi_1\eta_2\right)\\ &+ b_0\beta_0\chi_2 + b_1\beta_1\chi_2 + b_2\beta_0\chi_3 + \frac{1}{2}b_3\left(\beta_2\chi_2 + \beta_0\chi_4 + \beta_0\chi_2\eta_2\right),\end{aligned}$$

$$\begin{aligned}\hbar_3 =& \frac{1}{2}\left(a_0\beta_3 + a_1\beta_4 + a_2\beta_3\chi_1 + \frac{1}{2}a_3\left(\beta_5 + \beta_3\chi_2 + \beta_3\eta_2\right)\right)\\ &+ \frac{1}{2}\left(a_0\beta_1\chi_2 + a_1\beta_2\chi_2 + a_2\beta_1\chi_3 + \frac{1}{2}a_3\left(\beta_3\chi_2 + \beta_1\chi_4 + \beta_1\chi_2\eta_2\right)\right)\\ &+ \frac{1}{2}\left(a_0\beta_1\eta_2 + a_1\beta_2\eta_2 + a_2\beta_1\chi_1\eta_2 + \frac{1}{2}a_3\left(\beta_3\eta_2 + \beta_1\chi_2\eta_2 + \beta_1\eta_4\right)\right)\\ &+ \frac{1}{2}\left(b_0\beta_2\chi_1 + b_1\beta_3\chi_1 + b_2\beta_2\chi_2 + \frac{1}{2}b_3\left(\beta_4\chi_1 + \beta_2\chi_3 + \beta_2\chi_1\eta_2\right)\right)\end{aligned}$$

Brought down

Continued

$$+\frac{1}{2}\left(b_0\beta_0\chi_3+b_1\beta_1\chi_3+b_2\beta_0\chi_4+\frac{1}{2}b_3\left(\beta_2\chi_3+\beta_0\chi_5+\beta_0\chi_3\eta_2\right)\right)$$
$$+\frac{1}{2}\left(b_0\beta_0\chi_1\eta_2+b_1\beta_1\chi_1\eta_2+b_2\beta_0\chi_2\eta_2+\frac{1}{2}b_3\left(\beta_2\chi_1\eta_2+\beta_0\chi_3\eta_2+\beta_0\chi_1\eta_4\right)\right).$$

By now, we have determined all parameters in the local asymptotic solution to the Boltzmann equation on the cell interface $f(\mathbf{0},\boldsymbol{\xi},t)$.

5.3.2 *Evaluation of fluxes on the cell interface by M-GKFS*

Once the distribution function on the cell interface is obtained, the numerical fluxes of the Navier-Stokes equations can be computed by substituting Eq. (5.64) into Eq. (5.22), which yields

$$\begin{aligned}\overline{\mathbf{F}}&=\left\langle\xi_1\overline{\boldsymbol{\varphi}}_\alpha f(\mathbf{0},\boldsymbol{\xi},t)\right\rangle\\&=\left\langle\xi_1\overline{\boldsymbol{\varphi}}_\alpha g(\mathbf{0},\boldsymbol{\xi},t)(1-\tau B)\right\rangle\\&\quad-\tau\left[\left\langle\xi_1\overline{\boldsymbol{\varphi}}_\alpha g^L\left(a^L\xi_1+b^L\xi_2\right)\right\rangle_{>0}\right.\\&\quad\left.+\left\langle\xi_1\overline{\boldsymbol{\varphi}}_\alpha g^R\left(a^R\xi_1+b^R\xi_2\right)\right\rangle_{<0}\right].\end{aligned}\tag{5.71}$$

The detailed expression of Eq. (5.71) can be obtained by using the recurrence relationships shown in **Appendix A**. For the convenience of application, the explicit expressions of $\overline{\mathbf{F}}$ are given as forms:

$$F_\rho=\rho u_1,\tag{5.72}$$

$$F_{\rho u_1}=\rho\ell_1-\tau\left(\rho^L\lambda_1^L+\rho^R\lambda_1^R\right),\tag{5.73}$$

$$F_{\rho u_2}=\rho\ell_2-\tau\left(\rho^L\lambda_2^L+\rho^R\lambda_2^R\right),\tag{5.74}$$

$$F_{\rho E}=\rho\ell_3-\tau\left(\rho^L\lambda_3^L+\rho^R\lambda_3^R\right),\tag{5.75}$$

where

$$\ell_1 = \left(1 - \tau B_0\right)\gamma_2 - \tau\left(B_1\gamma_3 + B_2\gamma_2\chi_1 + \frac{1}{2}B_3\left(\gamma_4 + \gamma_2\chi_2 + \gamma_2\eta_2\right)\right),$$

$$\ell_2 = \left(1 - \tau B_0\right)\gamma_1\chi_1 - \tau\left(B_1\gamma_2\chi_1 + B_2\gamma_1\chi_2 + \frac{1}{2}B_3\left(\gamma_3\chi_1 + \gamma_1\chi_3 + \gamma_1\chi_1\eta_2\right)\right),$$

$$\begin{aligned}\ell_3 = &\frac{1}{2}\left[\left(1 - \tau B_0\right)\gamma_3 - \tau\left(B_1\gamma_4 + B_2\gamma_3\chi_1 + \frac{1}{2}B_3\left(\gamma_5 + \gamma_3\chi_2 + \gamma_3\eta_2\right)\right)\right] \\ &+ \frac{1}{2}\left[\left(1 - \tau B_0\right)\gamma_1\chi_2 - \tau\left(B_1\gamma_2\chi_2 + B_2\gamma_1\chi_3 + \frac{1}{2}B_3\left(\gamma_3\chi_2 + \gamma_1\chi_4 + \gamma_1\chi_2\eta_2\right)\right)\right] \\ &+ \frac{1}{2}\left[\left(1 - \tau B_0\right)\gamma_1\eta_2 - \tau\left(B_1\gamma_2\eta_2 + B_2\gamma_1\chi_1\eta_2 + \frac{1}{2}B_3\left(\gamma_3\eta_2 + \gamma_1\chi_2\eta_2 + \gamma_1\eta_4\right)\right)\right],\end{aligned}$$

$$\begin{aligned}\bar{\lambda}_1 = &\, a_0\beta_3 + a_1\beta_4 + a_2\beta_3\chi_1 + \frac{1}{2}a_3\left(\beta_5 + \beta_3\chi_2 + \beta_3\eta_2\right) \\ &+ b_0\beta_2\chi_1 + b_1\beta_3\chi_1 + b_2\beta_2\chi_2 + \frac{1}{2}b_3\left(\beta_4\chi_1 + \beta_2\chi_3 + \beta_2\chi_1\eta_2\right),\end{aligned}$$

$$\begin{aligned}\bar{\lambda}_2 = &\, a_0\beta_2\chi_1 + a_1\beta_3\chi_1 + a_2\beta_2\chi_2 + \frac{1}{2}a_3\left(\beta_4\chi_1 + \beta_2\chi_3 + \beta_2\chi_1\eta_2\right) \\ &+ b_0\beta_1\chi_2 + b_1\beta_2\chi_2 + b_2\beta_1\chi_3 + \frac{1}{2}b_3\left(\beta_3\chi_2 + \beta_1\chi_4 + \beta_1\chi_2\eta_2\right),\end{aligned}$$

$$\begin{aligned}\bar{\lambda}_3 = &\frac{1}{2}\left(a_0\beta_4 + a_1\beta_5 + a_2\beta_4\chi_1 + \frac{1}{2}a_3\left(\beta_6 + \beta_4\chi_2 + \beta_4\eta_2\right)\right) \\ &+ \frac{1}{2}\left(a_0\beta_2\chi_2 + a_1\beta_3\chi_2 + a_2\beta_2\chi_3 + \frac{1}{2}a_3\left(\beta_4\chi_2 + \beta_2\chi_4 + \beta_2\chi_2\eta_2\right)\right) \\ &+ \frac{1}{2}\left(a_0\beta_2\eta_2 + a_1\beta_3\eta_2 + a_2\beta_2\chi_1\eta_2 + \frac{1}{2}a_3\left(\beta_4\eta_2 + \beta_2\chi_2\eta_2 + \beta_2\eta_4\right)\right) \\ &+ \frac{1}{2}\left(b_0\beta_3\chi_1 + b_1\beta_4\chi_1 + b_2\beta_3\chi_2 + \frac{1}{2}b_3\left(\beta_5\chi_1 + \beta_3\chi_3 + \beta_3\chi_1\eta_2\right)\right) \\ &+ \frac{1}{2}\left(b_0\beta_1\chi_3 + b_1\beta_2\chi_3 + b_2\beta_1\chi_4 + \frac{1}{2}b_3\left(\beta_3\chi_3 + \beta_1\chi_5 + \beta_1\chi_3\eta_2\right)\right) \\ &+ \frac{1}{2}\left(b_0\beta_1\chi_1\eta_2 + b_1\beta_2\chi_1\eta_2 + b_2\beta_1\chi_2\eta_2 + \frac{1}{2}b_3\left(\beta_3\chi_1\eta_2 + \beta_1\chi_3\eta_2 + \beta_1\chi_1\eta_4\right)\right).\end{aligned}$$

Note that, in Eqs. (5.72)-(5.75), the flow variables without superscripts "L" and "R" are determined from the conservative variables on the cell interface, $\overline{\mathbf{W}}^{face}$. Comparatively, the expressions of numerical fluxes of the Navier-Stokes equations given by the M-GKFS are more concise than those of the gas kinetic BGK scheme (see Eq. (5.49)).

Like the gas kinetic BGK scheme, when there is a shock wave with a thickness in the order of the cell size, the numerical viscosity should be combined with the physical viscosity to interpret the pressure jump. Thus, Eq. (5.52) is also adopted in M-GKFS to calculate the collision time scale. In addition, the Prandtl number in M-GKFS is fixed at one since the BGK collision model and the Maxwellian distribution function are used. In order to adjust the Prandtl number to any realistic value, one convenient approach is to modify the heat flux with a variable Prandtl number, which has been shown in Eq. (5.51). Being different from the gas kinetic BGK scheme, the heat flux in M-GKFS is computed by

$$q = \left\langle \frac{1}{2}\left(\xi_1 - u_1^{face}\right)\left(\sum_{i=1}^{D}\left(\xi_i - u_i^{face}\right)^2 + \sum_{j=1}^{K}\zeta_j^2\right) f(\mathbf{0}, \boldsymbol{\xi}, t)\right\rangle. \tag{5.76}$$

For the 2D case, the above equation can be expanded as

$$\begin{aligned} q &= \left\langle \frac{1}{2}\left(\xi_1 - u_1\right)\left(\left(\xi_1 - u_1\right)^2 + \left(\xi_2 - u_2\right)^2 + \sum_{j=1}^{K}\zeta_j^2\right) f(\mathbf{0}, \boldsymbol{\xi}, t)\right\rangle \\ &= \left\langle \frac{1}{2}\left(\xi_1 - u_1\right)\left(\xi_1^2 + \xi_2^2 + \sum_{j=1}^{K}\zeta_j^2 + u_1^2 + u_2^2 - 2\xi_1 u_1 - 2\xi_2 u_2\right) f(\mathbf{0}, \boldsymbol{\xi}, t)\right\rangle. \end{aligned} \tag{5.77}$$

It can be found from Eq. (5.77) that q is a function of components of $\overline{\mathbf{F}}$ and the conservative variables on the cell interface, i.e.,

$$q = F_{\rho E} - u_1 F_{\rho u_1} - u_2 F_{\rho u_2} - u_1\left(\rho E - \rho u_1^2 - \rho u_2^2\right). \tag{5.78}$$

Note that, in Eqs. (5.77) and (5.78), the superscript "*face*" in ρ, u_1, u_2 and E has been omitted for simplicity. Since the conservative variables and numerical fluxes have been reconstructed, this modification requires little extra computational effort.

5.4 Sample Applications

Sun *et al.* [2015] have performed a series of tests for the M-GKFS from incompressible flow regime to hypersonic flow regime. It has been shown that the M-GKFS has comparable accuracy and stability as the gas kinetic BGK scheme while significantly improves the computational efficiency. Comparatively, the M-GKFS reduces more than 40% of the CPU time in the test case of shock-boundary layer interaction. For illustration purposes, three sample applications will be presented in this section. In the simulation, the Venkatakrishnan's limiter [Venkatakrishnan, 1995] or the van Leer's limiter [Van Leer, 1997] is used to calculate the conservative flow variables on two sides of the cell interface. For temporal discretization, the four-stage Runge-Kutta scheme or the Lower-Upper Symmetric-Gauss-Seidel (LU-SGS) method [Yoon and Jameson, 1988; Yoon and Kwak, 1991] is used.

5.4.1 *Couette flow*

The Couette flow is a benchmark test case for the viscous heat conduction. Its configuration is the flow between two infinite parallel plates with the distance of *H*. The bottom fixed wall is stationary and has a constant temperature T_0, while the top plate is moving at a speed u_0

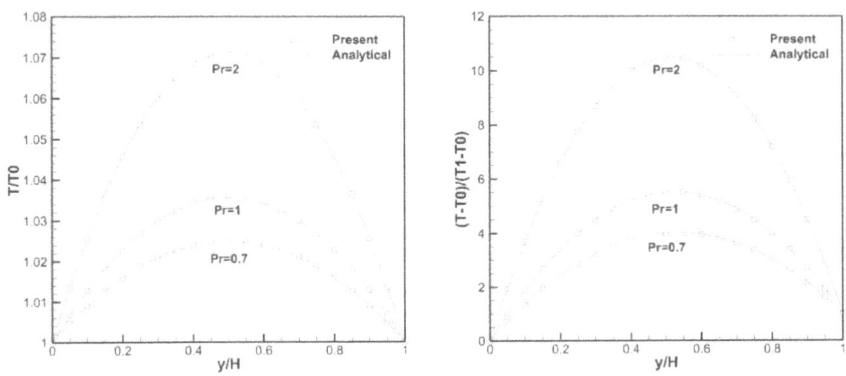

Figure 5.1. Comparison of temperature profiles for various Prandtl numbers with $T_1 = T_0$ (left) or $T_1 \neq T_0$ (right).

and maintained at the temperature of T_1. When assuming constant viscosity and heat conductivity, the temperature distribution at steady state can be derived as

$$T - T_0 = Pr\frac{u_0^2}{2c_p}\frac{y}{H}\left(1-\frac{y}{H}\right), \quad \text{when } T_0 = T_1, \tag{5.79}$$

$$\frac{T - T_0}{T_1 - T_0} = \frac{y}{H} + Pr\frac{Ec}{2}\frac{y}{H}\left(1-\frac{y}{H}\right), \quad \text{when } T_0 \neq T_1. \tag{5.80}$$

The distance of two plates H is set as 1 in this case and the mesh size of 20×40 is used in the simulation. The periodic boundary condition is applied at the inlet and the outlet while the isothermal no-slip boundary condition is imposed on the top and the bottom walls.

The condition of $T_1 = T_0$ and $u_0 = 1$ is considered first. Three test cases with different Prandtl numbers of $Pr = 0.7$, 1 and 2 are solved. The left figure in Figure 5.1 shows the temperature profiles along the vertical central line obtained by the M-GKFS and analytical solutions under different Prandtl numbers. It can be observed that all the present results match well with the analytical solutions. In addition, the condition of $T_1 \neq T_0$ is also simulated. The right figure of Figure 5.1 presents the solutions of Eckert number of $Ec = 40$ with different Prandtl numbers, in which the simulated results also agree well with the analytical solutions. These results validate the robustness of the Prandtl number correction technique used in M-GKFS.

5.4.2 *Hypersonic flow around a circular cylinder*

To investigate the capability of M-GKFS in simulating hypersonic flows with strong shock waves, the hypersonic flow over a circular cylinder is simulated. The inflow condition follows the experimental setups by Wieting [1987], in which the free-stream Mach number is $M_\infty = 8.03$, the Reynolds number is $Re = 1.835 \times 10^5$, the free-stream temperature is $T_\infty = 124.94K$ and the wall temperature is $T_w = 294.44K$. The mesh size adopted here is 160×160 and the cell Reynolds number reads

$Re_{cell} = \rho_\infty u_\infty \Delta r / \mu_\infty = 1.835$, where Δr is the mesh spacing of the first cell adjacent to the cylinder surface.

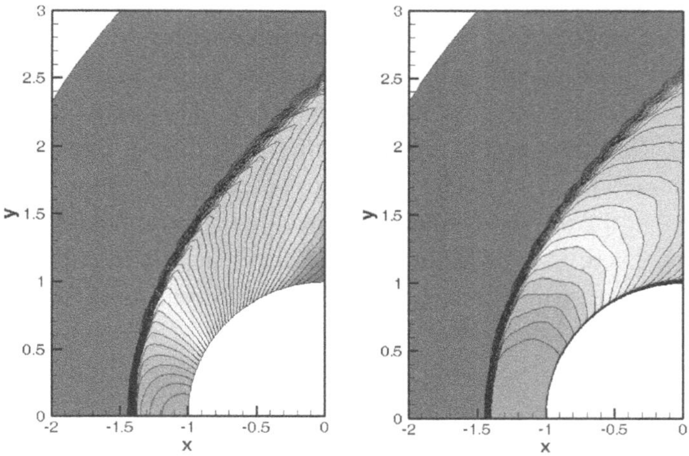

Figure 5.2. Pressure (left) and temperature (right) contours for hypersonic flow around a circular cylinder.

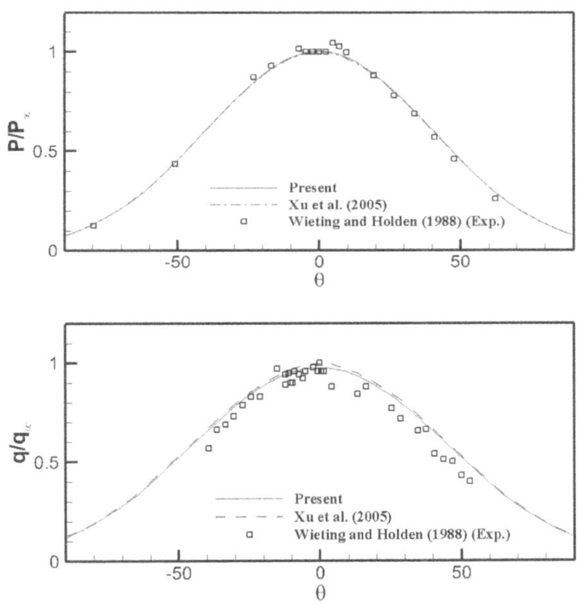

Figure 5.3. Comparison of pressure (upper) and heat flux (lower) distributions along the cylindrical surface for hypersonic flow around a circular cylinder.

Figure 5.2 shows the pressure and the temperature contours around the cylinder, indicating the absence of unphysical oscillations or "carbuncle phenomenon" in the present simulation. Figure 5.3 presents the pressure and the heat flux distributions along the cylindrical surface, which have been normalized by the pressure $p_0 = 0.9209$ and the heat flux $q_0 = 0.003655$ at the stagnation point. The experimental data [Wieting, 1987] and results given by the multi-dimensional gas kinetic BGK scheme [Xu *et al.*, 2005] are also included for comparison. It can be found that the present results match well with those obtained by Xu *et al.* [2005]. There is a slight deviation in the heat flux distribution at the stagnation point, which may be caused by different mesh sizes used in the simulations.

5.4.3 *DLR-F6 wing-body configuration*

The DLR-F6 wing-body configuration is a generic transport aircraft model from the second AIAA CFD drag prediction workshop (DPW II) [Laflin *et al.*, 2005]. The flow parameters used here include the free-stream Mach number of $M_\infty = 0.75$, the mean-chord based Reynolds number of $Re = 3\times10^6$ and the angle of attack $\alpha = 0.49^\circ$. The experimental data have been collected by the S2MA wind tunnel facility of ONERA in France in 1990s and provided by the DPW-II organizing committee [Laflin *et al.*, 2005]. Computations are carried out on the mesh provided by the NASA website*. Owing to the limitation of virtual memory, only the coarse mesh with 26 blocks and 2298880 cells are used. Figure 5.4 shows the pressure contours of the DLR-F6 wing-body configuration obtained by the M-GKFS, which indicates that smooth and stable solutions can be obtained by the proposed solver. Quantitative comparisons of pressure coefficient distributions at selected span-wise locations are then displayed in Figure 5.5. The results obtained by the M-GKFS show good agreement with the experimental data [Laflin *et al.*, 2005] and the numerical results of Vassberg *et al.* [2007] and Yang *et al.* [2014a]. To further verify the force coefficients, another test is conducted at the free stream condition of Mach number $M_\infty = 0.75$, Reynolds

---

* https://aiaa-dpw.larc.nasa.gov/Workshop2/workshop2.html

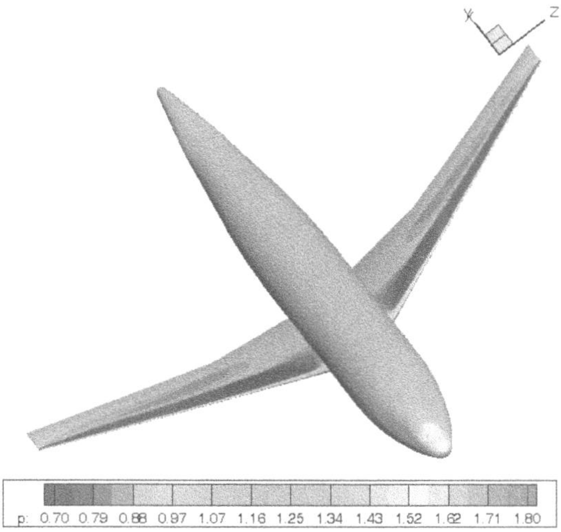

Figure 5.4. Pressure contours for DLR-F6 wing-body.

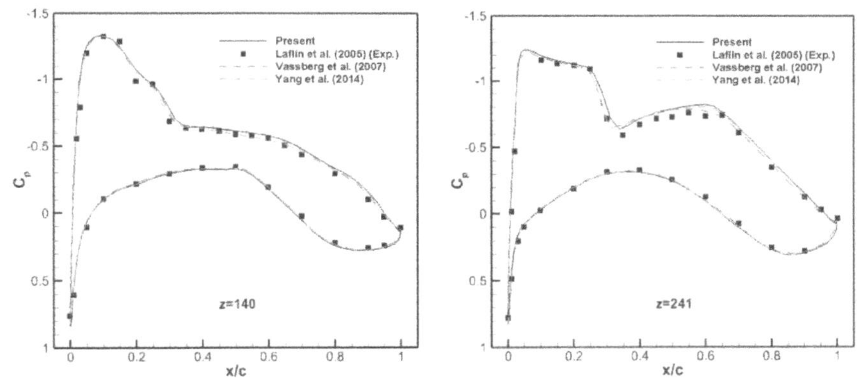

Figure 5.5. Comparison of pressure coefficient distributions at selected span-wise locations on DLR-F6 wing-body.

Table 5.1. Comparison of force coefficients for DLR-F6 wing-body configuration.

References	C_l	$C_{d,p}$	$C_{d,f}$	C_d	C_m
Vassberg *et al.* [2007]	0.51600	0.01502	0.01229	0.02731	-0.15280
Yang *et al.* [2014a]	0.52312	0.01554	0.00979	0.02533	-0.14988
Present	0.52470	0.01549	0.00947	0.02496	-0.16230

number $Re = 5\times10^6$ and angle of attack $\alpha = 0^\circ$. As shown in Table 5.1, the present force coefficients match well with the results of LBFS [Yang *et al.*, 2014a] and the reference data of Vassberg *et al.* [2007]. Note that, in this table, C_l, $C_{d,p}$, $C_{d,f}$, C_d and C_m are the lift coefficient, the pressure drag coefficient, the friction drag coefficient, the total drag coefficient and the moment coefficient, respectively.

5.5 Summary

In this chapter, the M-GKS and M-GKFS are introduced for solving flow problems in the continuum flow regime. In these two methods, the Navier-Stokes equations are discretized by the FVM and the numerical fluxes on the cell interface are reconstructed by the local solution to the Boltzmann equation with BGK collision model and the Maxwellian distribution function. In particular, the local integral solution to the Boltzmann equation is used in M-GKS, while the local asymptotic solution to the Boltzmann equation is utilized in M-GKFS. Since the local asymptotic solution to the Boltzmann equation is simpler than the local integral solution to the Boltzmann equation, the expressions of the reconstructed numerical fluxes given by the M-GKFS are more concise than those of M-GKS, which facilitates implementation and improves computational efficiency. Sample applications of M-GKFS to flow problems ranging from low to high speed regime demonstrate its effectiveness and accuracy.

Chapter 6

Simplified Distribution Function-Based Gas Kinetic Flux Solvers

In the previous chapter, we have described the Maxwellian function-based gas kinetic flux solver (M-GKFS) for simulation of inviscid and viscous flows [Sun *et al.*, 2015; Sun *et al.*, 2016; Sun *et al.*, 2017]. Like the conventional Maxwellian function-based gas kinetic scheme (M-GKS) [Xu, 2001; Tang, 2001; Xiong *et al.*, 2011; Kumar *et al.*, 2013], in M-GKFS, the Maxwellian distribution function is used as the equilibrium state. Since the integral domain of the particle velocity space for the Maxwellian distribution function is infinity, the error function and the exponential function are involved in the expressions of numerical fluxes for the M-GKFS, which makes the M-GKFS more complicated and less efficient than the conventional Navier-Stokes solvers [Toro, 2009; Blazek, 2015]. Thus, further simplification of M-GKFS is necessary to enrich its applicability in practical problems.

Recognizing the origin of the tediousness of M-GKFS, which is the Maxwellian distribution function, the most straightforward way of nurturing this method is to simplify its equilibrium distribution function. Along this direction, the circular function-based GKFS (C-GKFS) for 2D cases and the sphere function-based GKFS (S-GKFS) for 3D cases have been developed [Yang *et al.*, 2013b; Yang *et al.*, 2014b; Yang *et al.*, 2017a], which will be presented in detail in this chapter. In these solvers, the Maxwellian distribution function is simplified as the circular function and the sphere function by assuming that all particles in the particle velocity space are concentrated on a circle or on a spherical surface. And the integration domain in the particle velocity space is thus transformed from an infinite domain into a finite circle or spherical surface. Such simplification would also facilitate the local reconstruction of numerical fluxes on the cell interface from the asymptotic solution to the Boltzmann equation, and in the meantime, contribute to the explicitness and conciseness of resultant formulations. Given these characteristics, it is thus expected that the C-GKFS and S-GKFS will be more efficient than the M-GKFS.

In addition, the lattice Boltzmann model can also be regarded as one kind of simplification to the Maxwellian distribution function. By applying the compressible lattice Boltzmann model, the compressible lattice Boltzmann flux solver (LBFS) has also been developed [Yang *et al.*, 2013a; Yang *et al.*, 2016d; Yang *et al.*, 2016e]. This solver will be introduced in this chapter as well.

6.1 From Maxwellian Function to Circular and Sphere Functions

The continuous Boltzmann equation with the Bhatnagar-Gross-Krook (BGK) collision model [Bhatnagar *et al.*, 1954] reads

$$\frac{\partial f}{\partial t} + \boldsymbol{\xi} \cdot \nabla f = \frac{g - f}{\tau}, \tag{6.1}$$

where f is the gas distribution function that varies with the physical space $\mathbf{x} = (x_1, x_2, \ldots, x_D)$, the particle velocity space $(\boldsymbol{\xi}, \boldsymbol{\zeta})$, and time t. $\boldsymbol{\xi} = (\xi_1, \xi_2, \ldots, \xi_D)$ is the translational particle velocity in D-dimensional space and $\boldsymbol{\zeta} = (\zeta_1, \zeta_2, \ldots, \zeta_K)$ represents the rest components of the translational particle velocity (ξ_1, ξ_2, ξ_3) in 3-dimensional space and the rotational particle velocity, i.e., $K = 3 - D + N$, where N denotes the number of rotational degrees of freedom. g is the equilibrium state approached by f through particle collisions within a collision time scale τ. For the BGK model, the equilibrium state is a Maxwellian distribution function given by

$$g = g_M = \rho \left(\frac{\lambda}{\pi} \right)^{\frac{D+K}{2}} e^{-\lambda \left[\sum_{i=1}^{D} (\xi_i - u_i)^2 + \sum_{j=1}^{K} \zeta_j^2 \right]}. \tag{6.2}$$

Here, ρ is the density of the mean flow; u_i is the macroscopic flow velocity component in the ith direction of D-dimensional space; and $\lambda = m/(2kT) = 1/(2RT)$, where m, k, R and T are the molecular mass, the Boltzmann constant, the gas constant and the temperature, respectively. The specific heat ratio γ can be expressed by

$$\gamma = \frac{b+2}{b} = \frac{K+D+2}{K+D}, \tag{6.3}$$

where b is the total number of degrees of freedom of molecules. For monatomic gas and diatomic gas, b takes the value of 3 and 5, respectively.

Successful recovery of the Navier-Stokes equations from Eq. (6.1) requires that the Maxwellian distribution function (6.2) has to fulfill the following 7 moment relationships [Xu, 1998; Qu, 2009]:

$$\int g_M d\Xi = \rho, \tag{6.4a}$$

$$\int g_M \xi_\alpha d\Xi = \rho u_\alpha, \tag{6.4b}$$

$$\int g_M \left(\xi_\alpha \xi_\alpha + \sum_{j=1}^{K} \zeta_j^2 \right) d\Xi = \rho\left(u_\alpha u_\alpha + bRT\right), \tag{6.4c}$$

$$\int g_M \xi_\alpha \xi_\beta d\Xi = \rho u_\alpha u_\beta + p\delta_{\alpha\beta}, \tag{6.4d}$$

$$\int g_M \left(\xi_\alpha \xi_\alpha + \sum_{j=1}^{K} \zeta_j^2 \right) \xi_\beta d\Xi = \rho\left[u_\alpha u_\alpha + (b+2)RT\right]u_\beta, \tag{6.4e}$$

$$\int g_M \xi_\alpha \xi_\beta \xi_\chi d\Xi = p\left(u_\alpha \delta_{\beta\chi} + u_\beta \delta_{\chi\alpha} + u_\chi \delta_{\alpha\beta}\right) + \rho u_\alpha u_\beta u_\chi, \tag{6.4f}$$

$$\begin{aligned} &\int g_M \left(\xi_\chi \xi_\chi + \sum_{j=1}^{K} \zeta_j^2 \right) \xi_\alpha \xi_\beta d\Xi \\ &= \rho\left\{(b+2)R^2T^2\delta_{\alpha\beta} + \left[(b+4)u_\alpha u_\beta + u_\chi u_\chi \delta_{\alpha\beta}\right]RT + u_\chi u_\chi u_\alpha u_\beta\right\}, \end{aligned} \tag{6.4g}$$

where ξ_α, ξ_β, ξ_χ and u_α, u_β, u_χ are the components of the particle velocity and macroscopic flow velocity along the α -, β - and χ - directions, respectively. $d\Xi = d\xi_\alpha d\xi_\beta d\xi_\chi d\zeta_1 d\zeta_2 ... d\zeta_K$ is the volume element in the particle velocity space. And the integration takes place from $-\infty$ to ∞. In fact, Eqs. (6.4a)-(6.4c) are used to recover macroscopic quantities such as the density, the momentum and the energy, while Eqs. (6.4d)-(6.4g) are dedicated to recovering convective and diffusive fluxes in the momentum and the energy equations.

In view of the property of the exponential function, the Maxwellian distribution function in Eq. (6.2) can be considered as the product of two parts, i.e.,

$$g_M = g_{M1} g_{M2}, \tag{6.5}$$

where

$$g_{M1} = \rho \left(\frac{\lambda}{\pi} \right)^{\frac{D}{2}} e^{-\lambda \sum_{i=1}^{D} (\xi_i - u_i)^2}, \tag{6.6}$$

$$g_{M2} = \left(\frac{\lambda}{\pi} \right)^{\frac{K}{2}} e^{-\lambda \sum_{j=1}^{K} \zeta_j^2}. \tag{6.7}$$

Note that g_{M1} is only related to the translational particle velocity in D-dimensional space $\boldsymbol{\xi} = (\xi_1, \xi_2, ..., \xi_D)$ and g_{M2} is the function of $\boldsymbol{\zeta} = (\zeta_1, \zeta_2, ..., \zeta_K)$. Furthermore, the volume element $d\Xi$ can be expressed as $d\Xi = d\Xi_1 d\Xi_2$, where $d\Xi_1 = d\xi_\alpha d\xi_\beta d\xi_\chi$ and $d\Xi_2 = d\zeta_1 d\zeta_2 ... d\zeta_K$. Substituting Eqs. (6.5)-(6.7) into Eqs. (6.4a)-(6.4g) and noting that $\int g_{M2} d\Xi_2 = 1$, we have

$$\int g_M d\Xi = \int g_{M1} d\Xi_1, \tag{6.8a}$$

$$\int g_M \xi_\alpha d\Xi = \int g_{M1} \xi_\alpha d\Xi_1, \tag{6.8b}$$

$$\int g_M \left(\xi_\alpha \xi_\alpha + \sum_{j=1}^{K} \zeta_j^2 \right) d\Xi = \int g_{M1} \left(\xi_\alpha \xi_\alpha + 2e_p \right) d\Xi_1, \tag{6.8c}$$

$$\int g_M \xi_\alpha \xi_\beta d\Xi = \int g_{M1} \xi_\alpha \xi_\beta d\Xi_1 , \tag{6.8d}$$

$$\int g_M \left(\xi_\alpha \xi_\alpha + \sum_{j=1}^{K} \zeta_j^2 \right) \xi_\beta d\Xi = \int g_{M1} \left(\xi_\alpha \xi_\alpha + 2e_p \right) \xi_\beta d\Xi_1 , \tag{6.8e}$$

$$\int g_M \xi_\alpha \xi_\beta \xi_\chi d\Xi = \int g_{M1} \xi_\alpha \xi_\beta \xi_\chi d\Xi_1 , \tag{6.8f}$$

$$\int g_M \left(\xi_\chi \xi_\chi + \sum_{j=1}^{K} \zeta_j^2 \right) \xi_\alpha \xi_\beta d\Xi = \int g_{M1} \left(\xi_\chi \xi_\chi + 2e_p \right) \xi_\alpha \xi_\beta d\Xi_1 , \tag{6.8g}$$

where e_p is the particle potential energy defined as

$$e_p = \frac{1}{2} \int g_{M2} \sum_{j=1}^{K} \zeta_j^2 d\Xi_2 = \frac{K}{4\lambda} = \left[1 - \frac{D}{2}(\gamma - 1) \right] e . \tag{6.9}$$

Here, e is the potential energy of the mean flow, $e = p/[(\gamma-1)\rho]$, where p is the pressure. Note that in the derivation of Eq. (6.9), Eq. (6.3) and the expression of λ have been used. The moment relationships shown in Eqs. (6.8a)-(6.8g) serve as the basis of constructing gas kinetic flux solvers. In particular, the inviscid GKFS should satisfy Eqs. (6.8a)-(6.8e), while its viscous counterpart should be consistent with Eqs. (6.8a)-(6.8g).

The above Eqs. (6.8a)-(6.8g) and (6.9) reveal the independence between the particle potential energy and the translational particle velocity $\boldsymbol{\xi}$. Consequently, the original Maxwellian distribution function (Eq. (6.2)) can be simplified into Eq. (6.6) and the particle potential energy can be separately evaluated through Eq. (6.9). This conclusion is applicable not only for the GKFS, but also for various lattice Boltzmann models, which indicates that only one parameter is needed to represent the particle potential energy in lattice Boltzmann models. Based on this principle, the formulations of the particle potential energy proposed by Kataoka and Tsuthara [2004a; 2004b] and Dellar [2008] may not be physically robust since they are dependent on the translational particle velocity. On the contrary, recent progresses made by Qu *et al.* [2007a;

2007b] and Yang *et al.* [2012; 2013a] recognized such independency and developed several simpler lattice Boltzmann models.

Although significant simplification has been made by replacing the original Maxwellian distribution function g_M with g_{M1}, the integration is still performed over the infinite particle velocity space, due to which the asymptotic solution to the Boltzmann equation can not be applied straightforwardly. To resolve this issue, Shu and his coworkers [Yang *et al.*, 2014b; Yang *et al.*, 2017a] proposed a novel transformation of the distribution function to convert the infinite integral domain of the particle velocity space into a finite one. Based on this transformation, a series of simplified distribution functions can be developed. Before presenting these simplified distribution functions, the mean kinetic energy of particles is defined as

$$c^2 = \frac{\frac{1}{2}\int \sum_{i=1}^{D}(\xi_i - u_i)^2 g_{M1} d\Xi_1}{\frac{1}{2}\int g_{M1} d\Xi_1} = \frac{D}{2\lambda} = D(\gamma - 1)e. \tag{6.10}$$

Based on the mean kinetic energy, the circular function in the two-dimensional (2D) case and the sphere function in the three-dimensional (3D) case can be developed. Detailed derivations will be presented in the following sections.

6.1.1 *Two-dimensional simplified distribution: circular function*

In the 2D condition, the simplified Maxwellian distribution function (Eq. (6.6)) can be equivalently written as

$$g_{M1} = \rho\left(\frac{\lambda}{\pi}\right)^{\frac{D}{2}} e^{-\lambda\left[(\xi_1 - u_1)^2 + (\xi_2 - u_2)^2\right]}. \tag{6.11}$$

Letting $r^2 = (\xi_1 - u_1)^2 + (\xi_2 - u_2)^2$, the mass conservation relation recovered by Eq. (6.8a) in the polar coordinate system can be expressed as

$$\rho = \int g_{M1} d\Xi_1 = \int_0^{2\pi} \int_0^{\infty} \rho \left(\frac{\lambda}{\pi} \right)^{\frac{D}{2}} e^{-\lambda r^2} r dr d\theta . \tag{6.12}$$

Suppose that all particles in the particle velocity space are concentrated on a circle with the radius of c and centered at (u_1, u_2), i.e., $c^2 = (\xi_1 - u_1)^2 + (\xi_2 - u_2)^2$. Thus, the distribution function along the radial direction becomes a Dirac delta function, which further simplifies Eq. (6.12) as

$$\rho = \int_0^{2\pi} \int_0^{\infty} \rho \left(\frac{\lambda}{\pi} \right)^{\frac{D}{2}} e^{-\lambda r^2} \delta(r-c) r dr d\theta = 2\pi c \rho \left(\frac{\lambda}{\pi} \right)^{\frac{D}{2}} e^{-\lambda c^2} , \tag{6.13}$$

where $\delta(r-c)$ is the Dirac delta function that follows the standard definition of $\int_0^{\infty} f(r)\delta(r-c) dr = f(c)$ ($f(r)$ is a smooth function). From Eq. (6.13), we can get the 2D simplified distribution function as follows:

$$g_C = \begin{cases} \dfrac{\rho}{2\pi} & \text{if } (\xi_1 - u_1)^2 + (\xi_2 - u_2)^2 = c^2, \\ 0 & \text{otherwise.} \end{cases} \tag{6.14}$$

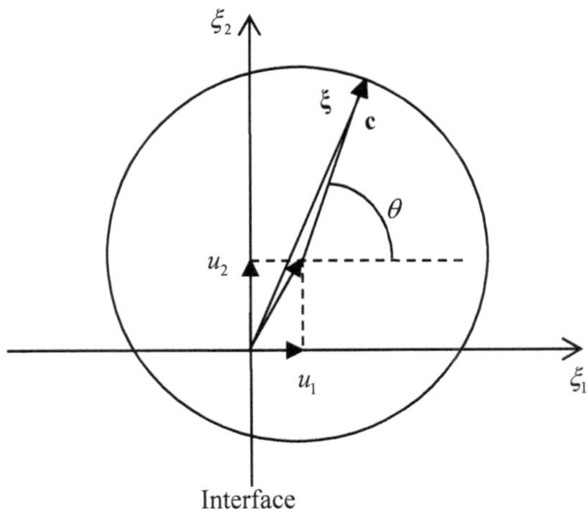

Figure 6.1. Sketch of the circular function.

As a result, Eq. (6.8a) can then be derived as

$$\rho = \int_0^{2\pi}\int_0^{\infty} \rho\left(\frac{\lambda}{\pi}\right)^{\frac{D}{2}} e^{-\lambda r^2}\delta(r-c) r dr d\theta = \int_0^{2\pi} g_C d\theta . \tag{6.15}$$

The 2D simplified distribution function is termed as "circular function" due to its distribution along a circle. As shown in Eq. (6.14) and Figure 6.1, on the circle, the particle velocity components in the Cartesian coordinate system can be written as

$$\xi_1 = u_1 + c\cos(\theta), \tag{6.16a}$$

$$\xi_2 = u_2 + c\sin(\theta), \tag{6.16b}$$

where θ is the angle between the radial direction of the circle and the normal direction of the cell interface.

With notations $c_1 = c\cos(\theta)$, $c_2 = c\sin(\theta)$ and $\xi_\alpha = u_\alpha + c_\alpha$, by substituting Eqs. (6.14) and (6.16a)-(6.16b) into the right-hand side of Eqs. (6.8a)-(6.8g), the following 7 moment relationships (from the zeroth order to the fourth order) are obtained:

$$\int_0^{2\pi} g_C d\theta = \rho, \tag{6.17a}$$

$$\int_0^{2\pi} g_C \xi_\alpha d\theta = \rho u_\alpha, \tag{6.17b}$$

$$\int_0^{2\pi} g_C \left(\xi_\alpha\xi_\alpha + 2e_p\right) d\theta = \rho\left(u_\alpha u_\alpha + bRT\right), \tag{6.17c}$$

$$\int_0^{2\pi} g_C \xi_\alpha \xi_\beta d\theta = \rho u_\alpha u_\beta + p\delta_{\alpha\beta}, \tag{6.17d}$$

$$\int_0^{2\pi} g_C \left(\xi_\alpha\xi_\alpha + 2e_p\right)\xi_\beta d\theta = \rho\left[u_\alpha u_\alpha + (b+2)RT\right]u_\beta, \tag{6.17e}$$

$$\int_0^{2\pi} g_C \xi_\alpha \xi_\beta \xi_\chi d\theta = p\left(u_\alpha\delta_{\beta\chi} + u_\beta\delta_{\chi\alpha} + u_\chi\delta_{\alpha\beta}\right) + \rho u_\alpha u_\beta u_\chi, \tag{6.17f}$$

$$\int_0^{2\pi} g_C\left(\xi_\chi\xi_\chi + 2e_p\right)\xi_\alpha\xi_\beta d\theta = \rho\left\{bR^2T^2\delta_{\alpha\beta} + \left[(b+4)u_\alpha u_\beta + u_\chi u_\chi\delta_{\alpha\beta}\right]RT + u_\chi u_\chi u_\alpha u_\beta\right\}. \quad (6.17g)$$

Comparing with Eqs. (6.4a)-(6.4g), the integral domain in Eqs. (6.17a)-(6.17g) is finite. This modification removes the numerical obstacles brought by the infinite integral domain and thus makes the application of the circular function be much easier than the original Maxwellian distribution function for reconstructing the GKFS by the asymptotic solution to the Boltzmann equation. Comparing the two sets of equations, one could find a slight difference between Eq. (6.17g) and Eq. (6.4g). Through the Chapman-Enskog expansion analysis (see **Section 6.1.3**), it reveals that this difference does not interfere with recovering the Navier-Stokes equations, but will yield a fixed value of γ for the Prandtl number. Such limitation could be removed by a simple modification of the energy flux, which will be elaborated in **Section 6.5**.

6.1.2 *Three-dimensional simplified distribution: sphere function*

For the 3D case, the simplified Maxwellian distribution function (Eq. (6.6)) can be expressed as

$$g_{M1} = \rho\left(\frac{\lambda}{\pi}\right)^{\frac{D}{2}} e^{-\lambda\left[(\xi_1-u_1)^2+(\xi_2-u_2)^2+(\xi_3-u_3)^2\right]}. \quad (6.18)$$

Similar to the derivation of the circular function, the 3D simplified distribution function can be obtained by transforming Eq. (6.8a) into the spherical coordinate system and assuming that all particles in the particle velocity space are concentrated on the surface of a sphere with the radius of c and centered at (u_1, u_2, u_3) [Yang *et al.*, 2015; Yang *et al.*, 2017a]. The obtained distribution function exhibits the characteristics of the Dirac delta function along the radial direction. Assuming $c^2 = (\xi_1 - u_1)^2 + (\xi_2 - u_2)^2 + (\xi_3 - u_3)^2$, the mass conservation relation in Eq. (6.8a) can be rewritten as

$$\begin{aligned}\rho &= \int_0^{2\pi}\int_0^{\pi}\int_0^{\infty} \rho\left(\frac{\lambda}{\pi}\right)^{\frac{D}{2}} e^{-\lambda r^2}\delta(r-c)r^2\sin(\varphi)drd\varphi d\theta \\ &= 4\pi c^2\rho\left(\frac{\lambda}{\pi}\right)^{\frac{D}{2}} e^{-\lambda c^2}.\end{aligned} \tag{6.19}$$

The above equation yields the following 3D simplified distribution function:

$$g_S = \begin{cases} \dfrac{\rho}{4\pi} & \text{if } (\xi_1 - u_1)^2 + (\xi_2 - u_2)^2 + (\xi_3 - u_3)^2 = c^2, \\ 0 & \text{otherwise.} \end{cases} \tag{6.20}$$

Since the 3D simplified distribution function is defined on the spherical surface, for simplicity, it is termed as "sphere function" hereafter. From Eq. (6.20) and Figure 6.2, the components of the particle velocity vector projected in the Cartesian coordinate system can be written as

$$\xi_1 = u_1 + c\sin(\varphi)\cos(\theta), \tag{6.21a}$$

$$\xi_2 = u_2 + c\sin(\varphi)\sin(\theta), \tag{6.21b}$$

$$\xi_3 = u_3 + c\cos(\varphi). \tag{6.21c}$$

The definition of angles θ and φ can refer to Figure 6.2.

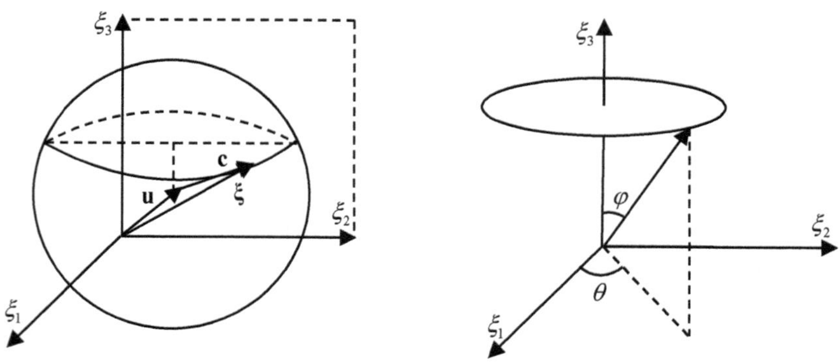

Figure 6.2. Sketch of the sphere function.

With the notations of $c_1 = c\sin(\varphi)\cos(\theta)$, $c_2 = c\sin(\varphi)\sin(\theta)$, $c_3 = c\cos(\varphi)$ and $\xi_\alpha = u_\alpha + c_\alpha$, by substituting Eqs. (6.20) and (6.21a)-(6.21c) into the right-hand side of Eqs. (6.8a)-(6.8g), the following 7 moment relationships can be obtained:

$$\int_0^{2\pi}\int_0^{\pi} g_S \sin(\varphi)\,d\varphi d\theta = \rho, \tag{6.22a}$$

$$\int_0^{2\pi}\int_0^{\pi} g_S \xi_\alpha \sin(\varphi)\,d\varphi d\theta = \rho u_\alpha, \tag{6.22b}$$

$$\int_0^{2\pi}\int_0^{\pi} g_S \left(\xi_\alpha\xi_\alpha + 2e_p\right)\sin(\varphi)\,d\varphi d\theta = \rho\left(u_\alpha u_\alpha + bRT\right), \tag{6.22c}$$

$$\int_0^{2\pi}\int_0^{\pi} g_S \xi_\alpha\xi_\beta \sin(\varphi)\,d\varphi d\theta = \rho u_\alpha u_\beta + p\delta_{\alpha\beta}, \tag{6.22d}$$

$$\begin{aligned}&\int_0^{2\pi}\int_0^{\pi} g_S \left(\xi_\alpha\xi_\alpha + 2e_p\right)\xi_\beta \sin(\varphi)\,d\varphi d\theta \\ &= \rho\left[u_\alpha u_\alpha + (b+2)RT\right]u_\beta,\end{aligned} \tag{6.22e}$$

$$\begin{aligned}&\int_0^{2\pi}\int_0^{\pi} g_S \xi_\alpha\xi_\beta\xi_\chi \sin(\varphi)\,d\varphi d\theta \\ &= p\left(u_\alpha\delta_{\beta\chi} + u_\beta\delta_{\chi\alpha} + u_\chi\delta_{\alpha\beta}\right) + \rho u_\alpha u_\beta u_\chi,\end{aligned} \tag{6.22f}$$

$$\begin{aligned}&\int_0^{2\pi}\int_0^{\pi} g_S \left(\xi_\chi\xi_\chi + 2e_p\right)\xi_\alpha\xi_\beta \sin(\varphi)\,d\varphi d\theta \\ &= \rho\left\{bR^2T^2\delta_{\alpha\beta} + \left[(b+4)u_\alpha u_\beta + u_\chi u_\chi\delta_{\alpha\beta}\right]RT + u_\chi u_\chi u_\alpha u_\beta\right\}.\end{aligned} \tag{6.22g}$$

The merits of the sphere function are very similar to those of the circular function. It transforms the infinite integral domain of the original Maxwellian distribution function into a finite one, which facilitates the implementation when reconstructing the GKFS by the asymptotic solution to the Boltzmann equation and could remarkably improve computational efficiency. However, detailed Chapman-Enskog expansion analysis would reveal that the Prandtl number of the recovered macroscopic governing equations is fixed at γ. Further modifications are needed to enrich its applicability in various thermal conditions.

6.1.3 *Chapman-Enskog expansion analysis to recover Euler/Navier-Stokes equations*

Within the continuum flow regime, both the Euler/Navier-Stokes equations and the Boltzmann equation (6.1) are mathematical models of the same physical problem. Their consistency could be revealed through the Chapman-Enskog expansion analysis in the continuous limit. In GKFS, the Euler/Navier-Stokes equations are globally resolved, while the numerical fluxes on the cell interface are locally reconstructed by the asymptotic solution to the Boltzmann equation, i.e., the distribution function truncated to the Navier-Stokes level on the cell interface. In this section, the Chapman-Enskog expansion analysis will be performed on the cell interface to establish the relationship between the distribution function with simplified equilibria and the numerical fluxes of the Euler/Navier-Stokes equations.

Since the moment relationships of the circular function and those of the sphere function are identical, the Boltzmann equation with sphere function is taken as an example to illustrate the Chapman-Enskog expansion analysis. For simplicity, Eq. (6.1) is rewritten as

$$f = f^{eq} - \tau Df, \tag{6.23}$$

where $D = (\partial_t + \xi \cdot \nabla)$ is the operator of substantial derivative and $f^{eq} = g$ is the equilibrium state. The expansion of Eq. (6.23) in terms of τ yields

$$f = f^{eq} - \tau Df^{eq} + O(\tau^2). \tag{6.24}$$

Here, the equilibrium state is taken as the sphere function, i.e., $f^{eq} = g_S$. To simplify the notation, the integration of an arbitrary function ϕ over the spherical surface is defined as

$$\langle \phi \rangle = \int_0^{2\pi} \int_0^{\pi} \phi \sin(\varphi) d\varphi d\theta, \tag{6.25}$$

where ϕ represents any function. To fulfill the compatibility condition, the non-equilibrium part of the distribution function should not contribute to the calculation of conservative flow variables of the Euler/Navier-Stokes equations, i.e.,

$$\left\langle \boldsymbol{\varphi}_{a}\left(f - f^{eq}\right)\right\rangle = 0, \tag{6.26}$$

where $\boldsymbol{\varphi}_{\alpha}$ stands for the moment vector given by

$$\boldsymbol{\varphi}_{\alpha} = \left(1, \boldsymbol{\xi}, \frac{1}{2}|\boldsymbol{\xi}|^{2} + e_{p}\right)^{T}. \tag{6.27}$$

In general conditions, the particle velocity $\boldsymbol{\xi}$ and moment $\boldsymbol{\varphi}_{a}$ are determined by the macroscopic flow variables and the angles θ and φ for the sphere function, as shown in Eqs. (6.21a)-(6.21c). However, since the Chapman-Enskog expansion analysis is carried out locally on the cell interface, $\boldsymbol{\varphi}_{a}$ can be viewed as a function of angles θ and φ only.

Substituting Eq. (6.24) into Eq. (6.26) gives

$$\partial_{t}\left\langle \boldsymbol{\varphi}_{a} f^{eq}\right\rangle + \nabla \cdot \left\langle \boldsymbol{\xi}\boldsymbol{\varphi}_{a} f^{eq}\right\rangle = O(\tau). \tag{6.28}$$

Applying the relationships in Eqs. (6.22a)-(6.22e), Eq. (6.28) can be rewritten as

$$\partial_{t}\rho + \nabla \cdot (\rho \mathbf{u}) = O(\tau), \tag{6.29a}$$

$$\partial_{t}(\rho \mathbf{u}) + \nabla \cdot (\rho \mathbf{u}\mathbf{u} + p\mathbf{I}) = O(\tau), \tag{6.29b}$$

$$\partial_{t}(\rho E) + \nabla \cdot \left[(\rho E + p)\mathbf{u}\right] = O(\tau), \tag{6.29c}$$

where $E = e + \mathbf{u} \cdot \mathbf{u}/2$ is the total energy and $\mathbf{I}$ is the unit tensor. It can be seen that, when truncated at $O(\tau)$, the Boltzmann equation can recover the Euler equations.

In order to derive the Navier-Stokes equations from the Boltzmann equation, higher-order terms should be considered. To begin with, let us apply the operator D to Eq. (6.24)

$$Df = Df^{eq} - D\left(\tau D f^{eq}\right) + O\left(\tau^{2}\right). \tag{6.30}$$

From Eq. (6.23), we have $Df = \left(f^{eq} - f\right)/\tau$. Furthermore, multiplying $\boldsymbol{\varphi}_a$ on both sides of Eq. (6.30) and performing integration over the spherical surface for the resultant equation yield

$$\begin{aligned} &\partial_t \left\langle \boldsymbol{\varphi}_a f^{eq} \right\rangle + \nabla \cdot \left\langle \boldsymbol{\xi} \boldsymbol{\varphi}_a f^{eq} \right\rangle \\ &= \partial_t \left\langle \boldsymbol{\varphi}_a \left(\tau D f^{eq}\right) \right\rangle + \nabla \cdot \left\langle \boldsymbol{\xi} \boldsymbol{\varphi}_a \left(\tau D f^{eq}\right) \right\rangle + O\left(\tau^2\right). \end{aligned} \tag{6.31}$$

Note that the compatibility condition has been used in the derivation of Eq. (6.31). Substituting Eq. (6.24) into the first term on the right-hand side of Eq. (6.31) and utilizing the compatibility condition, Eq. (6.31) can be reduced to

$$\partial_t \left\langle \boldsymbol{\varphi}_a f^{eq} \right\rangle + \nabla \cdot \left\langle \boldsymbol{\xi} \boldsymbol{\varphi}_a f^{eq} \right\rangle = \nabla \cdot \left\langle \boldsymbol{\xi} \boldsymbol{\varphi}_a \left(\tau D f^{eq}\right) \right\rangle + O\left(\tau^2\right). \tag{6.32}$$

Relevant components of the moment vector $\boldsymbol{\varphi}_a$ in Eq. (6.32) would yield the continuity, the momentum and the energy equations at the macroscopic scale.

The first component of $\boldsymbol{\varphi}_a$ is 1, which gives the following relationship when being substituted into Eq. (6.32):

$$\partial_t \left\langle f^{eq} \right\rangle + \nabla \cdot \left\langle \boldsymbol{\xi} f^{eq} \right\rangle = \nabla \cdot \left\langle \boldsymbol{\xi} \left(\tau D f^{eq}\right) \right\rangle + O\left(\tau^2\right). \tag{6.33}$$

Utilizing Eq. (6.24) and the compatibility condition (6.26) allows further simplification of the above equation into

$$\partial_t \left\langle f^{eq} \right\rangle + \nabla \cdot \left\langle \boldsymbol{\xi} f^{eq} \right\rangle = O\left(\tau^2\right). \tag{6.34}$$

Applying the relationships in Eqs. (6.22a)-(6.22b), Eq. (6.34) can be rewritten as

$$\partial_t \rho + \nabla \cdot \left(\rho \mathbf{u}\right) = O\left(\tau^2\right), \tag{6.35}$$

which recovers the continuity equation in the order of $O\left(\tau^2\right)$.

Considering the second component of $\boldsymbol{\varphi}_a$ in Eq. (6.32) gives

$$\begin{aligned}&\partial_t\left\langle \boldsymbol{\xi} f^{eq}\right\rangle+\nabla\cdot\left\langle \boldsymbol{\xi\xi} f^{eq}\right\rangle\\&=\nabla\cdot\left[\tau\left(\partial_t\left\langle \boldsymbol{\xi\xi} f^{eq}\right\rangle+\nabla\cdot\left\langle \boldsymbol{\xi\xi\xi} f^{eq}\right\rangle\right)\right]+O\left(\tau^2\right).\end{aligned} \tag{6.36}$$

Applying the relationships (6.22b), (6.22d) and (6.22f) and noting that

$$\begin{aligned}\partial_t\left\langle \boldsymbol{\xi\xi} f^{eq}\right\rangle&=\partial_t\left(\rho\mathbf{uu}+p\mathbf{I}\right)\\&=\partial_t\left(\rho\mathbf{u}\right)\mathbf{u}+\mathbf{u}\partial_t\left(\rho\mathbf{u}\right)-\mathbf{uu}\partial_t\rho+\left(\gamma-1\right)\partial_t\left(\rho e\right)\mathbf{I}\\&=-\nabla\cdot\left(\rho\mathbf{uuu}\right)-\left[\left(\nabla p\right)\mathbf{u}+\mathbf{u}\nabla p\right]\\&\quad-\nabla\cdot\left(p\mathbf{u}\right)\mathbf{I}-\left(\gamma-1\right)p\left(\nabla\cdot\mathbf{u}\right)\mathbf{I}+O\left(\tau\right),\end{aligned}$$

$$\begin{aligned}\nabla\cdot\left\langle \boldsymbol{\xi\xi\xi} f^{eq}\right\rangle&=\partial_\chi\left\langle \xi_\alpha\xi_\beta\xi_\chi f^{eq}\right\rangle\\&=\partial_\chi\left[p(u_\alpha\delta_{\beta\chi}+u_\beta\delta_{\chi\alpha}+u_\chi\delta_{\alpha\beta})+\rho u_\alpha u_\beta u_\chi\right]\\&=\nabla\cdot\left(\rho\mathbf{uuu}\right)+\nabla\cdot\left(p\mathbf{u}\right)\mathbf{I}\\&\quad+\left[\left(\nabla p\right)\mathbf{u}+\mathbf{u}\nabla p\right]+p\left[\nabla\mathbf{u}+\left(\nabla\mathbf{u}\right)^T\right],\end{aligned}$$

we have

$$\partial_t\left(\rho\mathbf{u}\right)+\nabla\cdot\left(\rho\mathbf{uu}+p\mathbf{I}\right)=\nabla\cdot\boldsymbol{\Pi}+O\left(\tau^2\right), \tag{6.37}$$

where

$$\boldsymbol{\Pi}=\mu\left[\left(\nabla\mathbf{u}+\left(\nabla\mathbf{u}\right)^T\right)-\frac{2}{3}(\nabla\cdot\mathbf{u})\mathbf{I}\right]+\mu_b\left(\nabla\cdot\mathbf{u}\right)\mathbf{I},$$

$$\mu=\tau p,$$

$$\mu_b=\left(\frac{5}{3}-\gamma\right)\mu.$$

Note that Eqs. (6.29a)-(6.29c) have been used for the calculation of $\partial_t\left(\rho\mathbf{uu}+p\mathbf{I}\right)$. Here, μ is the dynamic viscosity and μ_b is the bulk viscosity. It can be seen from Eq. (6.37) that, by applying the sphere function on the cell interface, the momentum equation can basically be recovered from the Boltzmann equation in the order of $O(\tau^2)$. However, one limitation of the recovered equation is that the bulk viscosity would vanish only for the monatomic gas which has the specific heat ratio of $\gamma = 5/3$.

Finally, we consider the last component of $\boldsymbol{\varphi}_a$ in Eq. (6.32), which yields

$$\begin{aligned}
&\partial_t \left\langle \left(\frac{1}{2}|\boldsymbol{\xi}|^2 + e_p\right) f^{eq} \right\rangle + \nabla \cdot \left\langle \left(\frac{1}{2}|\boldsymbol{\xi}|^2 + e_p\right) \boldsymbol{\xi} f^{eq} \right\rangle \\
&= \nabla \cdot \left[\tau \left(\partial_t \left\langle \left(\frac{1}{2}|\boldsymbol{\xi}|^2 + e_p\right) \boldsymbol{\xi} f^{eq} \right\rangle + \nabla \cdot \left\langle \left(\frac{1}{2}|\boldsymbol{\xi}|^2 + e_p\right) \boldsymbol{\xi}\boldsymbol{\xi} f^{eq} \right\rangle \right) \right] \\
&\quad + O(\tau^2).
\end{aligned} \tag{6.38}$$

Applying the relationships (6.22c), (6.22e) and (6.22g) and noting that

$$\begin{aligned}
\partial_t \left\langle \left(\frac{1}{2}|\boldsymbol{\xi}|^2 + e_p\right) \boldsymbol{\xi} f^{eq} \right\rangle &= \partial_t \left[(\rho E + p) \mathbf{u} \right] \\
&= -\nabla \cdot (\rho E \mathbf{u}\mathbf{u}) - 2\mathbf{u} \nabla \cdot (p\mathbf{u}) - E \nabla p - p\mathbf{u} \cdot \nabla \mathbf{u} \\
&\quad - (\gamma - 1) e \nabla p - (\gamma - 1) p \mathbf{u} \nabla \cdot \mathbf{u} + O(\tau),
\end{aligned}$$

$$\begin{aligned}
&\nabla \cdot \left\langle \left(\frac{1}{2}|\boldsymbol{\xi}|^2 + e_p\right) \boldsymbol{\xi}\boldsymbol{\xi} f^{eq} \right\rangle \\
&= \partial_\beta \left\langle \left(\frac{1}{2}\xi_\chi \xi_\chi + e_p\right) \xi_\alpha \xi_\beta f^{eq} \right\rangle \\
&= \partial_\beta \left\{ \frac{\rho}{2} \left[b R^2 T^2 \delta_{\alpha\beta} + \left[(b+4) u_\alpha u_\beta + u_\chi u_\chi \delta_{\alpha\beta} \right] RT + u_\chi u_\chi u_\alpha u_\beta \right] \right\} \\
&= \nabla (pE) + \nabla \cdot (\rho E \mathbf{u}\mathbf{u}) + 2\mathbf{u} \nabla \cdot (p\mathbf{u}) + p\mathbf{u} \cdot \left[\nabla \mathbf{u} + (\nabla \mathbf{u})^T \right],
\end{aligned}$$

we have

$$\begin{aligned}
&\partial_t (\rho E) + \nabla \cdot \left[(\rho E + p) \mathbf{u} \right] \\
&= \nabla \cdot (\kappa \nabla T) + \nabla \cdot (\mathbf{u} \cdot \boldsymbol{\Pi}) - \nabla \cdot \left[\mu/\rho \nabla p \right] + O(\tau^2).
\end{aligned} \tag{6.39}$$

Here, κ is the thermal conductivity, which is related to the collision time scale τ through

$$\kappa = \tau p c_V = \mu c_V . \tag{6.40}$$

And the Prandtl number thus yields

$$Pr = \frac{\mu c_p}{\kappa} = \frac{\mu c_p}{\mu c_v} = \gamma . \tag{6.41}$$

Note that Eqs. (6.29a)-(6.29c) have been used for the calculation of $\partial_t[(\rho E + p)\mathbf{u}]$. Overall, the energy equation can be recovered from the Boltzmann equation with the sphere function adopted on the cell interface in the order of $O(\tau^2)$. However, it is noteworthy that the resultant Prandtl number is fixed at γ and an error term of $\nabla \cdot [\mu/\rho \nabla p]$ emerges. Since this error term is actually a second-order derivative of the pressure, it is neglected in the developments of C-GKFS and S-GKFS.

From Eq. (6.32), it can be seen that the numerical fluxes $\mathbf{F}$ of the Navier-Stokes equations at the cell interface $\mathbf{x}$ and time t are actually computed by

$$\mathbf{F} = \langle \boldsymbol{\xi} \boldsymbol{\varphi}_a f \rangle \big|_{(\mathbf{x},t)} . \tag{6.42}$$

Note that $\mathbf{F}$ consists of the inviscid part $\mathbf{F}_c$ and the viscous part $\mathbf{F}_v$. $f(\mathbf{x},t)$ is the local solution to the Boltzmann equation with the sphere function adopted on the cell interface, which is given by

$$f(\mathbf{x},t) = f^{eq}(\mathbf{x},t) - \tau D f^{eq}(\mathbf{x},t) + O(\tau^2) . \tag{6.43}$$

In fact, as shown in the above Chapman-Enskog expansion analysis, the first term and the second term on the right-hand side of Eq. (6.43) contribute to the calculation of $\mathbf{F}_c$ and $\mathbf{F}_v$, respectively. By using the backward difference method, Eq. (6.43) can be approximated by

$$\begin{aligned} f(\mathbf{x},t) = & f^{eq}(\mathbf{x},t) + \tau_0 \left[f^{eq}(\mathbf{x} - \boldsymbol{\xi}\delta_t, t - \delta_t) - f^{eq}(\mathbf{x},t) \right] \\ & + O\left(\tau^2, \tau\delta_t, |\boldsymbol{\xi}|^2 \tau\delta_t\right). \end{aligned} \tag{6.44}$$

Furthermore, by substituting Eq. (6.44) into Eq. (6.42), we have

$$\mathbf{F} = \left\langle \boldsymbol{\xi}\boldsymbol{\varphi}_a f^{eq} \right\rangle\Big|_{(\mathbf{x},t)} + \tau_0 \left[\left\langle \boldsymbol{\xi}\boldsymbol{\varphi}_a f^{eq} \right\rangle\Big|_{(\mathbf{x}-\boldsymbol{\xi}\delta_t, t-\delta_t)} - \left\langle \boldsymbol{\xi}\boldsymbol{\varphi}_a f^{eq} \right\rangle\Big|_{(\mathbf{x},t)} \right] + O\left(\tau^2, \tau\delta_t, |\boldsymbol{\xi}|^2 \tau\delta_t\right), \tag{6.45}$$

where $\tau_0 = \tau/\delta_t$ is the dimensionless collision time scale and δ_t is the streaming time step. $|\boldsymbol{\xi}|\delta_t$ represents the streaming distance, which is in the order of the cell size, i.e., $|\boldsymbol{\xi}|\delta_t \propto \Delta x$. In addition, the collision time scale τ is usually smaller than the streaming time step δ_t in the continuous limit due to the small dynamic viscosity. As a result, the proposed strategy of evaluating numerical fluxes could expect the second-order of accuracy in both space and time. In the following sections, Eq. (6.45) will serve as the basis to reconstruct numerical fluxes of the Euler/Navier-Stokes equations for the C-GKFS, the S-GKFS and the compressible LBFS.

6.2 Circular Function-Based GKFS for Inviscid Flows

In **Section 6.1**, two simplified equilibrium distribution functions are derived from the Maxwellian distribution function, and connections between the local solution to the Boltzmann equation with simplified distribution functions and the numerical fluxes of macroscopic governing equations are established through the Chapman-Enskog expansion analysis. Such a strategy can be adopted in the reconstruction of inviscid fluxes in the Euler equations [Yang *et al.*, 2013b], which yields the C-GKFS to be presented in this section for simulation of 2D inviscid flows.

For the 2D case, the Euler equations can be written as

$$\frac{\partial \mathbf{W}}{\partial t} + \frac{\partial \mathbf{F}_{cx}}{\partial x} + \frac{\partial \mathbf{F}_{cy}}{\partial y} = \mathbf{0}, \tag{6.46}$$

where the conservative variable vector $\mathbf{W}$ and the convective flux vectors $\mathbf{F}_{cx}$ and $\mathbf{F}_{cy}$ are respectively given by

$$\mathbf{W}=\begin{bmatrix}\rho\\ \rho u\\ \rho v\\ \rho E\end{bmatrix},\ \mathbf{F}_{cx}=\begin{bmatrix}\rho u\\ \rho uu+p\\ \rho uv\\ (\rho E+p)u\end{bmatrix},\ \mathbf{F}_{cy}=\begin{bmatrix}\rho v\\ \rho vu\\ \rho vv+p\\ (\rho E+p)v\end{bmatrix}. \tag{6.47}$$

Here, $\mathbf{u}=(u,v)$ is the velocity vector expressed in the global Cartesian coordinate system. For the ease of solving flow problems on unstructured meshes with curved boundaries, the finite volume method (FVM) is usually utilized to discretize Eq. (6.46) into the following form:

$$\frac{\partial \mathbf{W}_i}{\partial t}=-\frac{1}{\Omega_i}\sum_{j=1}^{N_f}\mathbf{F}_{cnj}S_j\,, \tag{6.48}$$

where i is the index of the cell, Ω_i and N_f represent the volume and the number of faces of cell i, respectively. S_j denotes the area of the jth interface of the control volume i. $\mathbf{F}_{cn}$ is convective flux vector across the cell interface, which is given by

$$\mathbf{F}_{cn}=n_x\mathbf{F}_{cx}+n_y\mathbf{F}_{cy}=\begin{bmatrix}\rho u_1\\ \rho uu_1+pn_x\\ \rho vu_1+pn_y\\ (\rho E+p)u_1\end{bmatrix}, \tag{6.49}$$

where $\mathbf{n}=(n_x,n_y)$ denotes the unit normal vector of the cell interface; u_1 is the normal velocity component on the cell interface. In GKFS, the conservative variables are defined at cell centers. Thus, proper evaluation of the numerical flux vector $\mathbf{F}_{cn}$ on the cell interface would be essential to the solution of Eq. (6.48).

To facilitate analysis, a local coordinate system is defined on the cell interface with x_1-axis pointing to the normal direction and x_2-axis aligned with the tangential direction, which forms a right-hand coordinate system in the 2D case. The unit vectors of x_1-axis and x_2-axis are denoted as $\mathbf{n}_1=(n_{1x},n_{1y})$ and $\mathbf{n}_2=(n_{2x},n_{2y})$, respectively. In the local coordinate system, the vectors of the conservative variable and numerical flux can be expressed as

$$\overline{\mathbf{W}} = \begin{bmatrix} \rho \\ \rho u_1 \\ \rho u_2 \\ \rho E \end{bmatrix}, \quad \overline{\mathbf{F}}_c = \begin{bmatrix} \rho u_1 \\ \rho u_1 u_1 + p \\ \rho u_1 u_2 \\ (\rho E + p) u_1 \end{bmatrix}, \tag{6.50}$$

where $\mathbf{u} = (u_1, u_2)$ is the velocity vector expressed in the local coordinate system. These vectors can be projected to the global Cartesian coordinate system through the following formulations:

$$\mathbf{W} = \begin{bmatrix} \overline{\mathbf{W}}(1) \\ n_{1x}\overline{\mathbf{W}}(2) + n_{2x}\overline{\mathbf{W}}(3) \\ n_{1y}\overline{\mathbf{W}}(2) + n_{2y}\overline{\mathbf{W}}(3) \\ \overline{\mathbf{W}}(4) \end{bmatrix}, \quad \mathbf{F}_{cn} = \begin{bmatrix} \overline{\mathbf{F}}_c(1) \\ n_{1x}\overline{\mathbf{F}}_c(2) + n_{2x}\overline{\mathbf{F}}_c(3) \\ n_{1y}\overline{\mathbf{F}}_c(2) + n_{2y}\overline{\mathbf{F}}_c(3) \\ \overline{\mathbf{F}}_c(4) \end{bmatrix}. \tag{6.51}$$

Since $\mathbf{n}_1 = \mathbf{n} = (n_x, n_y)$ and $\mathbf{n}_2 = (-n_y, n_x)$, Eq. (6.51) can be rewritten as

$$\mathbf{W} = \begin{bmatrix} \overline{\mathbf{W}}(1) \\ n_x\overline{\mathbf{W}}(2) - n_y\overline{\mathbf{W}}(3) \\ n_x\overline{\mathbf{W}}(3) + n_y\overline{\mathbf{W}}(2) \\ \overline{\mathbf{W}}(4) \end{bmatrix}, \quad \mathbf{F}_{cn} = \begin{bmatrix} \overline{\mathbf{F}}_c(1) \\ n_x\overline{\mathbf{F}}_c(2) - n_y\overline{\mathbf{F}}_c(3) \\ n_x\overline{\mathbf{F}}_c(3) + n_y\overline{\mathbf{F}}_c(2) \\ \overline{\mathbf{F}}_c(4) \end{bmatrix}. \tag{6.52}$$

Here, $\overline{\mathbf{W}}(i)$ and $\overline{\mathbf{F}}_c(i)$, $i = 1,2,3,4$, are the components of $\overline{\mathbf{W}}$ and $\overline{\mathbf{F}}_c$, respectively. As shown in Eq. (6.52), the calculation of $\mathbf{F}_{cn}$ in the global Cartesian coordinate system is equivalent to evaluating $\overline{\mathbf{F}}_c$ in the local coordinate system.

According to the relationship between the distribution function and the macroscopic flux vector (see Eq. (6.45)), for the circular function, $\overline{\mathbf{F}}_c$ on the cell interface can be calculated by

$$\begin{aligned} \overline{\mathbf{F}}_c &= \int_0^{2\pi} \xi_1^{face} \overline{\boldsymbol{\varphi}}_\alpha^{face} g_C^{face} d\theta \\ &\quad + \tau_0 \left[\int_0^{2\pi} \xi_1^{cir} \overline{\boldsymbol{\varphi}}_\alpha^{cir} g_C^{cir} d\theta - \int_0^{2\pi} \xi_1^{face} \overline{\boldsymbol{\varphi}}_\alpha^{face} g_C^{face} d\theta \right] \\ &= \mathbf{F}_c^I + \tau_0 \left[\mathbf{F}_c^{II} - \mathbf{F}_c^I \right]. \end{aligned} \tag{6.53}$$

In Eq. (6.53), the superscript "*face*" denotes the location on the cell interface "$(\mathbf{0},t)$" and the superscript "*cir*" represents the location along the circle "$(-\boldsymbol{\xi}\delta_t, t-\delta_t)$". $\overline{\boldsymbol{\varphi}}_a$ is the moment vector for the circular function expressed in the local coordinate system, which is given by

$$\overline{\boldsymbol{\varphi}}_a = \left(1, \xi_1, \xi_2, \frac{1}{2}\left(\xi_1^2 + \xi_2^2\right) + e_p\right)^T. \tag{6.54}$$

Here, $\boldsymbol{\xi} = (\xi_1, \xi_2)$ is the particle velocity vector expressed in the local coordinate system. From Eq. (6.53), it can be observed that there are two parts of the flux vector on the cell interface. One is the flux vector attributed to the equilibrium distribution function and the moments on the cell interface, which is denoted as $\mathbf{F}_c^I$. The other is the flux vector attributed to the equilibrium distribution function and the moments at the surrounding points of the cell interface, which is referred as $\mathbf{F}_c^{II}$. It is known from the Chapman-Enskog expansion analysis that the first term on the right-hand side of Eq. (6.53) contributes to the inviscid fluxes while the second term corresponds to the viscous fluxes. For the inviscid flows considered in this section, the second term on the right-hand side of Eq. (6.53) can be viewed as the numerical dissipation and τ_0 is thus the weighting parameter of the numerical dissipation. The calculation of τ_0 will be introduced later.

The evaluations of equilibrium states and the moments depend on the conservative variables at the same physical locations. According to the moment relationships of the circular function, the conservative variables on the cell interface can be calculated by

$$\overline{\mathbf{W}}^{face} = \int_0^{2\pi} \overline{\boldsymbol{\varphi}}_\alpha^{face} g_C^{face} d\theta. \tag{6.55}$$

Substituting Eq. (6.44) into Eq. (6.55) and utilizing the compatibility condition (6.26), we have

$$\overline{\mathbf{W}}^{face} = \int_0^{2\pi} \overline{\boldsymbol{\varphi}}_\alpha^{face} g_C^{face} d\theta = \int_0^{2\pi} \overline{\boldsymbol{\varphi}}_\alpha^{cir} g_C^{cir} d\theta. \tag{6.56}$$

It can be seen from Eq. (6.56) that the conservative variables on the cell interface are determined by the equilibrium distribution function and the moments at the surrounding points of the cell interface. Therefore, the key to evaluate $\overline{\mathbf{F}}_c$ in C-GKFS is to calculate g_C^{cir} and $\overline{\boldsymbol{\varphi}}_\alpha^{cir}$.

Since this solver is restricted to inviscid flows, we can locally construct a Riemann problem with piecewise constant initial conditions over the cell interface. Then, the equilibrium distribution function and the moments at the surrounding points of the cell interface can be given according to the location of $-\boldsymbol{\xi}^+\delta_t$, where $\boldsymbol{\xi}^+$ is the predicted particle velocity on the cell interface given by

$$\begin{aligned}\boldsymbol{\xi}^+ &= \left(\xi_1^+, \xi_2^+\right) \\ &= \left(u_1^+ + c^+\cos(\theta), u_2^+ + c^+\sin(\theta)\right),\end{aligned} \tag{6.57}$$

where u_1^+, u_2^+ and c^+ are the predicted mean flow velocity components and the effective peculiar velocity on the cell interface, respectively. As indicated in the work of Yang *et al.* [2013b; 2014b], these quantities can be computed by Roe averages,

$$u_1^+ = \frac{u_1^L\sqrt{\rho^L} + u_1^R\sqrt{\rho^R}}{\sqrt{\rho^L} + \sqrt{\rho^R}}, \tag{6.58a}$$

$$u_2^+ = \frac{u_2^L\sqrt{\rho^L} + u_2^R\sqrt{\rho^R}}{\sqrt{\rho^L} + \sqrt{\rho^R}}, \tag{6.58b}$$

$$H^+ = \frac{H^L\sqrt{\rho^L} + H^R\sqrt{\rho^R}}{\sqrt{\rho^L} + \sqrt{\rho^R}}, \tag{6.58c}$$

$$c^+ = \sqrt{D\left(H^+ - \frac{1}{2}\left(\left(u_1^+\right)^2 + \left(u_2^+\right)^2\right)\right)\frac{(\gamma-1)}{\gamma}}, \tag{6.58d}$$

where H is the total enthalpy defined by $H = \left(u_1^2 + u_2^2\right)/2 + \gamma e$. Furthermore, since ξ_1^+ is perpendicular to the cell interface, its positive or negative sign poses a direct influence on the evaluation of g_C^{cir} and $\overline{\boldsymbol{\varphi}}_\alpha^{cir}$. Specifically,

$$g_C^{cir} = \begin{cases} g_C^L & \text{if } -\xi_1^+ \delta_t \le 0, \\ g_C^R & \text{if } -\xi_1^+ \delta_t > 0. \end{cases} \tag{6.59}$$

$$\overline{\boldsymbol{\varphi}}_\alpha^{cir} = \begin{cases} \overline{\boldsymbol{\varphi}}_\alpha^L & \text{if } -\xi_1^+ \delta_t \le 0, \\ \overline{\boldsymbol{\varphi}}_\alpha^R & \text{if } -\xi_1^+ \delta_t > 0. \end{cases} \tag{6.60}$$

Here, g_C^L and g_C^R are the piecewise constant distributions for the equilibrium state on the left and the right sides of the cell interface. $\overline{\boldsymbol{\varphi}}_\alpha^L$ and $\overline{\boldsymbol{\varphi}}_\alpha^R$ are the moments on two sides of the cell interface. g_C^L, $\overline{\boldsymbol{\varphi}}_\alpha^L$ and g_C^R, $\overline{\boldsymbol{\varphi}}_\alpha^R$ can be respectively calculated by the conservative variables on the left and the right sides of the cell interface $\mathbf{W}^L$ and $\mathbf{W}^R$. In addition, since the streaming time step δ_t is always positive, Eqs. (6.59) and (6.60) can be rewritten as

$$g_C^{cir} = \begin{cases} g_C^L & \text{if } \xi_1^+ \ge 0, \\ g_C^R & \text{if } \xi_1^+ < 0. \end{cases} \tag{6.61}$$

$$\overline{\boldsymbol{\varphi}}_\alpha^{cir} = \begin{cases} \overline{\boldsymbol{\varphi}}_\alpha^L & \text{if } \xi_1^+ \ge 0, \\ \overline{\boldsymbol{\varphi}}_\alpha^R & \text{if } \xi_1^+ < 0. \end{cases} \tag{6.62}$$

Substituting Eqs. (6.61) and (6.62) into Eq. (6.56), we have

$$\overline{\mathbf{W}}^{face} = \int_{\xi_1^+ \ge 0} \overline{\boldsymbol{\varphi}}_\alpha^L g_C^L d\theta + \int_{\xi_1^+ < 0} \overline{\boldsymbol{\varphi}}_\alpha^R g_C^R d\theta . \tag{6.63}$$

In the following part, we will show detailed expressions of Eq. (6.63) according to specific values of ξ_1^+ on the cell interface.

Case 1: $u_1^+ > c^+$, i.e., ξ_1^+ is always larger than 0. Thus, the integral of Eq. (6.63) yields

$$\overline{\mathbf{W}}^{face} = \int_0^{2\pi} \overline{\boldsymbol{\varphi}}_\alpha^L g_C^L d\theta = \begin{bmatrix} \rho \\ \rho u_1 \\ \rho u_2 \\ \frac{1}{2}\rho\left(u_1^2 + u_2^2 + c^2 + 2e_p\right) \end{bmatrix}^L . \tag{6.64}$$

In this case, the conservative variables on the cell interface are fully determined by those on the left side of the cell interface.

Case 2: $u_1^+ < -c^+$, i.e., ξ_1^+ is always less than 0. The conservative variables on the cell interface are equivalent to those on the right side of the cell interface, i.e.,

$$\overline{\mathbf{W}}^{face} = \int_0^{2\pi} \overline{\boldsymbol{\varphi}}_\alpha^R g_C^R d\theta = \begin{bmatrix} \rho \\ \rho u_1 \\ \rho u_2 \\ \frac{1}{2}\rho\left(u_1^2 + u_2^2 + c^2 + 2e_p\right) \end{bmatrix}^R . \tag{6.65}$$

Case 3: $-c^+ \le u_1^+ \le c^+$. In this case, the conservative variables on both sides of the cell interface contribute to the calculation of conservative variables on the cell interface. Eq. (6.63) can be specified as

$$\overline{\mathbf{W}}^{face} = \int_{-\theta_0}^{\theta_0} \overline{\boldsymbol{\varphi}}_\alpha^L g_C^L d\theta + \int_{\theta_0}^{2\pi-\theta_0} \overline{\boldsymbol{\varphi}}_\alpha^R g_C^R d\theta, \tag{6.66}$$

where $\theta_0 = \arccos\left(-u_1^+/c^+\right)$. By setting $\theta_1 = \theta_0$ and $\theta_2 = \pi - \theta_0$, the explicit form of Eq. (6.66) can be written as

$$\overline{\mathbf{W}}^{face} = \begin{bmatrix} \dfrac{\rho\theta_1}{\pi} \\ \dfrac{\rho\left(u_1\theta_1 + c\sin\left(\theta_1\right)\right)}{\pi} \\ \dfrac{\rho u_2\theta_1}{\pi} \\ \dfrac{\rho\left(W_{e1}\left(\theta_1\right) + 2u_1 c\sin\left(\theta_1\right)\right)}{2\pi} \end{bmatrix}^L + \begin{bmatrix} \dfrac{\rho\theta_2}{\pi} \\ \dfrac{\rho\left(u_1\theta_2 - c\sin\left(\theta_1\right)\right)}{\pi} \\ \dfrac{\rho u_2\theta_2}{\pi} \\ \dfrac{\rho\left(W_{e1}\left(\theta_2\right) - 2u_1 c\sin\left(\theta_1\right)\right)}{2\pi} \end{bmatrix}^R, \tag{6.67}$$

where

$$W_{e1}(\theta) = \theta\left(u_1^2 + u_2^2 + c^2 + 2e_p\right).$$

Once the conservative variables on the cell interface are obtained, g_C^{face} and $\overline{\boldsymbol{\varphi}}_\alpha^{face}$ can be constructed by applying Eqs. (6.14) and (6.54), respectively. Finally, the flux vector attributed to the equilibrium distribution function and the moments on the cell interface can be calculated by substituting g_C^{face} and $\overline{\boldsymbol{\varphi}}_\alpha^{face}$ into the integral of $\mathbf{F}_c^I$. Alternatively, a more efficient way is to directly substitute $\overline{\mathbf{W}}^{face}$ into the expression of the inviscid flux vector. As such, $\mathbf{F}_c^I$ can be expressed by

$$\mathbf{F}_c^I = \begin{bmatrix} \rho u_1 \\ \rho u_1^2 + \frac{1}{2}\rho c^2 \\ \rho u_1 u_2 \\ \frac{1}{2}\rho u_1\left(u_1^2 + u_2^2 + 2c^2 + 2e_p\right) \end{bmatrix}^{face}. \tag{6.68}$$

Now, the second issue is to compute the flux vector attributed to the equilibrium distribution function and the moments at the surrounding points of the cell interface. Since g_C^{cir} and $\overline{\boldsymbol{\varphi}}_\alpha^{cir}$ have been determined by Eqs. (6.61) and (6.62), $\mathbf{F}_c^{II}$ can be computed straightforwardly. Like the derivation process of $\overline{\mathbf{W}}^{face}$, there are three cases to be considered for the calculation of $\mathbf{F}_c^{II}$.

Case 1: $u_1^+ > c^+$. In this case, the flux vector on the cell interface is determined by the conservative variables on the left side of the cell interface, i.e.,

$$\mathbf{F}_c^{II} = \int_0^{2\pi} \xi_1^L \overline{\boldsymbol{\varphi}}_\alpha^L g_C^L d\theta = \begin{bmatrix} \rho u_1 \\ \rho u_1^2 + \frac{1}{2}\rho c^2 \\ \rho u_1 u_2 \\ \frac{1}{2}\rho u_1\left(u_1^2 + u_2^2 + 2c^2 + 2e_p\right) \end{bmatrix}^{L}. \tag{6.69}$$

Case 2: $u_1^+ < -c^+$. In this case, the flux vector on the cell interface is evaluated by the conservative variables on the right side of the cell interface, i.e.,

$$\mathbf{F}_c^{II} = \int_0^{2\pi} \xi_1^R \overline{\boldsymbol{\varphi}}_\alpha^R g_C^R d\theta = \begin{bmatrix} \rho u_1 \\ \rho u_1^2 + \frac{1}{2}\rho c^2 \\ \rho u_1 u_2 \\ \frac{1}{2}\rho u_1\left(u_1^2 + u_2^2 + 2c^2 + 2e_p\right) \end{bmatrix}^R . \quad (6.70)$$

Case 3: $-c^+ \le u_1^+ \le c^+$. According to the definition of ξ_1^+, $\mathbf{F}_c^{II}$ can be calculated by

$$\mathbf{F}_c^{II} = \int_{-\theta_0}^{\theta_0} \xi_1^L \overline{\boldsymbol{\varphi}}_\alpha^L g_C^L d\theta + \int_{\theta_0}^{2\pi-\theta_0} \xi_1^R \overline{\boldsymbol{\varphi}}_\alpha^R g_C^R d\theta$$

$$= \begin{bmatrix} \dfrac{\rho\left(u_1\theta_1 + c\sin(\theta_1)\right)}{\pi} \\ \dfrac{\rho\left(2u_1^2\theta_1 + c^2\theta_1 + F_{mx}(\theta_1)\right)}{2\pi} \\ \dfrac{\rho u_2\left(u_1\theta_1 + c\sin(\theta_1)\right)}{\pi} \\ \dfrac{\rho\left(F_{e1}(\theta_1) + F_{e2}(\theta_1)\right)}{2\pi} \end{bmatrix}^L + \begin{bmatrix} \dfrac{\rho\left(u_1\theta_2 - c\sin(\theta_1)\right)}{\pi} \\ \dfrac{\rho\left(2u_1^2\theta_2 + c^2\theta_2 - F_{mx}(\theta_1)\right)}{2\pi} \\ \dfrac{\rho u_2\left(u_1\theta_2 - c\sin(\theta_1)\right)}{\pi} \\ \dfrac{\rho\left(F_{e1}(\theta_2) - F_{e2}(\theta_1)\right)}{2\pi} \end{bmatrix}^R , \quad (6.71)$$

where

$$F_{mx}(\theta) = \left(4u_1 c + c^2\cos(\theta)\right)\sin(\theta),$$

$$F_{e1}(\theta) = \left(2u_1 c^2 + u_1^3 + u_1 u_2^2 + 2u_1 e_p\right)\theta,$$

$$F_{e2}(\theta) = \left(3cu_1^2 + cu_2^2 + c^3 + 2ce_p + u_1 c^2\cos(\theta)\right)\sin(\theta).$$

As mentioned earlier, τ_0 can be regarded as the weighting coefficient of the numerical dissipation in C-GKFS for inviscid flows. $\tau_0 = 0$ means that no numerical dissipation is introduced into the reconstruction of inviscid fluxes, while $\tau_0 = 1$ indicates that the entire non-equilibrium distribution function is included in the calculation of inviscid fluxes and relatively large numerical dissipation is contained. A switch function can be introduced here to adjust the value of τ_0 as per needs. It is expected that the switch function is zero in smooth regions such as in the boundary layer, while approaching to one around the strong shock wave. Given these criteria, one possible form of the switch function could be

$$\tau_0 = \tanh\left(C \frac{\left| p^L - p^R \right|}{p^L + p^R} \right), \tag{6.72}$$

where tanh(x) is the hyperbolic tangent function, p^L and p^R are the pressure on the left and the right side of the cell interface. C is an amplification factor which usually takes the value of 10-100 in simulations. From Eq. (6.72), it can be seen that the value of τ_0 will tend to one only in the regions where significant pressure jump exists over the interface.

On the other hand, to well simulate compressible flows with strong shock waves, the pressure in surrounding cells should also be considered in evaluating the switch function. Inspired by the construction of the limiter function by Barth and Jespersen [1989], a slight modification is made on Eq. (6.72). The new form of the switch function is defined as

$$\tau_0 = \max\left\{\tau^L, \tau^R\right\}, \tag{6.73}$$

$$\tau^L = \max_{i=1,N_{fL}} \left\{\tau_i\right\}, \tag{6.74}$$

$$\tau^R = \max_{i=1,N_{fR}} \left\{\tau_i\right\}, \tag{6.75}$$

where, τ^L and τ^R are the maximum values of the switch function at the left and the right control volumes. N_{fL} and N_{fR} are the numbers of surfaces of the left and the right control volumes, respectively. τ_i in Eqs. (6.74) and (6.75) is the original switch function calculated by Eq. (6.72).

By substituting $\mathbf{F}_c^I$, $\mathbf{F}_c^{II}$ and τ_0 into Eq. (6.53), we can obtain the full expression of C-GKFS for simulation of inviscid flows. The calculation procedure of inviscid fluxes on the cell interface is summarized as below:

(a) For each cell interface, use interpolation given in Eq. (5.35) to obtain the macroscopic flow variables on the left and the right sides of the cell interface, $\mathbf{W}^L$ and $\mathbf{W}^R$;
(b) Use Roe averages in Eqs. (6.58a)-(6.58d) to calculate u_1^+, u_2^+ and c^+ on the cell interface;
(c) Calculate the conservative variables on the cell interface $\overline{\mathbf{W}}^{face}$ from Eqs. (6.64), (6.65) and (6.67), and further compute the flux $\mathbf{F}_c^I$ from Eq. (6.68);
(d) Compute $\mathbf{F}_c^{II}$ by using Eqs. (6.69)-(6.71);
(e) Calculate τ_0 from Eqs. (6.73)-(6.75);
(f) Substitute $\mathbf{F}_c^I$, $\mathbf{F}_c^{II}$ and τ_0 into Eq. (6.53) to obtain the full expression of inviscid fluxes.

Different from the M-GKFS presented in the last chapter, neither the error function nor the exponential function is involved in the present scheme. As a result, the C-GKFS is expected to be more efficient than the M-GKFS, which will be validated by the numerical examples presented in **Section 6.6**. In addition, it can be seen from Eqs. (6.64), (6.65), (6.67) and Eqs. (6.69)-(6.71) that the conservative variables and fluxes on the cell interface are fully determined by the flow variables on the left and the right sides of the cell interface according to the ratio of u_1^+ to c^+. And the ratio of u_1^+ to c^+ is proportional to the local Mach number on the cell interface, i.e., $u_1^+/c^+ = Ma\sqrt{\gamma/D}$. From this point of view, the C-GKFS may have a similar mechanism to the Mach number-based flux splitting scheme for solving the Euler equations, such as the Van Leer's scheme [Van Leer, 1977; Van Leer, 1979; Van Leer, 1997] and the advection upstream splitting method (AUSM) [Liou and Steffen Jr, 1993; Liou, 2001].

6.3 Sphere Function-Based GKFS for Inviscid Flows

In this section, the derivation is extended to the 3D scenario. The sphere function forges the basis here to reconstruct the numerical fluxes on the cell interface with the local solution to the Boltzmann equation. And the S-GKFS for simulation of 3D inviscid flows is developed. To begin with, the 3D Euler equations read

$$\frac{\partial \mathbf{W}}{\partial t}+\frac{\partial \mathbf{F}_{cx}}{\partial x}+\frac{\partial \mathbf{F}_{cy}}{\partial y}+\frac{\partial \mathbf{F}_{cz}}{\partial z}=\mathbf{0}, \tag{6.76}$$

$$\mathbf{W}=\begin{bmatrix}\rho \\ \rho u \\ \rho v \\ \rho w \\ \rho E\end{bmatrix}, \tag{6.77}$$

$$\mathbf{F}_{cx}=\begin{bmatrix}\rho u \\ \rho uu+p \\ \rho uv \\ \rho uw \\ (\rho E+p)u\end{bmatrix},\ \mathbf{F}_{cy}=\begin{bmatrix}\rho v \\ \rho vu \\ \rho vv+p \\ \rho vw \\ (\rho E+p)v\end{bmatrix},\ \mathbf{F}_{cz}=\begin{bmatrix}\rho w \\ \rho wu \\ \rho wv \\ \rho ww+p \\ (\rho E+p)w\end{bmatrix}. \tag{6.78}$$

Here, $\mathbf{u}=(u,v,w)$ is the velocity vector expressed in the global Cartesian coordinate system. Eq. (6.76) can be discretized by FVM into:

$$\frac{\partial \mathbf{W}_i}{\partial t}=-\frac{1}{\Omega_i}\sum_{j=1}^{N_f}\mathbf{F}_{cnj}S_j, \tag{6.79}$$

$$\mathbf{F}_{cn}=n_{1x}\mathbf{F}_{cx}+n_{1y}\mathbf{F}_{cy}+n_{1z}\mathbf{F}_{cz}=\begin{bmatrix}\rho u_1 \\ \rho uu_1+pn_{1x} \\ \rho vu_1+pn_{1y} \\ \rho wu_1+pn_{1z} \\ (\rho E+p)u_1\end{bmatrix}, \tag{6.80}$$

where $\mathbf{n}_1=(n_{1x},n_{1y},n_{1z})$ denotes the unit normal vector of the cell interface.

For the convenience of derivation, the local coordinate system defined at the cell interface is introduced into the S-GKFS. In the local coordinate system, x_1-axis is taken as the normal direction pointing outwards of the cell interface of the control volume, x_2-axis and x_3-axis are chosen as two tangential directions of the cell interface, which are mutually orthogonal and form a right-hand coordinate system. For simplicity, x_2-axis can be set as parallel one of the edges of the cell interface and x_3-axis is then determined by using the right-hand rule. Like the direction 1, the unit vectors in the directions 2 and 3 of the local coordinate system are respectively denoted as $\mathbf{n}_2 = \left(n_{2x}, n_{2y}, n_{2z}\right)$ and $\mathbf{n}_3 = \left(n_{3x}, n_{3y}, n_{3z}\right)$. The conservative variables and fluxes expressed in the local coordinate system can be written as

$$\overline{\mathbf{W}} = \begin{bmatrix} \rho \\ \rho u_1 \\ \rho u_2 \\ \rho u_3 \\ \rho E \end{bmatrix}, \quad \overline{\mathbf{F}}_c = \begin{bmatrix} \rho u_1 \\ \rho u_1 u_1 + p \\ \rho u_1 u_2 \\ \rho u_1 u_3 \\ (\rho E + p) u_1 \end{bmatrix}, \tag{6.81}$$

where

$$u_1 = n_{1x} u + n_{1y} v + n_{1z} w, \tag{6.82a}$$

$$u_2 = n_{2x} u + n_{2y} v + n_{2z} w, \tag{6.82b}$$

$$u_3 = n_{3x} u + n_{3y} v + n_{3z} w. \tag{6.82c}$$

$\mathbf{u} = \left(u_1, u_2, u_3\right)$ is the velocity vector expressed in the local coordinate system. By applying the coordinate transformation from the local coordinate system to the global Cartesian coordinate system, the following relationships of conservative variables and fluxes expressed in two coordinate systems are obtained:

$$\mathbf{W} = \begin{bmatrix} \overline{\mathbf{W}}(1) \\ n_{1x}\overline{\mathbf{W}}(2) + n_{2x}\overline{\mathbf{W}}(3) + n_{3x}\overline{\mathbf{W}}(4) \\ n_{1y}\overline{\mathbf{W}}(2) + n_{2y}\overline{\mathbf{W}}(3) + n_{3y}\overline{\mathbf{W}}(4) \\ n_{1z}\overline{\mathbf{W}}(2) + n_{2z}\overline{\mathbf{W}}(3) + n_{3z}\overline{\mathbf{W}}(4) \\ \overline{\mathbf{W}}(5) \end{bmatrix}, \tag{6.83}$$

$$\mathbf{F}_{cn} = \begin{bmatrix} \overline{\mathbf{F}}_c(1) \\ n_{1x}\overline{\mathbf{F}}_c(2) + n_{2x}\overline{\mathbf{F}}_c(3) + n_{3x}\overline{\mathbf{F}}_c(4) \\ n_{1y}\overline{\mathbf{F}}_c(2) + n_{2y}\overline{\mathbf{F}}_c(3) + n_{3y}\overline{\mathbf{F}}_c(4) \\ n_{1z}\overline{\mathbf{F}}_c(2) + n_{2z}\overline{\mathbf{F}}_c(3) + n_{3z}\overline{\mathbf{F}}_c(4) \\ \overline{\mathbf{F}}_c(5) \end{bmatrix}. \tag{6.84}$$

Here, $\overline{\mathbf{W}}(i)$ and $\overline{\mathbf{F}}_c(i)$, $i = 1,2,...,5$, are the components of $\overline{\mathbf{W}}$ and $\overline{\mathbf{F}}_c$, respectively. As shown in Eq. (6.84), the calculation of $\mathbf{F}_{cn}$ in the global Cartesian coordinate system is equivalent to evaluating $\mathbf{F}_c$ in the local coordinate system.

Similar to the derivation of C-GKFS, the conservative variables and numerical fluxes of S-GKFS on the cell interface can be written as

$$\begin{aligned} \overline{\mathbf{W}}^{face} &= \int_0^{2\pi}\int_0^{\pi} \overline{\boldsymbol{\varphi}}_\alpha^{face} g_S^{face} \sin(\varphi) d\varphi d\theta \\ &= \int_0^{2\pi}\int_0^{\pi} \overline{\boldsymbol{\varphi}}_\alpha^{sph} g_S^{sph} \sin(\varphi) d\varphi d\theta, \end{aligned} \tag{6.85}$$

$$\begin{aligned} \overline{\mathbf{F}}_c &= \int_0^{2\pi}\int_0^{\pi} \xi_1^{face} \overline{\boldsymbol{\varphi}}_\alpha^{face} g_S^{face} \sin(\varphi) d\varphi d\theta \\ &\quad + \tau_0 \left[\int_0^{2\pi}\int_0^{\pi} \xi_1^{sph} \overline{\boldsymbol{\varphi}}_\alpha^{sph} g_S^{sph} \sin(\varphi) d\varphi d\theta \right. \\ &\quad \left. - \int_0^{2\pi}\int_0^{\pi} \xi_1^{face} \overline{\boldsymbol{\varphi}}_\alpha^{face} g_S^{face} \sin(\varphi) d\varphi d\theta \right] \\ &= \mathbf{F}_c^I + \tau_0 \left[\mathbf{F}_c^{II} - \mathbf{F}_c^I \right]. \end{aligned} \tag{6.86}$$

In Eqs. (6.85) and (6.86), the superscript " $face$ " denotes the location on the cell interface " $(\mathbf{0},t)$ " and the superscript " sph " represents the location on the spherical surface "$(-\xi\delta_t, t-\delta_t)$". In addition, $\mathbf{F}_c^I$ denotes the flux vector attributed to the equilibrium distribution function and the moments on the cell interface and $\mathbf{F}_c^{II}$ represents the flux vector attributed to the equilibrium distribution function and the moments on the spherical surface. $\overline{\boldsymbol{\varphi}}_a$ is the moment vector for the sphere function expressed in the local coordinate system, which is given by

$$\overline{\boldsymbol{\varphi}}_a = \left(1, \xi_1, \xi_2, \xi_3, \frac{1}{2}\left(\xi_1^2 + \xi_2^2 + \xi_3^2\right) + e_p\right)^T. \tag{6.87}$$

$\boldsymbol{\xi} = (\xi_1, \xi_2, \xi_3)$ is the particle velocity vector expressed in the local coordinate system. In order to calculate $\overline{\mathbf{W}}^{face}$ and $\overline{\mathbf{F}}_c$, the predicted particle velocity on the cell interface should be determined in advance. For the sphere function, it is given by

$$\begin{aligned}\boldsymbol{\xi}^+ &= \left(\xi_1^+, \xi_2^+, \xi_3^+\right) \\ &= \left(u_1^+ + c^+ \sin(\varphi)\cos(\theta), u_2^+ + c^+ \sin(\varphi)\sin(\theta), u_3^+ + c^+ \cos(\varphi)\right).\end{aligned} \tag{6.88}$$

Like the C-GKFS, the predicted mean flow velocities u_1^+, u_2^+ and u_3^+ and the effective peculiar velocity c^+ can also be calculated by Roe averages. In fact, the derivation processes of $\overline{\mathbf{W}}^{face}$, $\mathbf{F}_c^I$ and $\mathbf{F}_c^{II}$ are the same as those of C-GKFS. In the following, all the expressions of conservative variables and inviscid fluxes on the cell interface will be presented directly.

Firstly, the expressions of conservative variables on the cell interface $\overline{\mathbf{W}}^{face}$ and fluxes attributed to the equilibrium distribution function and the moments on the cell interface $\mathbf{F}_c^I$ are shown. There are three cases to be considered for the calculation of conservative variables on the cell interface.

Case 1: $u_1^+ > c^+$. In this case, the conservative variables on the cell interface are determined by the conservative variables on the left side of the cell interface, i.e.,

$$\overline{\mathbf{W}}^{face} = \int_0^{2\pi}\int_0^{\pi} \overline{\boldsymbol{\varphi}}_\alpha^L g_S^L \sin(\varphi)\, d\varphi d\theta = \begin{bmatrix} \rho \\ \rho u_1 \\ \rho u_2 \\ \rho u_3 \\ \frac{1}{2}\rho\left(u_1^2 + u_2^2 + u_3^2 + c^2 + 2e_p\right) \end{bmatrix}^L. \tag{6.89}$$

Case 2: $u_1^+ < -c^+$. In this case, the conservative variables on the cell interface are evaluated by the conservative variables on the right side of the cell interface, i.e.,

$$\overline{\mathbf{W}}^{face} = \int_0^{2\pi}\int_0^{\pi} \overline{\boldsymbol{\varphi}}_\alpha^R g_S^R \sin(\varphi) d\varphi d\theta = \begin{bmatrix} \rho \\ \rho u_1 \\ \rho u_2 \\ \rho u_3 \\ \frac{1}{2}\rho\left(u_1^2 + u_2^2 + u_3^2 + c^2 + 2e_p\right) \end{bmatrix}^R. \tag{6.90}$$

Case 3: $-c^+ \le u_1^+ \le c^+$. In this case, the conservative variables on both sides of the cell interface contribute to the calculation of conservative variables on the cell interface.

$$\overline{\mathbf{W}}^{face} = \int_{-\theta_0}^{\theta_0}\int_0^{\pi} \overline{\boldsymbol{\varphi}}_\alpha^L g_S^L \sin(\varphi) d\varphi d\theta + \int_{\theta_0}^{2\pi-\theta_0}\int_0^{\pi} \overline{\boldsymbol{\varphi}}_\alpha^R g_S^R \sin(\varphi) d\varphi d\theta$$
$$= \begin{bmatrix} \frac{\rho\theta_1}{\pi} \\ \frac{\rho\left(4u_1\theta_1 + c\pi\sin(\theta_1)\right)}{4\pi} \\ \frac{\rho u_2\theta_1}{\pi} \\ \frac{\rho u_3\theta_1}{\pi} \\ \frac{\rho\left(W_{e1}(\theta_1) + u_1 c\pi\sin(\theta_1)\right)}{4\pi} \end{bmatrix}^L + \begin{bmatrix} \frac{\rho\theta_2}{\pi} \\ \frac{\rho\left(4u_1\theta_2 - c\pi\sin(\theta_1)\right)}{4\pi} \\ \frac{\rho u_2\theta_2}{\pi} \\ \frac{\rho u_3\theta_2}{\pi} \\ \frac{\rho\left(W_{e1}(\theta_2) - u_1 c\pi\sin(\theta_1)\right)}{4\pi} \end{bmatrix}^R, \tag{6.91}$$

where

$$W_{e1}(\theta) = 2\theta\left(u_1^2 + u_2^2 + u_3^2 + c^2 + 2e_p\right).$$

The definitions of θ_0, θ_1 and θ_2 are the same as those below Eq. (6.66).

By substituting $\overline{\mathbf{W}}^{face}$ into Eq. (6.81), the flux vector attributed to the equilibrium distribution function and the moments on the cell interface can be computed by

$$\mathbf{F}_c^I = \begin{bmatrix} \rho u_1 \\ \rho u_1^2 + \rho c^2/3 \\ \rho u_1 u_2 \\ \rho u_1 u_3 \\ \rho u_1\left(3u_1^2 + 3u_2^2 + 3u_3^2 + 5c^2 + 6e_p\right)/6 \end{bmatrix}^{face}. \tag{6.92}$$

Secondly, the expressions of the flux vector attributed to equilibrium distribution function and the moments at the surrounding points of the cell interface $\mathbf{F}_c^{II}$ are shown below.

Case 1: $u_1^+ > c^+$.

$$\mathbf{F}_c^{II} = \int_0^{2\pi}\int_0^{\pi} \xi_1^L \overline{\boldsymbol{\varphi}}_\alpha^L g_S^L \sin(\varphi)\, d\varphi d\theta$$
$$= \begin{bmatrix} \rho u_1 \\ \rho u_1^2 + \rho c^2/3 \\ \rho u_1 u_2 \\ \rho u_1 u_3 \\ \rho u_1\left(3u_1^2 + 3u_2^2 + 3u_3^2 + 5c^2 + 6e_p\right)/6 \end{bmatrix}^{L}. \tag{6.93}$$

Case 2: $u_1^+ < -c^+$.

$$\mathbf{F}_c^{II} = \int_0^{2\pi}\int_0^{\pi} \xi_1^R \overline{\boldsymbol{\varphi}}_\alpha^R g_S^R \sin(\varphi)\, d\varphi d\theta$$
$$= \begin{bmatrix} \rho u_1 \\ \rho u_1^2 + \rho c^2/3 \\ \rho u_1 u_2 \\ \rho u_1 u_3 \\ \rho u_1\left(3u_1^2 + 3u_2^2 + 3u_3^2 + 5c^2 + 6e_p\right)/6 \end{bmatrix}^{R}. \tag{6.94}$$

Case 3: $-c^{+} \le u_1^{+} \le c^{+}$.

$$\begin{aligned}\mathbf{F}_c^{II} &= \int_{-\theta_0}^{\theta_0} \int_0^{\pi} \xi_1^L \overline{\boldsymbol{\varphi}}_\alpha^L g_S^L \sin(\varphi)\, d\varphi d\theta \\ &+ \int_{\theta_0}^{2\pi-\theta_0} \int_0^{\pi} \xi_1^R \overline{\boldsymbol{\varphi}}_\alpha^R g_S^R \sin(\varphi)\, d\varphi d\theta \\ &= \mathbf{F}_c^{II,L} + \mathbf{F}_c^{II,R},\end{aligned} \tag{6.95}$$

where

$$\mathbf{F}_c^{II,L} = \begin{bmatrix} \dfrac{\rho\left(4u_1\theta_1 + c\pi\sin(\theta_1)\right)}{4\pi} \\ \dfrac{\rho\left(6u_1^2\theta_1 + 2c^2\theta_1 + F_{mx}(\theta_1)\right)}{6\pi} \\ \dfrac{\rho u_2\left(4u_1\theta_1 + c\pi\sin(\theta_1)\right)}{4\pi} \\ \dfrac{\rho u_3\left(4u_1\theta_1 + c\pi\sin(\theta_1)\right)}{4\pi} \\ \dfrac{\rho\left(F_{e1}(\theta_1) + F_{e2}(\theta_1)\right)}{24\pi} \end{bmatrix}^L,$$

$$\mathbf{F}_c^{II,R} = \begin{bmatrix} \dfrac{\rho\left(4u_1\theta_2 - c\pi\sin(\theta_1)\right)}{4\pi} \\ \dfrac{\rho\left(6u_1^2\theta_2 + 2c^2\theta_2 - F_{mx}(\theta_1)\right)}{6\pi} \\ \dfrac{\rho u_2\left(4u_1\theta_2 - c\pi\sin(\theta_1)\right)}{4\pi} \\ \dfrac{\rho u_3\left(4u_1\theta_2 - c\pi\sin(\theta_1)\right)}{4\pi} \\ \dfrac{\rho\left(F_{e1}(\theta_2) - F_{e2}(\theta_1)\right)}{24\pi} \end{bmatrix}^R,$$

$$F_{mx}(\theta) = \left(3u_1 c\pi + 2c^2 \cos(\theta)\right)\sin(\theta),$$

$$F_{e1}(\theta) = \left(20u_1 c^2 + 12u_1^3 + 12u_1 u_2^2 + 12u_1 u_3^2 + 24u_1 e_p\right)\theta,$$

$$F_{e2}(\theta) = \left(9cu_1^2\pi + 3cu_2^2\pi + 3cu_3^2\pi + 3c^3\pi \right.$$
$$\left. +6ce_p\pi + 8u_1 c^2 \cos(\theta)\right)\sin(\theta).$$

6.4 D1Q4 Model-Based LBFS for Inviscid Flows

Since the compressible inviscid flows are intended to simulate, the compressible lattice Boltzmann model can also be utilized to reconstruct the numerical fluxes of the Euler equations in the framework of LBFS. In recent years, a great deal of effort has been devoted to develop lattice Boltzmann models for simulation of compressible flows and some interesting models have been developed. For example, Qu *et al.* [2007a; 2007b] replaced the Maxwellian distribution function with a circular function, and then used the circular function to develop some 1D and 2D models. By using the proposed models, some supersonic flows with strong shock waves were simulated successfully. However, in these models, the lattice velocities are artificially specified. For simulation of transonic flows and hypersonic flows, the lattice velocities may be quite different. As reported in the work of Yang *et al.* [2013a], an unexpected spurious numerical oscillation is observed in the vicinity of the sonic line for simulation of hypersonic flows. Besides, by introducing the internal energy distribution function to take in charge of the temperature field, Li *et al.* [2007] presented a 2D coupled double-distribution-function model for simulation of viscous compressible flows. In this model, the Prandtl number and the specific heat ratio can be adjusted flexibly. Also aiming at the simulation of compressible flows, Yang *et al.* [2013a] proposed a platform to develop non-free parameter lattice Boltzmann models from the moment relationships. In the platform, the equilibrium distribution functions are derived from the moment relationships (from the zeroth order to the third order) which are associated with the conservative flow variables and fluxes, while the lattice velocities are obtained from the appended higher order moment relationships. In this section, the non-free parameter D1Q4 model (D1Q4 model for short) will be applied to establish the LBFS for compressible inviscid flows.

The distribution of discrete lattice velocities for the D1Q4 model is shown in Figure 6.3, which contains four equilibrium distribution functions g_1, g_2, g_3, g_4 and two lattice velocities d_1, d_2. Both g_1, g_2, g_3, g_4 and d_1, d_2 are derived from the moment relationships and given by

$$
\begin{aligned}
g_1 &= \frac{\rho\left(-d_1 d_2^2 - d_2^2 u + d_1 u^2 + d_1 c^2 + u^3 + 3uc^2\right)}{2d_1\left(d_1^2 - d_2^2\right)}, \\
g_2 &= \frac{\rho\left(-d_1 d_2^2 + d_2^2 u + d_1 u^2 + d_1 c^2 - u^3 - 3uc^2\right)}{2d_1\left(d_1^2 - d_2^2\right)}, \\
g_3 &= \frac{\rho\left(d_1^2 d_2 + d_1^2 u - d_2 u^2 - d_2 c^2 - u^3 - 3uc^2\right)}{2d_2\left(d_1^2 - d_2^2\right)}, \\
g_4 &= \frac{\rho\left(d_1^2 d_2 - d_1^2 u - d_2 u^2 - d_2 c^2 + u^3 + 3uc^2\right)}{2d_2\left(d_1^2 - d_2^2\right)}.
\end{aligned} \tag{6.96}
$$

$$
\begin{aligned}
d_1 &= \sqrt{u^2 + 3c^2 - \sqrt{4u^2c^2 + 6c^4}}, \\
d_2 &= \sqrt{u^2 + 3c^2 + \sqrt{4u^2c^2 + 6c^4}}.
\end{aligned} \tag{6.97}
$$

where, u is the velocity of the mean flow for the 1D case. For multi-dimensional cases, this lattice Boltzmann model should be applied along the normal direction to the cell interface, i.e., the velocity u should be replaced with the normal velocity u_1. c represents the peculiar velocity of particles defined as $c = \sqrt{Dp/\rho}$, D denotes the space dimension of

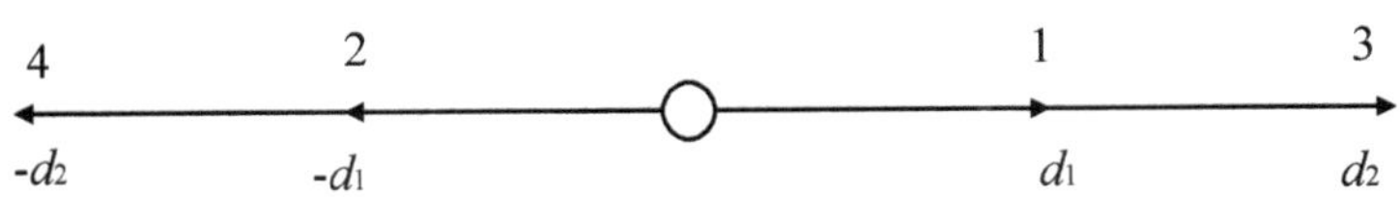

Figure 6.3. Lattice velocity distribution for the non-free parameter D1Q4 model.

lattice Boltzmann model. For the D1Q4 model, $D=1$ should be taken.

Here, we consider the 3D Euler equations for general cases. With FVM discretization, the 3D Euler equations can be written as Eq. (6.79). After introducing a local coordinate system defined on the cell interface, the calculation of $\mathbf{F}_{cn}$ in the global Cartesian coordinate system is equivalent to evaluating $\overline{\mathbf{F}}_c$ in the local coordinate system. However, since the D1Q4 model is applied along the normal direction to the cell interface, only the conservative variables and numerical fluxes attributed to the normal velocity can be calculated from the D1Q4 model. Similar to the derivation of C-GKFS, these conservative variables and numerical fluxes on the cell interface can be computed by

$$\overline{\mathbf{W}}_{D1Q4}^{face} = \sum_{i=1}^{4} \overline{\boldsymbol{\varphi}}_{\alpha}^{face} g_i^{face} = \sum_{i=1}^{4} \overline{\boldsymbol{\varphi}}_{\alpha}^{surd} g_i^{surd}, \tag{6.98}$$

$$\begin{aligned} \overline{\mathbf{F}}_{D1Q4} &= \sum_{i=1}^{4} \xi_i^{face} \overline{\boldsymbol{\varphi}}_{\alpha}^{face} g_i^{face} \\ &\quad + \tau_0 \left[\sum_{i=1}^{4} \xi_i^{surd} \overline{\boldsymbol{\varphi}}_{\alpha}^{surd} g_i^{surd} - \sum_{i=1}^{4} \xi_i^{face} \overline{\boldsymbol{\varphi}}_{\alpha}^{face} g_i^{face} \right] \\ &= \mathbf{F}_{D1Q4}^{I} + \tau_0 \left[\mathbf{F}_{D1Q4}^{II} - \mathbf{F}_{D1Q4}^{I} \right]. \end{aligned} \tag{6.99}$$

The subscript "*D1Q4*" is used to define the variables calculated from the local solution to the lattice Boltzmann equation (LBE) with the D1Q4 model. The superscript "$face$" denotes the location of the midpoint of the cell interface "$(\mathbf{0},t)$" and the superscript "$surd$" represents the location of surrounding streaming nodes of the cell interface "$(-\xi_i\delta_t, t-\delta_t)$". In addition, $\mathbf{F}_{D1Q4}^{I}$ denotes the flux vector attributed to the equilibrium distribution function and the moments on the cell interface and $\mathbf{F}_{D1Q4}^{II}$ represents the flux vector attributed to the equilibrium distribution function and the moments at the surrounding streaming nodes of the cell interface. $\overline{\boldsymbol{\varphi}}_a$ is the moment vector for the D1Q4 model expressed in the local coordinate system and is given by

$$\overline{\boldsymbol{\varphi}}_a = \left(1, \xi, \frac{1}{2}\xi^2 + e_p\right)^T, \tag{6.100}$$

where $\xi = \{\xi_1, \xi_2, \xi_3, \xi_4\} = \{d_1, -d_1, d_2, -d_2\}$ is the lattice velocity, as shown in Figure 6.3.

Since the 1D model is utilized, both $\overline{\mathbf{W}}_{D1Q4}^{face}$ and $\overline{\mathbf{F}}_{D1Q4}$ have only three components, which correspond to the conservative variables and fluxes of the 1D Euler equation. They have connections to the conservative variables and fluxes of the 3D Euler equations expressed in the local coordinate system as follows:

$$\overline{\mathbf{W}}^{face} = \begin{bmatrix} \overline{\mathbf{W}}_{D1Q4}^{face}(1) \\ \overline{\mathbf{W}}_{D1Q4}^{face}(2) \\ (\rho u_2)^{face} \\ (\rho u_3)^{face} \\ \overline{\mathbf{W}}_{D1Q4}^{face}(3) + \rho^{face}\left[\left(u_2^{face}\right)^2 + \left(u_3^{face}\right)^2\right]/2 \end{bmatrix}, \quad (6.101)$$

$$\overline{\mathbf{F}}_c = \begin{bmatrix} \overline{\mathbf{F}}_{D1Q4}(1) \\ \overline{\mathbf{F}}_{D1Q4}(2) \\ \rho u_1 u_2 \\ \rho u_1 u_3 \\ \overline{\mathbf{F}}_{D1Q4}(3) + \rho u_1\left(u_2^2 + u_3^2\right)/2 \end{bmatrix}. \quad (6.102)$$

It can be seen from Eqs. (6.101) and (6.102) that the contribution of tangential velocities to $\overline{\mathbf{W}}^{face}$ and $\overline{\mathbf{F}}_c$ should be calculated approximately. In addition, according to Eq. (6.98), the conservative variables and the equilibrium distribution function on the cell interface are determined by the equilibrium distribution function and the moments at the surrounding streaming nodes of the cell interface. Therefore, the key to evaluate $\overline{\mathbf{F}}_{D1Q4}$ in LBFS is to calculate g_i^{surd} and $\overline{\boldsymbol{\varphi}}_\alpha^{surd}$.

By assuming a local Riemann problem with a piecewise constant distribution of the equilibrium distribution function and conservative variables being formed over the cell interface, the equilibrium distribution function and the moments at the surrounding streaming nodes of the cell interface can be given by

$$g_i^{surd} = \begin{cases} g_i^L & \text{if } -\xi_i \delta_t \le 0, \\ g_i^R & \text{if } -\xi_i \delta_t > 0. \end{cases} \tag{6.103}$$

$$\overline{\boldsymbol{\varphi}}_\alpha^{surd} = \begin{cases} \overline{\boldsymbol{\varphi}}_\alpha^L & \text{if } -\xi_i \delta_t \le 0, \\ \overline{\boldsymbol{\varphi}}_\alpha^R & \text{if } -\xi_i \delta_t > 0. \end{cases} \tag{6.104}$$

Since the streaming time step δ_t is always positive, Eqs. (6.103) and (6.104) can be rewritten as

$$g_i^{surd} = \begin{cases} g_i^L & \text{if } i = 1,3, \\ g_i^R & \text{if } i = 2,4. \end{cases} \tag{6.105}$$

$$\overline{\boldsymbol{\varphi}}_\alpha^{surd} = \begin{cases} \overline{\boldsymbol{\varphi}}_\alpha^L & \text{if } i = 1,3, \\ \overline{\boldsymbol{\varphi}}_\alpha^R & \text{if } i = 2,4. \end{cases} \tag{6.106}$$

Substituting Eqs. (6.105) and (6.106) into Eq. (6.98), we have

$$\overline{\mathbf{W}}_{D1Q4}^{face} = \sum_{i=1,3} \overline{\boldsymbol{\varphi}}_\alpha^L g_i^L + \sum_{i=2,4} \overline{\boldsymbol{\varphi}}_\alpha^R g_i^R . \tag{6.107}$$

From Eq. (6.107), we can obtain the density, the normal velocity and the pressure on the cell interface in a straightforward way. The tangential velocities on the cell interface can be treated as the passive scalar quantities and approximated by

$$(\rho u_2)^{face} = \sum_{i=1,3} u_2^L g_i^L + \sum_{i=2,4} u_2^R g_i^R , \tag{6.108}$$

$$(\rho u_3)^{face} = \sum_{i=1,3} u_3^L g_i^L + \sum_{i=2,4} u_3^R g_i^R . \tag{6.109}$$

By substituting all the primitive variables on the cell interface into Eq. (6.81), the flux vector attributed to the equilibrium distribution function and the moments as well as the tangential velocities on the cell interface $\mathbf{F}_c^I$ can then be computed by

$$\mathbf{F}_c^I = \begin{bmatrix} \rho u_1 \\ \rho u_1 u_1 + p \\ \rho u_1 u_2 \\ \rho u_1 u_3 \\ (\rho E + p) u_1 \end{bmatrix}^{face} . \tag{6.110}$$

In addition, the flux vector attributed to equilibrium distribution function and the moments at the surrounding streaming nodes of the cell interface $\mathbf{F}_{D1Q4}^{II}$ can be computed by

$$\mathbf{F}_{D1Q4}^{II} = \sum_{i=1,3} \xi_i^L \overline{\boldsymbol{\varphi}}_\alpha^L g_i^L + \sum_{i=2,4} \xi_i^R \overline{\boldsymbol{\varphi}}_\alpha^R g_i^R . \tag{6.111}$$

And the contribution of tangential velocities to the momentum flux in the tangential directions and the energy flux can be approximated by

$$(\rho u_1 u_2)^{II} = \sum_{i=1,3} u_2^L \xi_i^L g_i^L + \sum_{i=2,4} u_2^R \xi_i^R g_i^R , \tag{6.112}$$

$$(\rho u_1 u_3)^{II} = \sum_{i=1,3} u_3^L \xi_i^L g_i^L + \sum_{i=2,4} u_3^R \xi_i^R g_i^R , \tag{6.113}$$

$$\begin{aligned} \left(\rho u_1 \left(u_2^2 + u_3^2\right)/2\right)^{II} &= \frac{1}{2} \sum_{i=1,3} \left(u_2^2 + u_3^2\right)^L \xi_i^L g_i^L \\ &+ \frac{1}{2} \sum_{i=2,4} \left(u_2^2 + u_3^2\right)^R \xi_i^R g_i^R . \end{aligned} \tag{6.114}$$

By substituting Eqs. (6.111)-(6.114) into Eq. (6.102), we can obtain $\mathbf{F}_c^{II}$ for the 3D Euler equations. Finally, the full expression of LBFS for simulation of 3D inviscid flows can be obtained by further substituting $\mathbf{F}_c^I$, $\mathbf{F}_c^{II}$ and τ_0 (see Eqs. (6.73)-(6.75)) into Eq. (6.86).

6.5 Circular Function-Based GKFS for Viscous Flows

The C-GKFS and S-GKFS presented in **Sections 6.2** and **6.3** aim to improve the computational efficiency of the conventional M-GKS and

M-GKFS in simulating inviscid flows. The improvement is achieved by removing the error function and the exponential form of equilibrium distribution functions in the conventional M-GKS and M-GKFS. On the other hand, it should be noted that the above C-GKFS and S-GKFS both treat the non-equilibrium part of the distribution function as the numerical dissipations, and are thus constrained to inviscid flows. To resolve viscous flow problems, the non-equilibrium term has to be evaluated appropriately and the collision time scale should be determined by the dynamic viscosity. In this section, the non-equilibrium part of the distribution function is approximated by the difference of equilibrium distribution functions on the cell interface and its surrounding points located at $-\xi\delta_t$. Since the circular function is defined along a circle rather than over the infinite domain, the interpolation points are distributed on the circle around the cell interface. And the equilibrium distribution function on the cell interface can be determined by the compatibility condition.

The 2D viscous flow can be interpreted by the Navier-Stokes equations

$$\frac{\partial \mathbf{W}}{\partial t}+\frac{\partial \mathbf{F}_{cx}}{\partial x}+\frac{\partial \mathbf{F}_{cy}}{\partial y}=\frac{\partial \mathbf{F}_{vx}}{\partial x}+\frac{\partial \mathbf{F}_{vy}}{\partial y}, \tag{6.115}$$

where the conservative variable vector $\mathbf{W}$, convective flux vectors $\mathbf{F}_{cx}$ and $\mathbf{F}_{cy}$, and viscous flux vectors $\mathbf{F}_{vx}$ and $\mathbf{F}_{vy}$ are given by

$$\mathbf{W}=\begin{bmatrix}\rho\\ \rho u\\ \rho v\\ \rho E\end{bmatrix}, \tag{6.116}$$

$$\mathbf{F}_{cx}=\begin{bmatrix}\rho u\\ \rho uu+p\\ \rho uv\\ (\rho E+p)u\end{bmatrix},\quad \mathbf{F}_{cy}=\begin{bmatrix}\rho v\\ \rho vu\\ \rho vv+p\\ (\rho E+p)v\end{bmatrix}, \tag{6.117}$$

$$\mathbf{F}_{vx} = \begin{bmatrix} 0 \\ \tau_{xx} \\ \tau_{yx} \\ \Theta_x \end{bmatrix}, \ \mathbf{F}_{vy} = \begin{bmatrix} 0 \\ \tau_{xy} \\ \tau_{yy} \\ \Theta_y \end{bmatrix}, \tag{6.118}$$

where τ_{ij} denotes the components of viscous stress tensor and Θ_i represents the term describing the work done by viscous stress and the heat conduction in the fluid. Detailed expressions of τ_{ij} and Θ_i can refer to Eqs. (1.6) and (1.17), respectively. The FVM discretization of Eq. (6.115) yields

$$\frac{\partial \mathbf{W}_i}{\partial t} = -\frac{1}{\Omega_i} \sum_{j=1}^{N_f} \mathbf{F}_{nj} S_j, \tag{6.119}$$

$$\mathbf{F}_n = n_x \left(\mathbf{F}_{cx} - \mathbf{F}_{vx} \right) + n_y \left(\mathbf{F}_{cy} - \mathbf{F}_{vy} \right), \tag{6.120}$$

where $\mathbf{F}_n$ is the numerical flux vector on the cell interface, which consists of the inviscid part and the viscous part.

Similar to the previous derivation for inviscid flows, a local coordinate system is defined on the cell interface with x_1-axis pointing to the normal direction and x_2-axis aligned with the tangential direction. In the local coordinate system, the conservative variable vector and the flux vector can be expressed as

$$\overline{\mathbf{W}} = \begin{bmatrix} \rho \\ \rho u_1 \\ \rho u_2 \\ \rho E \end{bmatrix}, \tag{6.121}$$

$$\overline{\mathbf{F}} = \begin{bmatrix} \rho u_1 \\ \rho u_1 u_1 + p \\ \rho u_1 u_2 \\ (\rho E + p) u_1 \end{bmatrix} - \begin{bmatrix} 0 \\ n_x \left(n_x \tau_{xx} + n_y \tau_{xy} \right) + n_y \left(n_x \tau_{yx} + n_y \tau_{yy} \right) \\ n_x \left(n_x \tau_{yx} + n_y \tau_{yy} \right) - n_y \left(n_x \tau_{xx} + n_y \tau_{xy} \right) \\ n_x \Theta_x + n_y \Theta_y \end{bmatrix}. \tag{6.122}$$

By applying the coordinate transformation from the local coordinate system to the global Cartesian coordinate system, the following relationships are obtained:

$$\mathbf{W} = \begin{bmatrix} \overline{\mathbf{W}}(1) \\ n_x \overline{\mathbf{W}}(2) - n_y \overline{\mathbf{W}}(3) \\ n_x \overline{\mathbf{W}}(3) + n_y \overline{\mathbf{W}}(2) \\ \overline{\mathbf{W}}(4) \end{bmatrix}, \ \mathbf{F}_n = \begin{bmatrix} \overline{\mathbf{F}}(1) \\ n_x \overline{\mathbf{F}}(2) - n_y \overline{\mathbf{F}}(3) \\ n_x \overline{\mathbf{F}}(3) + n_y \overline{\mathbf{F}}(2) \\ \overline{\mathbf{F}}(4) \end{bmatrix}. \quad (6.123)$$

As shown in Eq. (6.123), the calculation of $\mathbf{F}_n$ in the global Cartesian coordinate system is equivalent to evaluating $\overline{\mathbf{F}}$ in the local coordinate system. However, one should note that the Prandtl number corresponding to $\overline{\mathbf{F}}(4)$ evaluated by the C-GKFS is equal to the specific heat ratio. A correction to the heat flux is required to adjust to other Prandtl numbers.

In the following text, we denote "$(\mathbf{0},t)$" by the superscript "$face$" and "$(-\xi\delta_t, t-\delta_t)$" by the superscript "$cir$". According to the relationship between the distribution function and the macroscopic fluxes (see Eq. (6.45)), for the circular function, $\overline{\mathbf{F}}$ at the cell interface can be calculated by

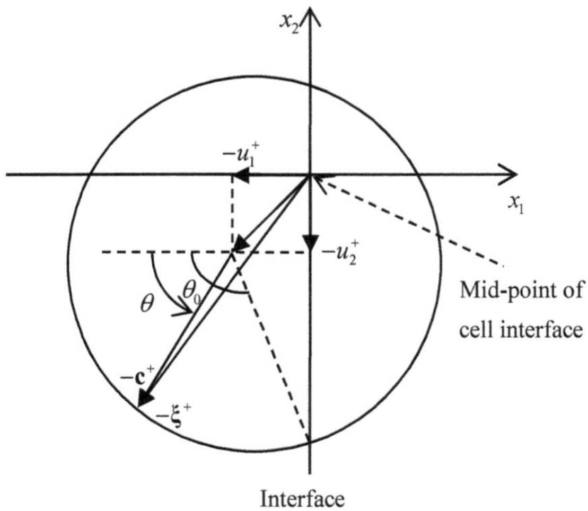

Figure 6.4. Sketch of the circular function in physical space at the cell interface ($\delta_t = 1$ for illustration).

$$\begin{aligned}\overline{\mathbf{F}} &= \int_0^{2\pi} \xi_1^{face} \overline{\boldsymbol{\varphi}}_\alpha^{face} g_C^{face} d\theta \\ &+ \tau_0 \left[\int_0^{2\pi} \xi_1^{cir} \overline{\boldsymbol{\varphi}}_\alpha^{cir} g_C^{cir} d\theta - \int_0^{2\pi} \xi_1^{face} \overline{\boldsymbol{\varphi}}_\alpha^{face} g_C^{face} d\theta \right] \\ &= \mathbf{F}^I + \tau_0 \left[\mathbf{F}^{II} - \mathbf{F}^I \right].\end{aligned} \tag{6.124}$$

As shown in Eq. (6.37), the dimensionless collision time scale is given by

$$\tau_0 = \frac{\tau}{\delta_t} = \frac{\mu}{p\delta_t}. \tag{6.125}$$

It can be seen from Eq. (6.124) that the C-GKFS for viscous flows also contains two parts of the flux vector on the cell interface. One is the flux vector attributed to the equilibrium distribution function and the moments on the cell interface, which is denoted as $\mathbf{F}^I$. The other is the flux vector attributed to the equilibrium distribution function and the moments along the circle, which is referred as $\mathbf{F}^{II}$. It is noteworthy that the piecewise constant distribution of the distribution function and conservative variables over the interface used in C-GKFS for inviscid flows would not be applicable to the viscous flows. Instead, the proposed C-GKFS for viscous flows assumes a linear distribution for the distribution function and conservative variables over the cell interface.

To calculate $\mathbf{F}^I$ and $\mathbf{F}^{II}$, the equilibrium distribution functions and the moments on the cell interface and its surrounding points should be determined first. By replacing the Maxwellian distribution function with the circular function as the equilibrium state, the integration is performed along the circle rather than over the infinite domain of $-\infty$ to $+\infty$. In other words, the equilibrium distribution functions at the surrounding points of the cell interface are distributed along the circle for the C-GKFS, as shown in Figure 6.4. This characteristic not only facilitates the interpolation algorithm, but also improves the computational efficiency as compared with the conventional M-GKS or M-GKFS.

Similar to the C-GKFS for inviscid flows, the calculation of flux vectors $\mathbf{F}^I$ and $\mathbf{F}^{II}$ relies on the equilibrium distribution function and the moments at the surrounding points of the cell interface (g_C^{cir} and

$\overline{\boldsymbol{\varphi}}_{\alpha}^{cir}$), while g_C^{cir} and $\overline{\boldsymbol{\varphi}}_{\alpha}^{cir}$ are both functions of conservative variables as shown in Eqs. (6.14) and (6.54). So, the key issue is to determine the conservative variables on the circle. Since the linear distribution for the distribution function and conservative variables is assumed in C-GKFS for simulation of viscous flows, the second-order interpolation should be used to calculate the conservative variables on the circle, which gives

$$\mathbf{W}^{cir} = \begin{cases} \mathbf{W}^L - \nabla \mathbf{W}^L \cdot \boldsymbol{\xi}^+ \delta_t, & \text{if } \xi_1^+ \geq 0, \\ \mathbf{W}^R - \nabla \mathbf{W}^R \cdot \boldsymbol{\xi}^+ \delta_t, & \text{if } \xi_1^+ < 0. \end{cases} \tag{6.126}$$

where $\mathbf{W}^L$, $\mathbf{W}^R$ and $\mathbf{W}^{cir}$ are the conservative variables on the left and the right sides of the cell interface and on the circle, respectively. $\nabla \mathbf{W}^L$ and $\nabla \mathbf{W}^R$ are the first-order derivatives of conservative variables at the left and the right cells around the cell interface. $\boldsymbol{\xi}^+$ is the predicted particle velocity on the cell interface, which is computed by Eq. (6.57).

Once the conservative variables along the circle are obtained, g_C^{cir} and $\overline{\boldsymbol{\varphi}}_{\alpha}^{cir}$ can be calculated straightforwardly. Taking $\xi_1^+ \geq 0$ as an example, with Eq. (6.126), the components of moments and the circular function along the circle can be computed by

$$\begin{aligned} \xi_1^{cir,L} = u_1^L &- \frac{\partial u_1^L}{\partial x_1}\left(u_1^+ + c^+ \cos(\theta)\right)\delta_t \\ &- \frac{\partial u_1^L}{\partial x_2}\left(u_2^+ + c^+ \sin(\theta)\right)\delta_t + c^+ \cos(\theta), \end{aligned} \tag{6.127a}$$

$$\begin{aligned} \xi_2^{cir,L} = u_2^L &- \frac{\partial u_2^L}{\partial x_1}\left(u_1^+ + c^+ \cos(\theta)\right)\delta_t \\ &- \frac{\partial u_2^L}{\partial x_2}\left(u_2^+ + c^+ \sin(\theta)\right)\delta_t + c^+ \sin(\theta), \end{aligned} \tag{6.127b}$$

$$\begin{aligned} e_p^{cir,L} = e_p^L &- \frac{\partial e_p^L}{\partial x_1}\left(u_1^+ + c^+ \cos(\theta)\right)\delta_t \\ &- \frac{\partial e_p^L}{\partial x_2}\left(u_2^+ + c^+ \sin(\theta)\right)\delta_t, \end{aligned} \tag{6.127c}$$

$$g_C^{cir,L} = g_C^L - \frac{\partial g_C^L}{\partial x_1}\left(u_1^+ + c^+ \cos(\theta)\right)\delta_t \\ - \frac{\partial g_C^L}{\partial x_2}\left(u_2^+ + c^+ \sin(\theta)\right)\delta_t . \tag{6.127d}$$

To facilitate the derivation of Eq. (6.124), we can rearrange Eqs. (6.127a)-(6.127d) into the following forms:

$$\xi_1^{cir,L} = a_0^L + a_1^L r + a_2^L s , \tag{6.128a}$$

$$\xi_2^{cir,L} = b_0^L + b_1^L r + b_2^L s , \tag{6.128b}$$

$$e_p^{cir,L} = e_0^L + e_1^L r + e_2^L s , \tag{6.128c}$$

$$g_C^{cir,L} = g_0^L + g_1^L r + g_2^L s , \tag{6.128d}$$

where

$$r = \cos(\theta), \; s = \sin(\theta),$$

$$a_0^L = u_1^L - \frac{\partial u_1^L}{\partial x_1} u_1^+ \delta_t - \frac{\partial u_1^L}{\partial x_2} u_2^+ \delta_t , \; a_1^L = c^+ - \frac{\partial u_1^L}{\partial x_1} c^+ \delta_t , \; a_2^L = -\frac{\partial u_1^L}{\partial x_2} c^+ \delta_t .$$

$$b_0^L = u_2^L - \frac{\partial u_2^L}{\partial x_1} u_1^+ \delta_t - \frac{\partial u_2^L}{\partial x_2} u_2^+ \delta_t , \; b_1^L = -\frac{\partial u_2^L}{\partial x_1} c^+ \delta_t , \; b_2^L = c^+ - \frac{\partial u_2^L}{\partial x_2} c^+ \delta_t .$$

$$e_0^L = e_p^L - \frac{\partial e_p^L}{\partial x_1} u_1^+ \delta_t - \frac{\partial e_p^L}{\partial x_2} u_2^+ \delta_t , \; e_1^L = -\frac{\partial e_p^L}{\partial x_1} c^+ \delta_t , \; e_2^L = -\frac{\partial e_p^L}{\partial x_2} c^+ \delta_t .$$

$$g_0^L = g_C^L - \frac{\partial g_C^L}{\partial x_1} u_1^+ \delta_t - \frac{\partial g_C^L}{\partial x_2} u_2^+ \delta_t , \; g_1^L = -\frac{\partial g_C^L}{\partial x_1} c^+ \delta_t , \; g_2^L = -\frac{\partial g_C^L}{\partial x_2} c^+ \delta_t .$$

Note that all the above coefficients are local constants except r and s. For the case of $\xi_1^+ < 0$, the components of moments and the circular function along the circle can be calculated by simply replacing the superscript " L " in Eqs. (6.127a)-(6.127d) and Eqs. (6.128a)-(6.128d) with the superscript " R ".

The detailed expressions of numerical fluxes of C-GKFS for viscous flows are the linear combination of a set of factors, which are the following functions of r and s :

$$h_{i,j}^{L} = \int_{-\theta_0}^{\theta_0} r^i s^j d\theta, \tag{6.129a}$$

$$h_{i,j}^{R} = \int_{\theta_0}^{2\pi-\theta_0} r^i s^j d\theta, \tag{6.129b}$$

where $\theta_0 = \arccos\left(-u_1^+ / c^+\right)$. As shown in Eqs. (6.129a)-(6.129b) and Figure 6.4, $h_{i,j}^{L}$ and $h_{i,j}^{R}$ are the integrals along the left circular arc and the right circular arc of the cell interface, respectively. It is easy to prove that $h_{i,j}^{L}$ and $h_{i,j}^{R}$ are precisely zero if j is an odd number. In addition, in the calculation of numerical fluxes of C-GKFS for viscous flows, only the factors with $i + j \le 4$ are required. In the following, the expressions of $h_{i,j}^{L}$ and $h_{i,j}^{R}$ are provided according to the relative location of the circle with respect to the cell interface.

Case 1: $u_1^+ > c^+$, i.e., the circle locates on the left side of the cell interface. Thus, $h_{i,j}^{L}$ and $h_{i,j}^{R}$ become

$$h_{i,j}^{L} = \int_{0}^{2\pi} r^i s^j d\theta, \tag{6.130a}$$

$$h_{i,j}^{R} = 0. \tag{6.130b}$$

For the case of $i + j \le 4$, the non-zero factors are as follows:

$$h_{0,0}^{L} = 2\pi, \; h_{2,0}^{L} = \pi, \; h_{0,2}^{L} = \pi, \; h_{4,0}^{L} = 3\pi/4,$$

$$h_{2,2}^{L} = \pi/4, \; h_{0,4}^{L} = 3\pi/4.$$

Case 2: $u_1^+ < -c^+$, i.e., the circle locates on the right side of the cell interface. As a result, $h_{i,j}^{L}$ and $h_{i,j}^{R}$ yield

$$h_{i,j}^{L} = 0, \tag{6.131a}$$

$$h_{i,j}^R = \int_0^{2\pi} r^i s^j d\theta \,. \tag{6.131b}$$

For the case of $i + j \le 4$, the non-zero factors are given by

$$h_{0,0}^R = 2\pi \,,\ h_{2,0}^R = \pi \,,\ h_{0,2}^R = \pi \,,\ h_{4,0}^R = 3\pi / 4 \,,$$

$$h_{2,2}^R = \pi / 4 \,,\ h_{0,4}^R = 3\pi / 4 \,.$$

Case 3: $-c^+ \le u_1^+ \le c^+$. In this case, the circle spans across the cell interface. Therefore, the non-zero elements of $h_{i,j}^L$ and $h_{i,j}^R$ with $i + j \le 4$ can be expressed by

$$h_{0,0}^L = 2\theta_1,\ h_{1,0}^L = 2\sin(\theta_1),$$

$$h_{2,0}^L = \cos(\theta_1)\sin(\theta_1) + \theta_1,\ h_{0,2}^L = -\cos(\theta_1)\sin(\theta_1) + \theta_1,$$

$$h_{3,0}^L = (2/3)(\cos(\theta_1))^2 \sin(\theta_1) + (4/3)\sin(\theta_1),$$

$$h_{1,2}^L = (2/3)(\sin(\theta_1))^3,$$

$$h_{4,0}^L = (1/2)(\cos(\theta_1))^3 \sin(\theta_1) + (3/4)\cos(\theta_1)\sin(\theta_1) + (3/4)\theta_1,$$

$$h_{2,2}^L = -(1/2)(\cos(\theta_1))^3 \sin(\theta_1) + (1/4)\cos(\theta_1)\sin(\theta_1) + (1/4)\theta_1,$$

$$h_{0,4}^L = -(1/2)(\sin(\theta_1))^3 \cos(\theta_1) - (3/4)\cos(\theta_1)\sin(\theta_1) + (3/4)\theta_1 \,.$$

$$h_{0,0}^R = 2\theta_2 \,,\ h_{1,0}^R = -2\sin(\theta_2),$$

$$h_{2,0}^R = \cos(\theta_2)\sin(\theta_2) + \theta_2,\ h_{0,2}^R = -\cos(\theta_2)\sin(\theta_2) + \theta_2 \,,$$

$$h_{3,0}^R = -(2/3)(\cos(\theta_2))^2 \sin(\theta_2) - (4/3)\sin(\theta_2),$$

$$h_{1,2}^R = -(2/3)(\sin(\theta_2))^3,$$

$$h_{4,0}^R = (1/2)(\cos(\theta_2))^3 \sin(\theta_2) + (3/4)\cos(\theta_2)\sin(\theta_2) + (3/4)\theta_2,$$

$$h_{2,2}^R = -(1/2)(\cos(\theta_2))^3 \sin(\theta_2) + (1/4)\cos(\theta_2)\sin(\theta_2) + (1/4)\theta_2,$$

$$h_{0,4}^R = -(1/2)(\sin(\theta_2))^3 \cos(\theta_2) - (3/4)\cos(\theta_2)\sin(\theta_2) + (3/4)\theta_2.$$

where the definitions of θ_1 and θ_2 are the same as those in Eq. (6.67).

To calculate the numerical fluxes attributed to the equilibrium distribution function and the moments on the cell interface, the conservative variables at the same physical location should be determined first. The compatibility condition gives

$$\overline{\mathbf{W}}^{face} = \int_0^{2\pi} \overline{\boldsymbol{\varphi}}_\alpha^{face} g_C^{face} d\theta = \int_0^{2\pi} \overline{\boldsymbol{\varphi}}_\alpha^{cir} g_C^{cir} d\theta. \tag{6.132}$$

Substituting Eqs. (6.128a)-(6.128d) into Eq. (6.132) and utilizing the definition of integrals of r and s, the explicit expressions of Eq. (6.132) can be written as

$$\overline{\mathbf{W}}^{face}(1) = [g_0 h_{0,0} + g_1 h_{1,0}]^L + [g_0 h_{0,0} + g_1 h_{1,0}]^R, \tag{6.133a}$$

$$\begin{aligned}\overline{\mathbf{W}}^{face}(2) = &[g_0 a_0 h_{0,0} + (g_0 a_1 + g_1 a_0) h_{1,0} + g_1 a_1 h_{2,0} + g_2 a_2 h_{0,2}]^L \\ &+ [g_0 a_0 h_{0,0} + (g_0 a_1 + g_1 a_0) h_{1,0} + g_1 a_1 h_{2,0} + g_2 a_2 h_{0,2}]^R,\end{aligned} \tag{6.133b}$$

$$\begin{aligned}\overline{\mathbf{W}}^{face}(3) = &[g_0 b_0 h_{0,0} + (g_0 b_1 + g_1 b_0) h_{1,0} + g_1 b_1 h_{2,0} + g_2 b_2 h_{0,2}]^L \\ &+ [g_0 b_0 h_{0,0} + (g_0 b_1 + g_1 b_0) h_{1,0} + g_1 b_1 h_{2,0} + g_2 b_2 h_{0,2}]^R,\end{aligned} \tag{6.133c}$$

$$\begin{aligned}\overline{\mathbf{W}}^{face}(4) = &[g_0 \chi_0 h_{0,0} + (g_0 \chi_1 + g_1 \chi_0) h_{1,0} + (g_0 \chi_3 + g_1 \chi_1) h_{2,0} \\ &+ (g_0 \chi_5 + g_2 \chi_2) h_{0,2} + g_1 \chi_3 h_{3,0} + (g_1 \chi_5 + g_2 \chi_4) h_{1,2}]^L \\ &+ [g_0 \chi_0 h_{0,0} + (g_0 \chi_1 + g_1 \chi_0) h_{1,0} + (g_0 \chi_3 + g_1 \chi_1) h_{2,0} \\ &+ (g_0 \chi_5 + g_2 \chi_2) h_{0,2} + g_1 \chi_3 h_{3,0} + (g_1 \chi_5 + g_2 \chi_4) h_{1,2}]^R,\end{aligned} \tag{6.133d}$$

where

$$\chi_0 = \frac{1}{2}\left(a_0^2 + b_0^2\right) + e_0,\ \chi_1 = a_0 a_1 + b_0 b_1 + e_1,\ \chi_2 = a_0 a_2 + b_0 b_2 + e_2,$$

$$\chi_3 = \frac{1}{2}\left(a_1^2 + b_1^2\right),\ \chi_4 = a_1 a_2 + b_1 b_2,\ \chi_5 = \frac{1}{2}\left(a_2^2 + b_2^2\right).$$

Once the conservative variables on the cell interface $\overline{\mathbf{W}}^{face}$ are obtained, g_C^{face} and $\overline{\boldsymbol{\varphi}}_\alpha^{face}$ can be constructed by applying Eqs. (6.14) and (6.54). Finally, the fluxes contributed by g_C^{face} and $\overline{\boldsymbol{\varphi}}_\alpha^{face}$ can be computed by the integral of Eq. (6.124). An alternative but simpler way to compute $\mathbf{F}^I$ is to substitute the conservative variables $\overline{\mathbf{W}}^{face}$ directly into the expression of inviscid fluxes. As a result, $\mathbf{F}^I$ can be calculated by

$$\mathbf{F}^I = \begin{bmatrix} \rho u_1 \\ \rho u_1 u_1 + p \\ \rho u_1 u_2 \\ (\rho E + p) u_1 \end{bmatrix}^{face}. \tag{6.134}$$

$\mathbf{F}^{II}$ is the flux vector related to the equilibrium distribution function and the moments on the circle, namely g_C^{cir} and $\overline{\boldsymbol{\varphi}}_\alpha^{cir}$. According to Eq. (6.124), we can get the expression of $\mathbf{F}^{II}$ as

$$\mathbf{F}^{II} = \int_0^{2\pi} \xi_1^{cir} \overline{\boldsymbol{\varphi}}_\alpha^{cir} g_C^{cir} d\theta. \tag{6.135}$$

Substituting Eqs. (6.128a)-(6.128d) into Eq. (6.135) and utilizing the definition of integrals of r and s, we can get

$$\mathbf{F}^{II}(1) = [\beta_0]^L + [\beta_0]^R, \tag{6.136a}$$

$$\mathbf{F}^{II}(2) = [a_0\beta_0 + a_1\beta_1 + a_2\beta_2]^L + [a_0\beta_0 + a_1\beta_1 + a_2\beta_2]^R, \tag{6.136b}$$

$$\mathbf{F}^{II}(3) = [b_0\beta_0 + b_1\beta_1 + b_2\beta_2]^L + [b_0\beta_0 + b_1\beta_1 + b_2\beta_2]^R, \tag{6.136c}$$

$$\begin{aligned}\mathbf{F}^{II}(4) = {} & [\chi_0\beta_0 + \chi_1\beta_1 + \chi_2\beta_2 + \chi_3\beta_3 + \chi_4\beta_4 + \chi_5\beta_5]^L \\ & + [\chi_0\beta_0 + \chi_1\beta_1 + \chi_2\beta_2 + \chi_3\beta_3 + \chi_4\beta_4 + \chi_5\beta_5]^R,\end{aligned} \tag{6.136d}$$

where

$$\beta_0 = \alpha_0 h_{0,0} + \alpha_1 h_{1,0} + \alpha_3 h_{2,0} + \alpha_5 h_{0,2},$$

$$\beta_1 = \alpha_0 h_{1,0} + \alpha_1 h_{2,0} + \alpha_3 h_{3,0} + \alpha_5 h_{1,2},$$

$$\beta_2 = \alpha_2 h_{0,2} + \alpha_4 h_{1,2},\ \beta_3 = \alpha_0 h_{2,0} + \alpha_1 h_{3,0} + \alpha_3 h_{4,0} + \alpha_5 h_{2,2},$$

$$\beta_4 = \alpha_2 h_{1,2} + \alpha_4 h_{2,2},\ \beta_5 = \alpha_0 h_{0,2} + \alpha_1 h_{1,2} + \alpha_3 h_{2,2} + \alpha_5 h_{0,4},$$

and

$$\alpha_0 = a_0 g_0,\ \alpha_1 = a_0 g_1 + a_1 g_0,\ \alpha_2 = a_0 g_2 + a_2 g_0,$$

$$\alpha_3 = a_1 g_1,\ \alpha_4 = a_1 g_2 + a_2 g_1,\ \alpha_5 = a_2 g_2.$$

The definitions of $\chi_0, \chi_1, \ldots, \chi_5$ are the same as those below Eq. (6.133d).

In the reconstruction of numerical fluxes from the local solution to the Boltzmann equation, the streaming process is involved, which is associated with the streaming time step δ_t. The principle for the choice of δ_t is that the location of $\left(-\xi^+ \delta_t, t-\delta_t\right)$ must be within the two cells sharing the interface in order to avoid extrapolation. A simple way to determine δ_t is given by

$$\delta_t = \frac{0.2 \times \min\{\Delta l, \Delta r\}}{\left(\max\{u_1^+, u_2^+\} + c^+\right)}, \tag{6.137}$$

where Δl and Δr are the shortest edges of the left and the right cells around the cell interface, respectively. In this way, the constraint of non-extrapolation can be satisfied for any non-excessively distorted grid.

Theoretically, the dimensionless collision time scale τ_0 should be determined by Eq. (6.125) since the dissipation of the Navier-Stokes equations is solely determined by the physical viscosity. However, when encountering severe cases such as shock waves, extra numerical dissipation is necessary to ensure convergence. Similar to the idea of the Jameson-Schmidt-Turkel (JST) scheme [Jameson *et al.*, 1981], Xu [2001] presented a simple but effective way to incorporate the numerical dissipation into the conventional M-GKS, which is also adopted in the present C-GKFS for viscous flows. In this approach, the dimensionless collision time scale τ_0 is computed by

$$\tau_0 = \frac{\mu}{p\delta_t} + \frac{\left|p^L - p^R\right|}{\left|p^L + p^R\right|}\frac{\Delta t}{\delta_t}, \tag{6.138}$$

where Δt is the real time step to solve Eq. (6.119). The second part of Eq. (6.138) corresponds to the numerical viscosity, which takes effect in the presence of shock waves.

As mentioned before, the Prandtl number of C-GKFS for viscous flows is fixed at γ. In order to solve problems with different Prandtl numbers, a convenient approach is to make a correction on the energy flux. First, the heat flux on the cell interface can be expressed as

$$q = -k\frac{\partial T}{\partial x_1} = -\frac{\gamma\mu}{\left[1 - D(\gamma - 1)/2\right]Pr}\frac{\partial e_p}{\partial x_1}. \tag{6.139}$$

Since $\partial e_p^L / \partial x_1$ and $\partial e_p^R / \partial x_1$ have been calculated in Eq. (6.127c), $\partial e_p / \partial x_1$ in the above equation can be taken as the simple average of those from two sides of the cell interface. With Eq. (6.139), the corrected energy flux can then be calculated by

$$\overline{\mathbf{F}}^{correct}(4) = \overline{\mathbf{F}}(4) + \left(1 - \frac{Pr}{\gamma}\right)q. \tag{6.140}$$

Such modification is simple and effective with very little extra computational efforts.

The above derivation reveals two major improvements made by the C-GKFS for viscous flows as compared with the conventional M-GKS. One is that the Maxwellian distribution function in the conventional M-GKS is replaced by a simple circular function in C-GKFS, which reduces integrals over the infinity domain of the particle velocity space to integrals along a circle. Such simplification not only facilitates the implementation, but also reduces the computational effort. The other improvement is that the non-equilibrium part of the distribution function is calculated by a simpler way in C-GKFS. It is computed by the difference of equilibrium distribution functions on the cell interface and its surrounding points. Since the equilibrium state is defined by the circular function in C-GKFS, interpolation points are distributed along a circle, which facilitates the mathematical interpolation. These improvements allow explicit and concise expressions of conservative variables and numerical fluxes on the cell interface. As a comparison, as mentioned in the work of Tang [2012], for the viscous flow simulated by the conventional M-GKS, a large number of coefficients related to the conservative variables and their derivatives on each cell interface and time step need to be calculated, which makes the evaluation of numerical fluxes by the conventional M-GKS inefficient.

6.6 Sample Applications

The C-GKFS, S-GKSF and LBFS introduced in this chapter have been validated in the literature by simulating various flow problems [Zhou *et al.*, 2017; Zhou *et al.*, 2018; Meng *et al.*, 2018; Yang *et al.*, 2018a]. In this section, three sample applications will be presented to illustrate the accuracy and efficiency of these solvers. In these applications, the conservative variables on the left and the right sides of the cell interface are calculated by the following interpolations:

$$\begin{aligned}\mathbf{W}^{L} &= \mathbf{W}_{c}^{L} + \Psi^{L}\left(\mathbf{x}_{b} - \mathbf{x}_{c}^{L}\right)\cdot\nabla\mathbf{W}^{L},\\ \mathbf{W}^{R} &= \mathbf{W}_{c}^{R} + \Psi^{R}\left(\mathbf{x}_{b} - \mathbf{x}_{c}^{R}\right)\cdot\nabla\mathbf{W}^{R}.\end{aligned} \tag{6.141}$$

Here, $\mathbf{W}_c^L$ and $\mathbf{W}_c^R$ are respectively the conservative flow variables at centers of the left and the right cells. $\mathbf{x}_c^L$, $\mathbf{x}_c^R$ and $\mathbf{x}_b$ are the coordinates of the left cell center, the right cell center and the midpoint of the cell interface, respectively. Ψ^L and Ψ^R are the limiter functions utilized in the left and the right cells. The gradients of conservative variables are calculated by the Green-Gauss theorem,

$$\nabla\mathbf{W}_i = \frac{1}{\Omega_i}\sum_{j=1}^{N_f}\frac{1}{2}\left(\mathbf{W}_i + \mathbf{W}_j\right)\mathbf{n}_{ij}S_{ij}\,, \tag{6.142}$$

where $\mathbf{n}_{ij}$ denotes the unit normal vector and S_{ij} represents the area of the cell interface shared by the cell i and cell j, respectively. The Venkatakrishnan's limiter [Venkatakrishnan, 1995] is chosen in the simulation, which gives

$$\Psi_i = \min_{j=1,N_f} \begin{cases} \dfrac{1}{\Delta_2}\left[\dfrac{\left(\Delta_{1,\max}^2+\varepsilon^2\right)\Delta_2+2\Delta_2^2\Delta_{1,\max}}{\Delta_{1,\max}^2+2\Delta_2^2+\Delta_{1,\max}\Delta_2+\varepsilon^2}\right], & \text{for } \Delta_2>0, \\ \dfrac{1}{\Delta_2}\left[\dfrac{\left(\Delta_{1,\min}^2+\varepsilon^2\right)\Delta_2+2\Delta_2^2\Delta_{1,\min}}{\Delta_{1,\min}^2+2\Delta_2^2+\Delta_{1,\min}\Delta_2+\varepsilon^2}\right], & \text{for } \Delta_2<0, \\ 1, & \text{for } \Delta_2=0. \end{cases} \quad (6.143)$$

where

$$\Delta_{1,\max} = \mathbf{W}_{\max} - \mathbf{W}_i,\ \Delta_{1,\min} = \mathbf{W}_{\min} - \mathbf{W}_i,\ \Delta_2 = \nabla \mathbf{W}_i \cdot \mathbf{r},\ \varepsilon^2 = \left(K_0 \Delta h\right)^3.$$

$\mathbf{W}_{\max}$ and $\mathbf{W}_{\min}$ are the maximum and the minimum values of conservative variables of all surrounding cells including the cell i itself. $\mathbf{r}$ denotes the vector pointing from the cell center to the midpoint of the cell interface. Δh stands for the square-root of the area of the control volume for the 2D case. K_0 is a constant, which is usually taken as 5.

In addition, the three-stage Runge-Kutta scheme is applied as the temporal discretization strategy for Eqs. (6.48), (6.79) and (6.119). Overall, Eqs. (6.48), (6.79) and (6.119) can be written as

$$\frac{d\mathbf{W}_i}{dt} = \mathbf{R}\left(\mathbf{W}_i\right), \quad (6.144)$$

where $\mathbf{R}\left(\mathbf{W}_i\right)$ is the residual. The discretization of Eq. (6.144) using the three-stage Runge-Kutta scheme can be written as

$$\mathbf{W}_i^{(1)} = \mathbf{W}_i^n + \Delta t_i \mathbf{R}\left(\mathbf{W}_i^n\right), \quad (6.145a)$$

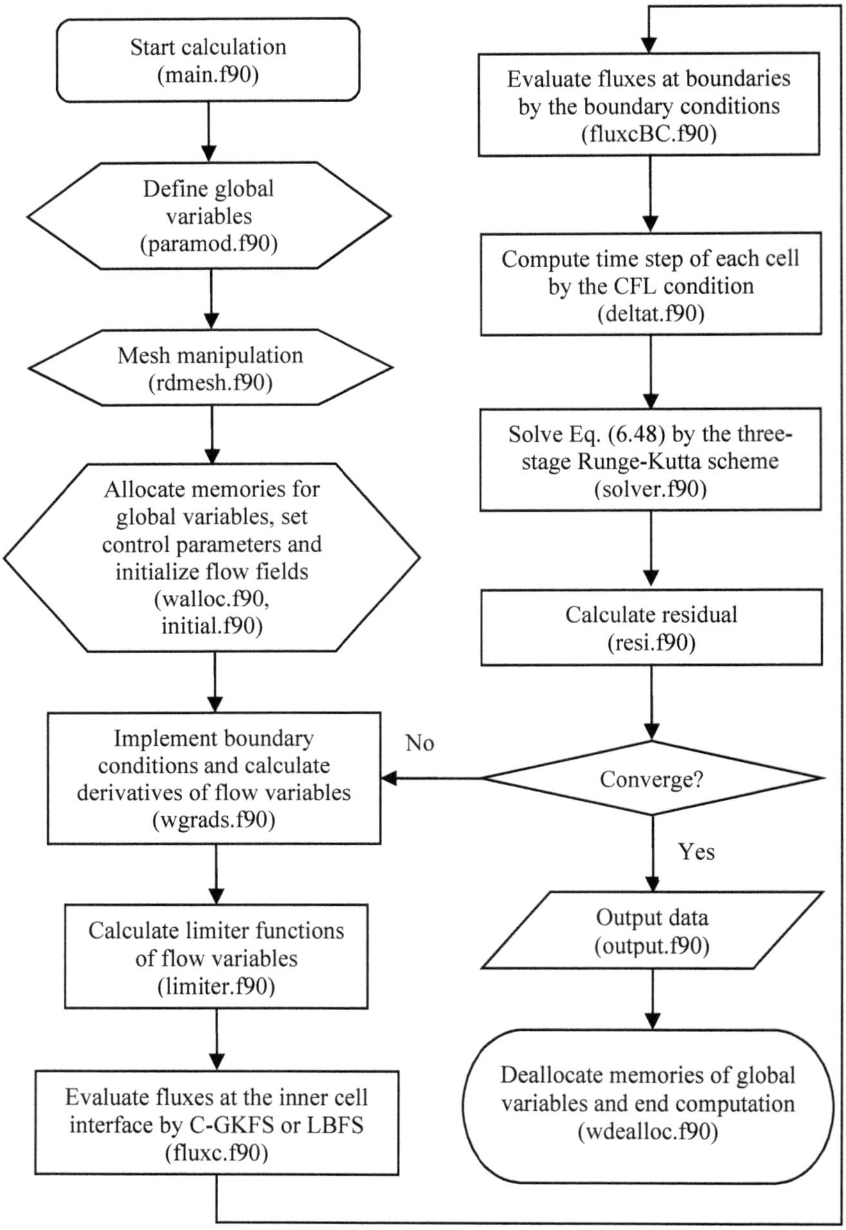

Figure 6.5. Flowchart of C-GKFS and LBFS for simulation of inviscid compressible flows.

$$\mathbf{W}_i^{(2)} = (3/4)\mathbf{W}_i^n + (1/4)\left[\mathbf{W}_i^{(1)} + \Delta t_i \mathbf{R}\left(\mathbf{W}_i^{(1)}\right)\right], \tag{6.145b}$$

$$\mathbf{W}_i^{n+1} = (1/3)\mathbf{W}_i^n + (2/3)\left[\mathbf{W}_i^{(2)} + \Delta t_i \mathbf{R}\left(\mathbf{W}_i^{(2)}\right)\right], \tag{6.145c}$$

where $\mathbf{W}_i^n$ and $\mathbf{W}_i^{n+1}$ are the conservative variables at the current time level and the new time level, respectively. Δt_i is the time step for cell i. For simulation of inviscid flows, Δt_i can be calculated by

$$\Delta t_i = \sigma \frac{\Omega_i}{\left(\hat{\Lambda}_c\right)_i}, \tag{6.146}$$

$$\left(\hat{\Lambda}_c\right)_i = \sum_{j=1}^{N_f} \left(\mathbf{u}_{ij} \cdot \mathbf{n}_{ij} + c_{ij}\right) S_{ij}, \tag{6.147}$$

where $\mathbf{u}_{ij}$ and c_{ij} are respectively the velocity and the sound speed on the cell interface, which are obtained by the average of those at the left and the right cells. σ is the Courant-Friedrichs-Lewy (CFL) number. The computational efficiency of C-GKFS/S-GKFS will be compared quantitatively with that of M-GKFS shown in the last chapter.

For the convenience of application, the flowchart of C-GKFS and LBFS for simulation of inviscid compressible flows is shown in Figure 6.5 and the sample code for solving transonic flow around RAE2822 airfoil is attached in **Code C**. The computational mesh can be generated by the commercial software POINTWISE with COBALT format and then transformed into a mesh input file by **Code B**.

6.6.1 *Two-dimensional inviscid flow*

The transonic flow around RAE2822 airfoil is firstly presented to access the performance of C-GKFS and LBFS in simulating 2D inviscid flows. In this case, the free-stream Mach number is taken as 0.75 and the angle of attack is chosen as 3 degrees. A C-type grid with 256×65 cells is used. Figure 6.6 shows the pressure contours computed by the C-GKFS and M-GKFS. As shown in this figure, the shock wave is well captured by both schemes without unphysical oscillations. Figure 6.7 depicts the pressure coefficient distributions along the surface of the airfoil calculated by the C-GKFS, M-GKFS and LBFS. Also displayed in this

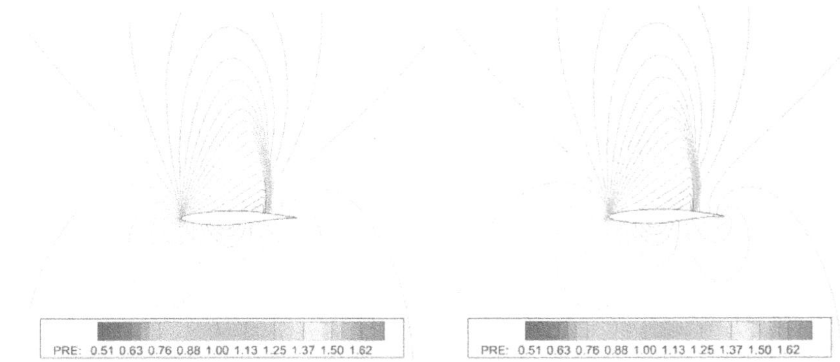

Figure 6.6. Pressure contours of flow around RAE2822 airfoil computed by the C-GKFS (left) and M-GKFS (right).

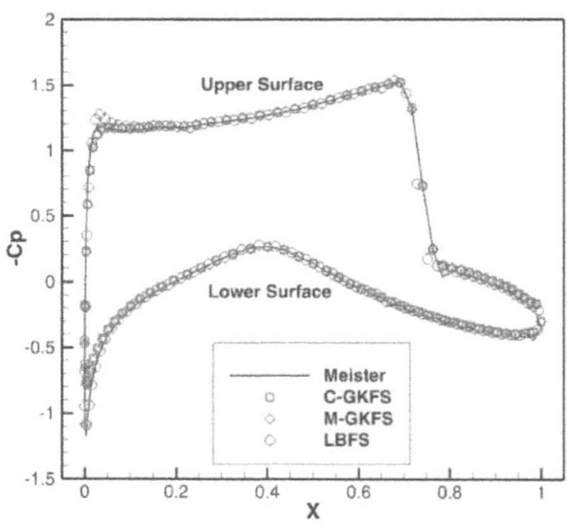

Figure 6.7. Comparison of pressure coefficient distributions on the surface of RAE2822 airfoil.

Table 6.1. Comparison of computational time (seconds) for different schemes.

Test Cases	C-GKFS/S-GKFS	M-GKFS
RAE2822 airfoil [Yang *et al.*, 2013b]	924.6	1314.5
ONERA M6 wing [Yang *et al.*, 2017a]	13705.2	26707.2

figure is the result computed on the same grid by solving the Euler equations with the JST scheme [Jameson, 1981]. Clearly, the results of C-GKFS, M-GKFS and LBFS match very well with each other and are in good agreement with the reference data. In addition, the lift coefficient (C_l) obtained by the C-GKFS, M-GKFS and LBFS are respectively 1.0436, 1.0461 and 1.0382, which agree well with the reference data of 1.0395 obtained by the JST scheme. Quantitative comparison of the computational efficiency indicates that the C-GKFS is more efficient than the M-GKFS, as shown in Table 6.1.

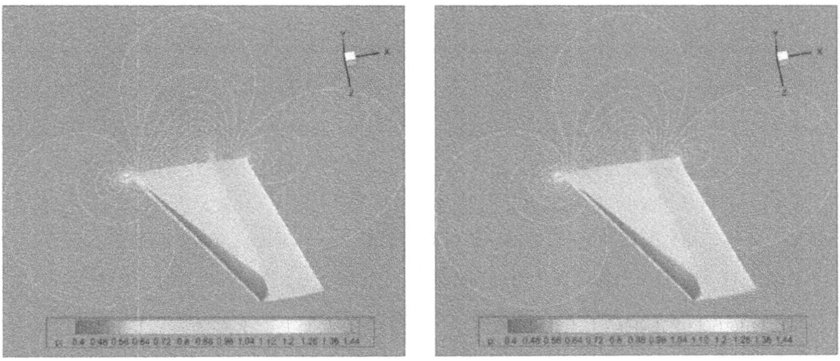

Figure 6.8. Pressure contours of flow around ONERA M6 wing computed by the S-GKFS (left) and M-GKFS (right).

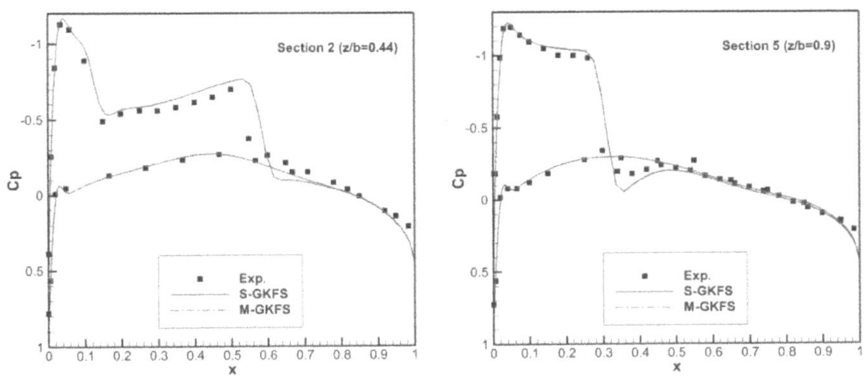

Figure 6.9. Comparison of pressure coefficient distributions at two selected sections on ONERA M6 wing.

6.6.2 *Three-dimensional inviscid flow*

To investigate the performance of S-GKFS in simulating 3D inviscid flows, the test of the transonic flow around ONERA M6 wing is carried out. In this simulation, the free-stream Mach number is taken as 0.8395 and the angle of attack is set as 3.06 degrees. The computational mesh provided by the NASA website*, which contains 294912 hexahedral cells, is utilized. The pressure contours obtained by the S-GKFS and M-GKFS are displayed in Figure 6.8. The "λ" shape of the shock wave on the upper surface of the wing is well recovered by both schemes, which is in agreement with the result from the WIND solver [Mani *et al.*, 1997]. Figure 6.9 compares the pressure coefficient distributions at two selected span-wise locations obtained by the S-GKFS, M-GKFS and experimental measurement [Schmitt and Charpin, 1979]. As shown in this figure, the results of S-GKFS and M-GKFS are basically the same and they generally agree well with the experimental data. The minor difference between the simulation results and the experimental data could be attributed to the inviscid assumption in the computation. In the meantime, the comparison of the computational time shown in Table 6.1 demonstrates our expectation that the S-GKFS bears higher numerical efficiency than the M-GKFS.

6.6.3 *Two-dimensional viscous flow*

In this case, the interaction of an oblique shock wave with a laminar boundary layer is considered. The shock hits the boundary layer on the wall at x=1m with an incident angle of 32.6 degrees. The Mach number of the shock wave is equal to 2 and the Reynolds number of the upstream flow is 2.96×10^5. The temperature of the free stream is taken as 117K and the adiabatic boundary condition is implemented on the bottom wall. The dynamic viscosity is computed by the Sutherland's law [Sutherland, 1893]. And the non-uniform mesh with 100×120 cells is used. The pressure contours obtained by the C-GKFS and the Roe scheme [Kim *et al.*, 2002] are shown in Figure 6.10. The results of these two schemes are

---

* https://www.grc.nasa.gov/www/wind/valid/m6wing/m6wing

basically the same and flow details in the separation zone of the boundary layer are well captured. Figure 6.11 compares the pressure distribution and the distribution of the skin friction coefficient along with the flat plate computed by different schemes. Also presented in this figure are the experimental data [Hakkinen *et al.*, 1959] and the numerical results computed by the conventional M-GKS [Xu, 2001]. Basically, good agreements are achieved. Moreover, it was reported by Yang *et al.* [2014b] that the computational efficiency of C-GKFS is slightly higher than that of the Roe scheme for this example, while Chae *et al.* [2000] indicated that the conventional M-GKS is about 1.5 times slower than the conventional Navier-Stokes solver for simulation of 2D viscous flows. Therefore, the C-GKFS is genuinely more efficient than the conventional M-GKS in simulating viscous flows.

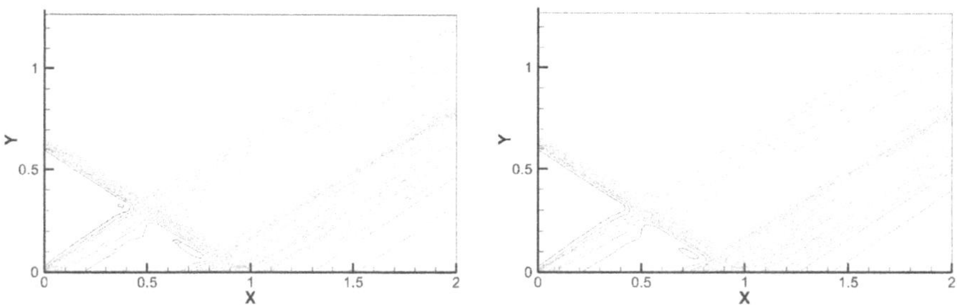

Figure 6.10. Pressure contours of shock-boundary layer interaction obtained by the C-GKFS (left) and the Roe scheme (right).

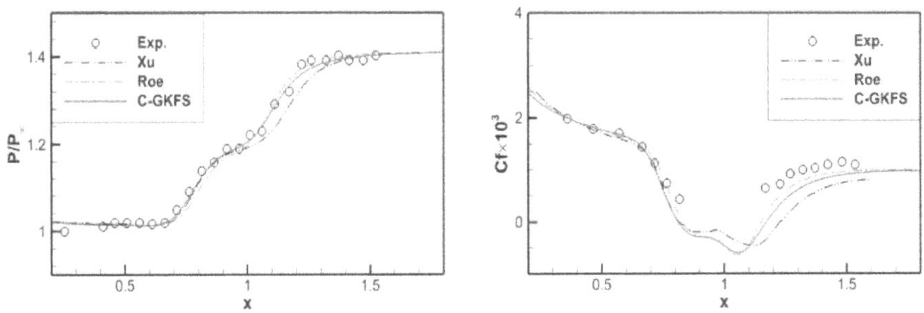

Figure 6.11. Pressure distribution (left) and skin friction coefficient distribution (right) along with flat plate for shock-boundary layer interaction.

6.7 Summary

This chapter presents two simplified distribution functions, namely, the circular function and the sphere function, to replace the Maxwellian distribution function in two- and three- dimensional conditions. The simplification is made on the basis of some assumptions. First, the simplified distribution function only considers the translational particle velocity in D-dimensional space, while the effects of rest components of the translational particle velocity in 3-dimensional space and the rotational particle velocity are reflected by the particle potential energy. Then, the simplified Maxwellian distribution function is further reduced to the circular function for the 2D case and the sphere function for the 3D case by assuming that all the particles are concentrated on a circle or on the spherical surface, respectively.

Based on the simplified distribution functions, various gas kinetic flux solvers are developed, aiming to resolve inviscid or viscous flow problems. In these solvers, the Euler/Navier-Stokes equations are discretized by the FVM and the numerical fluxes on the cell interface are reconstructed by the local asymptotic solution to the Boltzmann equation with the simplified distribution functions. This local solution involves the equilibrium part and the non-equilibrium part of the distribution function. Specifically, the equilibrium part is evaluated by the newly proposed circular function or sphere function, and the non-equilibrium part is approximated by the difference of two equilibrium distribution functions. The inviscid solvers assume a local Riemann problem with a piecewise constant distribution of the distribution function and conservative variables across the cell interface, and thus treat the non-equilibrium part of the distribution function as the numerical dissipation term. And the contribution of the dissipation term is weighted by the switch function which is affected by the pressure difference between two sides of the cell interface. The viscous solver, on the other hand, reflects the true physical viscosity through the non-equilibrium part of the distribution function; and the distribution function and conservative variables are assumed to follow the linear distribution over the cell interface.

The C-GKFS and S-GKFS presented in this chapter show advantages over the conventional M-GKS [Xu, 2001; Tang, 2001; Xiong *et al.*,

2011; Kumar *et al.*, 2013] and M-GKFS [Sun *et al.*, 2015; Sun *et al.*, 2016; Sun *et al.*, 2017] in two major aspects. By replacing the Maxwellian distribution function with a simple circular/sphere function, the integration domain is thus simplified from an infinite domain to the finite circle or spherical surface, which greatly reduces computational efforts. In the meantime, the novel approximation of the non-equilibrium distribution function allows explicit expression of the local solution to the Boltzmann equation, and thus facilitates the implementation of the solver. In addition, with the same strategy for approximation of the non-equilibrium distribution function and by replacing the Maxwellian distribution function with the D1Q4 model, the LBFS for compressible inviscid flows is also developed.

Sample applications of C-GKFS, S-GKFS and LBFS illustrate their effectiveness in simulating various inviscid or viscous flows. Good agreements are achieved between the C-GKFS, S-GKFS, LBFS and M-GKFS, which well validates the accuracy of the proposed solvers. Comparisons of computational time indicate that the C-GKFS and the S-GKFS are more efficient than the M-GKFS. It is thus believed that these simplified distribution function-based gas kinetic flux solvers have a great potential for practical applications.

Chapter 7

Discrete Gas Kinetic Flux Solvers

The simplification from the Maxwellian distribution function to the circular function in the two-dimensional (2D) case and the sphere function in the three-dimensional (3D) case greatly reduces the computational effort of the conventional gas kinetic schemes. And the corresponding circular function-based and sphere-function based gas kinetic flux solvers (C-GKFS and S-GKFS) are much more efficient than the Maxwellian function-based gas kinetic flux solver (M-GKFS). However, comparing with the conventional Navier-Stokes solvers, the formulations of C-GKFS for viscous flows are still complicated, which obstructs its wider application. Furthermore, the extension of S-GKFS to 3D viscous flows remains difficult. To resolve these issues, it was suggested that the integrations along the circle/spherical surface can be replaced by the quadrature algorithm [Yang *et al.*, 2016a; Yang *et al.*, 2016b]. The basic requirement here is to ensure the consistency with the moment relationships in C-GKFS and S-GKFS, so that the macroscopic variables, numerical fluxes, and the Navier-Stokes equations can be accurately recovered. This simplification yields the discrete gas kinetic flux solver (DGKFS), which will be presented in detail in this chapter.

7.1 Moment Relationships and Their Discretization Forms

The circular function and the sphere function derived in the previous chapter significantly simplifies the original Maxwellian distribution function. They replace the integral over the infinite particle velocity space by the line integral along the circle in 2D conditions and the surface integral over the sphere surface in 3D scenarios. However, analytical integration is still tedious from the numerical perspective. Simplifications can be made by using the quadrature algorithm in place of the analytical integral without violating the moment relationships of the distribution functions. To do that, the preliminary step is to construct the corresponding discrete velocity models.

7.1.1 *Two-dimensional discrete velocity model*

As indicated in the last chapter, in 2D conditions, the Maxwellian distribution function can be simplified to the following circular function:

$$g_C = \begin{cases} \dfrac{\rho}{2\pi} & \text{if } (\xi_1 - u_1)^2 + (\xi_2 - u_2)^2 = c^2, \\ 0 & \text{otherwise.} \end{cases} \tag{7.1}$$

where ρ is the density of the mean flow, $\boldsymbol{\xi} = (\xi_1, \xi_2)$ and $\mathbf{u} = (u_1, u_2)$ are the particle velocity vector and the macroscopic flow velocity vector. The square of radius c represents the mean kinetic energy of particles, which can be expressed as

$$c^2 = D(\gamma - 1)e. \tag{7.2}$$

Here, γ is the specific heat ratio, D is the number of dimensions ($D = 2$ for two-dimension). $e = p/[(\gamma - 1)\rho]$ is the potential energy of the mean flow, where p is the pressure.

The physical meaning of Eq. (7.1) is that all particles are concentrated on the circle. The velocity components of the particle on the circle can be written as follows in the Cartesian coordinate system:

$$\xi_1 = u_1 + c\cos(\theta), \tag{7.3a}$$

$$\xi_2 = u_2 + c\sin(\theta), \tag{7.3b}$$

where θ is the angle between the radial direction of the circle and the normal direction of the cell interface, which can be referred to Figure 6.1. With notations $c_1 = c\cos(\theta)$, $c_2 = c\sin(\theta)$ and $\xi_\alpha = u_\alpha + c_\alpha$, the following 7 moment relationships for the C-GKFS can be established to recover the Navier-Stokes equations:

$$\int_0^{2\pi} g_C d\theta = \rho , \tag{7.4a}$$

$$\int_0^{2\pi} g_C \xi_\alpha d\theta = \rho u_\alpha , \tag{7.4b}$$

$$\int_0^{2\pi} g_C \left(\xi_\alpha \xi_\alpha + 2e_p\right) d\theta = \rho\left(u_\alpha u_\alpha + bRT\right), \tag{7.4c}$$

$$\int_0^{2\pi} g_C \xi_\alpha \xi_\beta d\theta = \rho u_\alpha u_\beta + p\delta_{\alpha\beta} , \tag{7.4d}$$

$$\int_0^{2\pi} g_C \left(\xi_\alpha \xi_\alpha + 2e_p\right)\xi_\beta d\theta = \rho\left[u_\alpha u_\alpha + (b+2)RT\right]u_\beta , \tag{7.4e}$$

$$\int_0^{2\pi} g_C \xi_\alpha \xi_\beta \xi_\chi d\theta = p\left(u_\alpha \delta_{\beta\chi} + u_\beta \delta_{\chi\alpha} + u_\chi \delta_{\alpha\beta}\right) + \rho u_\alpha u_\beta u_\chi , \tag{7.4f}$$

$$\begin{aligned}&\int_0^{2\pi} g_C \left(\xi_\chi \xi_\chi + 2e_p\right)\xi_\alpha \xi_\beta d\theta \\ &= \rho\left\{bR^2T^2\delta_{\alpha\beta} + \left[(b+4)u_\alpha u_\beta + u_\chi u_\chi \delta_{\alpha\beta}\right]RT + u_\chi u_\chi u_\alpha u_\beta\right\},\end{aligned} \tag{7.4g}$$

where ξ_α, ξ_β, ξ_χ and u_α, u_β, u_χ are the particle velocities and macroscopic flow velocities in the α -, β - and χ - directions, respectively. R is the gas constant and T is the temperature of the mean flow. b is a constant related to γ, namely $b = 2/(\gamma-1)$. e_p is the potential energy of particles given by

$$e_p = \left[1 - \frac{D}{2}(\gamma-1)\right]e . \tag{7.5}$$

Eqs. (7.4a)-(7.4c) are used to recover macroscopic properties like the density, the momentum and the energy, while Eqs. (7.4d)-(7.4g) recover convective and diffusive fluxes in the momentum and energy equations.

Although the C-GKFS is simpler than the original M-GKFS, it is still a complex model, and its explicit formulations for viscous flows [Yang *et al.*, 2014b] are more complicated than the Navier-Stokes solvers. The essence of further simplification is to replace the analytical integral involved in C-GKFS. It is indicated in Eqs. (7.3a) and (7.3b) that only the particle velocity $\boldsymbol{\xi}$ is related to the angle θ. When the particle velocity $\boldsymbol{\xi}$ and the distribution function g_C are substituted into Eqs. (7.4a)-(7.4g), it brings the integrals of sine and cosine functions of θ. In fact, the substitution would use the following integrals,

$$\int_0^{2\pi} \sin^{i-1}(\theta)\cos^{j-1}(\theta)d\theta = A_{ij}, \quad 1 \le i+j \le 5, \tag{7.6}$$

where A_{ij} is the result of the integral equation (7.6), and both i and j are the positive integers. For $1 \le i+j \le 5$, we have

$$\mathbf{A} = \begin{bmatrix} 2\pi & 0 & \pi & 0 \\ 0 & 0 & 0 & \\ \pi & 0 & & \\ 0 & & & \end{bmatrix}. \tag{7.7}$$

To facilitate constructing the discrete velocity models, Eq. (7.6) is rewritten in the form of integral quadrature, i.e.,

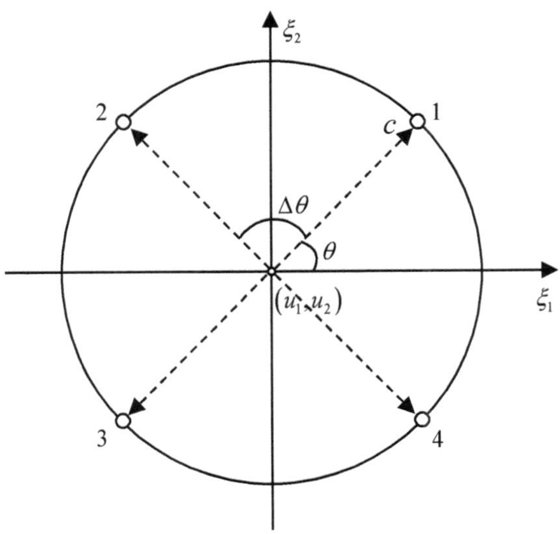

Figure 7.1. Schematic diagram of D2Q4 discrete velocity model.

$$\sum_{m=1}^{M} \sin^{i-1}(\theta_m)\cos^{j-1}(\theta_m)\Delta\theta = A_{ij}, \quad 1 \le i+j \le 5, \tag{7.8}$$

where $\Delta\theta$ is the angle between two adjacent discrete points shown in Figure 7.1, i.e., $\Delta\theta = 2\pi/M$. M is the number of discrete points on the circle. θ_m is the angle of the mth discrete point, which can be computed by

$$\theta_m = \left(m - \frac{1}{2}\right)\Delta\theta, \quad m = 1,2,\ldots,M. \tag{7.9}$$

This prompts us that, the discrete velocity models can be constructed by discretizing the angle θ into some discrete values θ_m as long as the relationship (7.8) can be exactly satisfied. Thus, Eq. (7.8) forms the framework of developing the discrete velocity models for simulation of 2D viscous flows.

Once the angle θ is discretized by Eq. (7.9), the corresponding discrete velocity model can be expressed as

$$g_m = \frac{\rho}{2\pi}, \quad m = 1,2,\ldots,M, \tag{7.10a}$$

$$\begin{aligned} \boldsymbol{\xi}_m &= \left(\xi_{m,1}, \xi_{m,2}\right) \\ &= \left(u_1 + c\cos(\theta_m), u_2 + c\sin(\theta_m)\right), \quad m = 1,2,\ldots,M. \end{aligned} \tag{7.10b}$$

It should be pointed out that Eqs. (7.10a) and (7.10b) are the general forms of the discrete velocity model, which may not satisfy Eq. (7.8). So, the remaining issue is to determine the value of M to fulfill the relationship (7.8). Setting $M = 4K$, where K is a positive integer, we can derive the following expressions:

$$\sum_{m=1}^{M} \sin^{0}(\theta_m)\cos^{0}(\theta_m)\Delta\theta = 2\pi, \tag{7.11a}$$

$$\sum_{m=1}^{M} \sin^{1}(\theta_m)\cos^{0}(\theta_m)\Delta\theta = 0, \tag{7.11b}$$

$$\sum_{m=1}^{M} \sin^0(\theta_m)\cos^1(\theta_m)\Delta\theta = 0, \tag{7.11c}$$

$$\sum_{m=1}^{M} \sin^2(\theta_m)\cos^0(\theta_m)\Delta\theta = \pi, \tag{7.11d}$$

$$\sum_{m=1}^{M} \sin^1(\theta_m)\cos^1(\theta_m)\Delta\theta = 0, \tag{7.11e}$$

$$\sum_{m=1}^{M} \sin^0(\theta_m)\cos^2(\theta_m)\Delta\theta = \pi, \tag{7.11f}$$

$$\sum_{m=1}^{M} \sin^3(\theta_m)\cos^0(\theta_m)\Delta\theta = 0, \tag{7.11g}$$

$$\sum_{m=1}^{M} \sin^2(\theta_m)\cos^1(\theta_m)\Delta\theta = 0, \tag{7.11h}$$

$$\sum_{m=1}^{M} \sin^1(\theta_m)\cos^2(\theta_m)\Delta\theta = 0, \tag{7.11i}$$

$$\sum_{m=1}^{M} \sin^0(\theta_m)\cos^3(\theta_m)\Delta\theta = 0. \tag{7.11j}$$

Eqs. (7.11a)-(7.11j) show that, with $M = 4K$, the discrete velocity model can exactly satisfy the 7 moment relationships in Eq. (7.8), which ensures correctly recovering the Navier-Stokes equations.

By choosing different values of K, different discrete velocity models can be established. For illustration, we can take $K = 1$ to develop a four-points discrete velocity model. The schematic diagram of this model is shown in Figure 7.1. The equilibrium distribution function and the discrete velocities of this model are given by

$$g_m = \frac{\rho}{2\pi}, \quad m = 1,2,3,4, \tag{7.12a}$$

$$\boldsymbol{\xi}_m = \begin{cases} \left(u_1 + \sqrt{2}/2c, u_2 + \sqrt{2}/2c\right), & m = 1, \\ \left(u_1 - \sqrt{2}/2c, u_2 + \sqrt{2}/2c\right), & m = 2, \\ \left(u_1 - \sqrt{2}/2c, u_2 - \sqrt{2}/2c\right), & m = 3, \\ \left(u_1 + \sqrt{2}/2c, u_2 - \sqrt{2}/2c\right), & m = 4. \end{cases} \tag{7.12b}$$

For simplicity, we name the two-dimensional four-points discrete velocity model as the D2Q4 model, where D denotes the dimension and Q represents the number of discrete velocities. Similarly, we can set $K = 2$ and $K = 3$ to derive the D2Q8 model and the D2Q12 model, respectively. In this chapter, we mainly focus on the D2Q4 model to investigate its performance in simulating viscous incompressible and compressible flows. At the same time, the D2Q8 and D2Q12 models are also presented to evaluate the influence of the number of discrete velocities on the computational accuracy and efficiency.

7.1.2 *Three-dimensional discrete velocity model*

For the 3D case, the Maxwellian distribution function can be simplified to the sphere function, which is expressed as

$$g_S = \begin{cases} \dfrac{\rho}{4\pi} & \text{if } (\xi_1 - u_1)^2 + (\xi_2 - u_2)^2 + (\xi_3 - u_3)^2 = c^2, \\ 0 & \text{otherwise.} \end{cases} \tag{7.13}$$

where $\boldsymbol{\xi} = (\xi_1, \xi_2, \xi_3)$ and $\mathbf{u} = (u_1, u_2, u_3)$ are the particle velocity vector and the macroscopic flow velocity vector in 3D conditions. The definition of c is the same as in Eq. (7.2), while the dimension should be set as $D = 3$ for the sphere function. From Eq. (7.13), the velocity components of the particle on the spherical surface can be written as follows in the Cartesian coordinate system:

$$\xi_1 = u_1 + c\sin(\varphi)\cos(\theta), \tag{7.14a}$$

$$\xi_2 = u_2 + c\sin(\varphi)\sin(\theta), \tag{7.14b}$$

$$\xi_3 = u_3 + c\cos(\varphi). \tag{7.14c}$$

The definitions of angles θ and φ can be referred to Figure 6.2. The moment relationships of the sphere function can be obtained by substituting Eqs. (7.13) and (7.14a)-(7.14c) into the integrals of moments for the Maxwellian distribution function. The details can be found in **Section 6.1.2**. With the notations of $c_1 = c\sin(\varphi)\cos(\theta)$, $c_2 = c\sin(\varphi)\sin(\theta)$, $c_3 = c\cos(\varphi)$ and $\xi_\alpha = u_\alpha + c_\alpha$, the following 7 moment relationships of the sphere function can be obtained:

$$\int_0^{2\pi}\int_0^{\pi} g_S \sin(\varphi)d\varphi d\theta = \rho, \tag{7.15a}$$

$$\int_0^{2\pi}\int_0^{\pi} g_S \xi_\alpha \sin(\varphi)d\varphi d\theta = \rho u_\alpha, \tag{7.15b}$$

$$\int_0^{2\pi}\int_0^{\pi} g_S (\xi_\alpha \xi_\alpha + 2e_p)\sin(\varphi)d\varphi d\theta = \rho(u_\alpha u_\alpha + bRT), \tag{7.15c}$$

$$\int_0^{2\pi}\int_0^{\pi} g_S \xi_\alpha \xi_\beta \sin(\varphi)d\varphi d\theta = \rho u_\alpha u_\beta + p\delta_{\alpha\beta}, \tag{7.15d}$$

$$\begin{aligned}&\int_0^{2\pi}\int_0^{\pi} g_S (\xi_\alpha \xi_\alpha + 2e_p)\xi_\beta \sin(\varphi)d\varphi d\theta \\ &= \rho\left[u_\alpha u_\alpha + (b+2)RT\right]u_\beta,\end{aligned} \tag{7.15e}$$

$$\begin{aligned}&\int_0^{2\pi}\int_0^{\pi} g_S \xi_\alpha \xi_\beta \xi_\chi \sin(\varphi)d\varphi d\theta \\ &= p(u_\alpha \delta_{\beta\chi} + u_\beta \delta_{\chi\alpha} + u_\chi \delta_{\alpha\beta}) + \rho u_\alpha u_\beta u_\chi,\end{aligned} \tag{7.15f}$$

$$\begin{aligned}&\int_0^{2\pi}\int_0^{\pi} g_S (\xi_\chi \xi_\chi + 2e_p)\xi_\alpha \xi_\beta \sin(\varphi)d\varphi d\theta \\ &= \rho\left\{bR^2T^2\delta_{\alpha\beta} + \left[(b+4)u_\alpha u_\beta + u_\chi u_\chi \delta_{\alpha\beta}\right]RT + u_\chi u_\chi u_\alpha u_\beta\right\}.\end{aligned} \tag{7.15g}$$

Note that the definition of the potential energy of particles e_p is the same as that in Eq. (7.5) with the dimension of $D = 3$. Like the circular function, Eqs. (7.15a)-(7.15c) are used to recover the fluid density, the momentum and the energy. Eqs. (7.15d) and (7.15e) are applied to retrieve the convective fluxes of momentum and energy equations. Eqs. (7.15f) and (7.15g) are utilized to recover the diffusive fluxes of momentum and energy equations.

As discussed above, it can be found that only the particle velocity ξ shown in Eqs. (7.14a)-(7.14c) is related to the angles θ and φ. And the distribution function g_S in Eq. (7.13) is irrelevant to these angles. Therefore, the relationships in Eqs. (7.15a)-(7.15g) only involve the integrals of sine and cosine functions, which can be generally written as

$$\int_0^{2\pi}\int_0^{\pi}\left[\sin(\varphi)\cos(\theta)\right]^{i-1}\left[\sin(\varphi)\sin(\theta)\right]^{j-1}\left[\cos(\varphi)\right]^{k-1}\sin(\varphi)d\varphi d\theta = A_{ijk}, \quad 1\le i+j+k\le 6, \tag{7.16}$$

where A_{ijk} is the result of the integral equation (7.16), and i, j and k are the positive integers. For $1\le i+j+k\le 6$, we have

$$A_{1,1,1}=4\pi,\ A_{1,3,1}=A_{3,1,1}=A_{1,1,3}=\frac{4}{3}\pi,\ \text{and } A_{ijk}=0 \text{ for other } i,\ j,\ k. \tag{7.17}$$

In order to construct the discrete velocity model from the sphere function, Eq. (7.16) should be rewritten into the quadrature form of

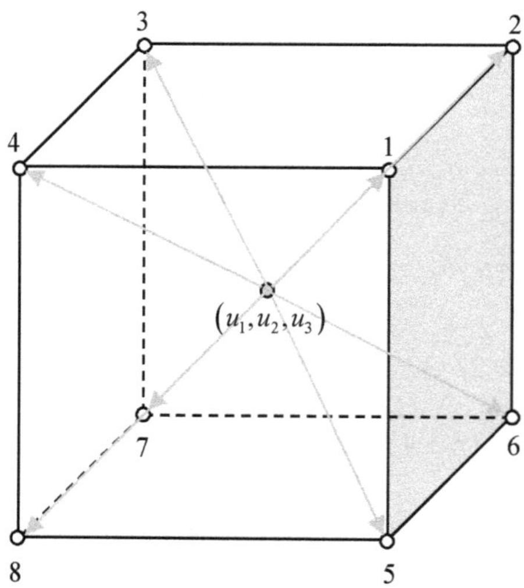

Figure 7.2. Schematic diagram of D3Q8 discrete velocity model.

$$\sum_{m=1}^{M}\sum_{n=1}^{N}\left[\sin(\varphi_n)\cos(\theta_m)\right]^{i-1}\left[\sin(\varphi_n)\sin(\theta_m)\right]^{j-1}\left[\cos(\varphi_n)\right]^{k-1}\Delta S$$
$$= A_{ijk}, \quad 1 \le i+j+k \le 6, \tag{7.18}$$

where ΔS is the area of the surface element on the spherical surface. M and N are respectively the numbers of discrete points of angle θ and angle φ. θ_m and φ_n are the angles of the discrete point. Eq. (7.18) forms the fundamental requirement for development of the 3D discrete velocity model from the sphere function.

As an example, let us consider the eight-points discrete approximation of the spherical surface. In this case, the discrete angles and the area of the surface element are given by

$$\theta_m = \left\{\frac{\pi}{4}, \frac{3\pi}{4}, \frac{5\pi}{4}, \frac{7\pi}{4}\right\}, \tag{7.19a}$$

$$\varphi_n = \left\{\arccos\left(\sqrt{3}/3\right), \arccos\left(-\sqrt{3}/3\right)\right\}, \tag{7.19b}$$

$$\Delta S = \frac{4\pi}{8}. \tag{7.19c}$$

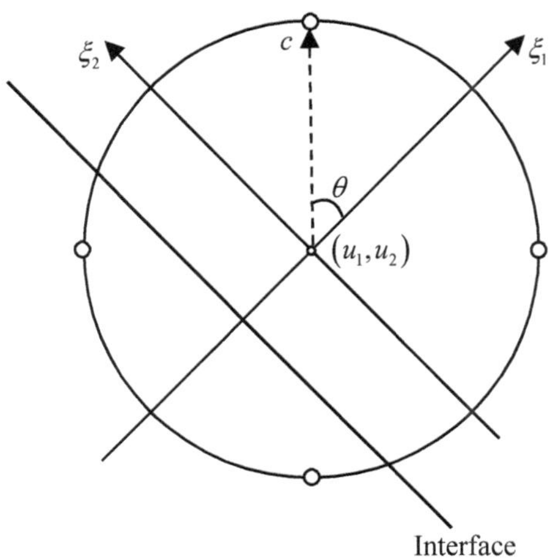

Figure 7.3. Schematic diagram of C-DGKFS with the D2Q4 model.

By substituting Eqs. (7.19a)-(7.19c) into the left-hand side of Eq. (7.18), it can be found that the discrete velocity model exactly satisfies the relationship (7.17). The schematic diagram of this model is shown in Figure 7.2. In particular, the equilibrium distribution function and discrete velocities of this model can be written as

$$g_q = \frac{\rho}{4\pi}, \quad q = 1, 2, \ldots, 8, \tag{7.20a}$$

$$\xi_q = \begin{cases} \left(u_1 + \sqrt{3}/3c, u_2 + \sqrt{3}/3c, u_3 + \sqrt{3}/3c\right), & q = 1, \\ \left(u_1 - \sqrt{3}/3c, u_2 + \sqrt{3}/3c, u_3 + \sqrt{3}/3c\right), & q = 2, \\ \left(u_1 - \sqrt{3}/3c, u_2 - \sqrt{3}/3c, u_3 + \sqrt{3}/3c\right), & q = 3, \\ \left(u_1 + \sqrt{3}/3c, u_2 - \sqrt{3}/3c, u_3 + \sqrt{3}/3c\right), & q = 4, \\ \left(u_1 + \sqrt{3}/3c, u_2 + \sqrt{3}/3c, u_3 - \sqrt{3}/3c\right), & q = 5, \\ \left(u_1 - \sqrt{3}/3c, u_2 + \sqrt{3}/3c, u_3 - \sqrt{3}/3c\right), & q = 6, \\ \left(u_1 - \sqrt{3}/3c, u_2 - \sqrt{3}/3c, u_3 - \sqrt{3}/3c\right), & q = 7, \\ \left(u_1 + \sqrt{3}/3c, u_2 - \sqrt{3}/3c, u_3 - \sqrt{3}/3c\right), & q = 8. \end{cases} \tag{7.20b}$$

where the subscript "q" denotes the qth discrete point, and $q = 4(n-1) + m$. For simplicity, we name the eight-points discrete velocity model as the D3Q8 model in this chapter.

7.2 Circular Function-Based DGKFS

In **Section 7.1.1**, several discrete velocity models are derived from the circular function; and the moment relationships remain unchanged. This property enables direct construction of the circular function-based discrete gas kinetic flux solver (C-DGKFS) from the discrete velocity models, which will be presented in this section. Similar to other flux solvers, the C-DGKFS is established on the base of the finite volume method (FVM). It aims to reconstruct the numerical fluxes on the cell interface.

For simplicity, the local coordinate system defined in Figure 7.3 is utilized, where x_1-axis points to the normal direction and the x_2-axis is aligned with the tangential direction. By employing discrete quadrature points, the connections between the discrete distribution functions f_m and the conservative variables and flux vector in 2D conditions can be written as follows:

$$\overline{\mathbf{W}} = \left(\rho, \rho u_1, \rho u_2, \rho E\right)^T = \sum_{m=1}^{M} \overline{\boldsymbol{\varphi}}_m f_m \Delta\theta, \tag{7.21}$$

$$\overline{\mathbf{F}} = \left(F_1, F_2, F_3, F_4\right)^T = \sum_{m=1}^{M} \xi_{m,1} \overline{\boldsymbol{\varphi}}_m f_m \Delta\theta, \tag{7.22}$$

where $\overline{\boldsymbol{\varphi}}_m$ is the moment vector for the discrete velocity models expressed in the local coordinate system, which is given by

$$\overline{\boldsymbol{\varphi}}_m = \left(1, \xi_{m,1}, \xi_{m,2}, \frac{1}{2}\left|\boldsymbol{\xi}_m\right|^2 + e_{p,m}\right)^T. \tag{7.23}$$

By applying the coordinate transformation from the local coordinate system to the global Cartesian coordinate system, the following relationships are obtained:

$$\mathbf{W} = \begin{bmatrix} \rho \\ \rho\left(u_1 n_x - u_2 n_y\right) \\ \rho\left(u_2 n_x + u_1 n_y\right) \\ \rho E \end{bmatrix}, \quad \mathbf{F}_n = \begin{bmatrix} F_1 \\ F_2 n_x - F_3 n_y \\ F_3 n_x + F_2 n_y \\ F_4 \end{bmatrix}, \tag{7.24}$$

where $\mathbf{n} = \left(n_x, n_y\right)$ denotes the unit normal vector of the cell interface. $E = e + \left|\mathbf{u}\right|^2/2$ is the total energy. Detailed expressions of $\mathbf{W}$ and $\mathbf{F}_n$ can be found in Eq. (6.116) and Eq. (6.120), respectively. Note that Eq. (7.21) is only used for calculation of conservative variables on the cell interface in C-DGKFS, the conservative variables at cell centers are updated by marching the Navier-Stokes equations in time.

With the FVM, the discrete form of the 2D Navier-Stokes equations reads

$$\frac{\partial \mathbf{W}_i}{\partial t} = -\frac{1}{\Omega_i}\sum_{j=1}^{N_f} \mathbf{F}_{nj} S_j , \tag{7.25}$$

where i is the index of the control volume, Ω_i and N_f represent the volume and the number of surfaces of the control volume i, respectively. S_j denotes the area of the jth interface of the control volume. To solve Eq. (7.25), the main task is to calculate the flux vector $\mathbf{F}_n$ on the cell interface. Furthermore, it can be seen from Eq. (7.24) that the calculation of $\mathbf{F}_n$ in the global Cartesian coordinate system is equivalent to evaluating $\overline{\mathbf{F}}$ in the local coordinate system by Eq. (7.22) and the key is to obtain the discrete distribution functions f_m and the moments $\overline{\boldsymbol{\varphi}}_m$.

Suppose that the cell interface is located at $\mathbf{x} = \mathbf{0}$. According to the Chapman-Enskog expansion analysis shown in **Section 6.1.3**, the distribution function on the cell interface consists of the equilibrium part and the non-equilibrium part, which can be written as

$$\begin{aligned} f_m(\mathbf{0},t) &= f_m^{eq}(\mathbf{0},t) + f_m^{neq}(\mathbf{0},t) \\ &= g_m(\mathbf{0},t) + \tau_0\left[g_m(-\boldsymbol{\xi}_m\delta_t, t-\delta_t) - g_m(\mathbf{0},t)\right] \\ &\quad + O\left(\tau^2, \tau\delta_t, |\boldsymbol{\xi}_m|^2 \tau\delta_t\right), \end{aligned} \tag{7.26}$$

where $g_m(\mathbf{0},t) = f_m^{eq}(\mathbf{0},t)$ is the equilibrium distribution function on the cell interface, and $g_m(-\boldsymbol{\xi}_m\delta_t, t-\delta_t)$ is the equilibrium distribution function on discrete points along the circle. τ_0 is the dimensionless collision time, which can be calculated by

$$\tau_0 = \frac{\tau}{\delta_t} = \frac{\mu}{p\delta_t}. \tag{7.27}$$

Here, μ denotes the dynamic viscosity, which is determined from either the Sutherland's law or the Reynolds number. δ_t is the streaming time step. To simplify the notation, the space and time locations of "$(\mathbf{0},t)$" and "$(-\boldsymbol{\xi}_m\delta_t, t-\delta_t)$" are respectively denoted by the superscripts "*face*" and "*cir*" in the following text. Substituting Eq. (7.26) into Eq. (7.22) and neglecting the second-order minor terms, we can get

$$\begin{aligned}\overline{\mathbf{F}} &= \sum_{m=1}^{M} \xi_{m,1}^{face} \overline{\boldsymbol{\varphi}}_m^{face} g_m^{face} \Delta\theta \\ &\quad + \tau_0 \left[\sum_{m=1}^{M} \xi_{m,1}^{cir} \overline{\boldsymbol{\varphi}}_m^{cir} g_m^{cir} \Delta\theta - \sum_{m=1}^{M} \xi_{m,1}^{face} \overline{\boldsymbol{\varphi}}_m^{face} g_m^{face} \Delta\theta \right] \\ &= \mathbf{F}^I + \tau_0 \left[\mathbf{F}^{II} - \mathbf{F}^I \right].\end{aligned} \tag{7.28}$$

This equation indicates that the numerical fluxes on the cell interface consist of two parts. One is the flux vector attributed to the equilibrium distribution functions and the moments on the cell interface, which is denoted as $\mathbf{F}^I$. The other part is the flux vector attributed to the equilibrium distribution functions and the moments on the circle, which is referred as $\mathbf{F}^{II}$.

The schematic diagram of C-DGKFS with the D2Q4 model is shown in Figure 7.3. To calculate the fluxes $\mathbf{F}^I$ and $\mathbf{F}^{II}$, $\overline{\boldsymbol{\varphi}}_m^{face}$, g_m^{face}, $\overline{\boldsymbol{\varphi}}_m^{cir}$ and g_m^{cir} should be computed in advance. In particular, $\overline{\boldsymbol{\varphi}}_m^{cir}$ and g_m^{cir} are associated with the conservative variables on the discrete points along the circle which can be obtained by the linear interpolation as follows:

$$\mathbf{W}_m^{cir} = \begin{cases} \mathbf{W}^L - \nabla\mathbf{W}^L \cdot \boldsymbol{\xi}_m^+ \delta_t, & \text{if } \xi_{m,1}^+ \geq 0, \\ \mathbf{W}^R - \nabla\mathbf{W}^R \cdot \boldsymbol{\xi}_m^+ \delta_t, & \text{if } \xi_{m,1}^+ < 0. \end{cases} \tag{7.29}$$

where $\mathbf{W}^L$, $\mathbf{W}^R$ and $\mathbf{W}_m^{cir}$ are the conservative variables on the left and the right sides of the cell interface and at the discrete points of the circle, respectively. $\nabla\mathbf{W}^L$ and $\nabla\mathbf{W}^R$ are the first-order derivatives of conservative variables at the left and the right cells around the cell interface. $\boldsymbol{\xi}_m^+$ is the predicted particle velocity on the cell interface given by

$$\boldsymbol{\xi}_m^+ = \left(\xi_{m,1}^+, \xi_{m,2}^+\right) = \left(u_1^+ + c^+ \cos(\theta_m), u_2^+ + c^+ \sin(\theta_m)\right), \tag{7.30}$$

where u_1^+, u_2^+ and c^+ are the predicted mean flow velocity components and the effective peculiar velocity on the cell interface, respectively. These quantities can be determined by the Roe averages, as shown in Eqs. (6.58a)-(6.58d).

Once the conservative variables at the discrete points along the circle are obtained, the particle velocities, the particle potential energy and the equilibrium distribution function can be calculated by

$$\xi_{m,1}^{cir} = u_{m,1}^{cir} + c^+ \cos(\theta_m), \tag{7.31}$$

$$\xi_{m,2}^{cir} = u_{m,2}^{cir} + c^+ \sin(\theta_m), \tag{7.32}$$

$$e_{p,m}^{cir} = \left[1 - \frac{D}{2}(\gamma - 1)\right] e_m^{cir}, \tag{7.33}$$

$$g_m^{cir} = \rho_m^{cir} / 2\pi, \tag{7.34}$$

where ρ_m^{cir}, e_m^{cir}, $u_{m,1}^{cir}$ and $u_{m,2}^{cir}$ are respectively the density, the potential energy, the normal velocity and the tangential velocity of the mean flow on the discrete points. According to the compatibility condition, the conservative variables on the cell interface can be computed by

$$\overline{\mathbf{W}}^{face} = \sum_{m=1}^{M} \overline{\boldsymbol{\varphi}}_m^{cir} g_m^{cir} \Delta\theta. \tag{7.35}$$

Substituting Eqs. (7.31)-(7.34) into Eq. (7.35), we have

$$\overline{\mathbf{W}}^{face} = \begin{bmatrix} \sum_{m=1}^{M} g_m^{cir} \Delta\theta \\ \sum_{m=1}^{M} \xi_{m,1}^{cir} g_m^{cir} \Delta\theta \\ \sum_{m=1}^{M} \xi_{m,2}^{cir} g_m^{cir} \Delta\theta \\ \sum_{m=1}^{M} \left(\frac{1}{2} \left| \boldsymbol{\xi}_m^{cir} \right|^2 + e_{p,m}^{cir} \right) g_m^{cir} \Delta\theta \end{bmatrix}. \tag{7.36}$$

Furthermore, $\overline{\boldsymbol{\varphi}}_m^{face}$ and g_m^{face} can be determined in a straightforward way by substituting $\overline{\mathbf{W}}^{face}$ into Eqs. (7.12a) and (7.12b).

Once the equilibrium distribution functions and the moments at the discrete points along the circle and on the cell interface are obtained, the fluxes $\mathbf{F}^I$ and $\mathbf{F}^{II}$ can be evaluated directly from Eq. (7.28). An alternative and simpler way to compute $\mathbf{F}^I$ is to substitute the conservative variables $\overline{\mathbf{W}}^{face}$ directly into the expression of the inviscid flux vector, which gives

$$\mathbf{F}^I = \begin{bmatrix} \rho u_1 \\ \rho u_1 u_1 + p \\ \rho u_1 u_2 \\ (\rho E + p) u_1 \end{bmatrix}^{face} . \tag{7.37}$$

And the flux vector $\mathbf{F}^{II}$ reads

$$\mathbf{F}^{II} = \sum_{m=1}^{M} \xi_{m,1}^{cir} \overline{\boldsymbol{\varphi}}_m^{cir} g_m^{cir} \Delta\theta = \begin{bmatrix} \sum_{m=1}^{M} \xi_{m,1}^{cir} g_m^{cir} \Delta\theta \\ \sum_{m=1}^{M} \xi_{m,1}^{cir} \xi_{m,1}^{cir} g_m^{cir} \Delta\theta \\ \sum_{m=1}^{M} \xi_{m,1}^{cir} \xi_{m,2}^{cir} g_m^{cir} \Delta\theta \\ \sum_{m=1}^{M} \xi_{m,1}^{cir} \left(\frac{1}{2} \left| \boldsymbol{\xi}_m^{cir} \right|^2 + e_{p,m}^{cir} \right) g_m^{cir} \Delta\theta \end{bmatrix} . \tag{7.38}$$

The total numerical fluxes on the cell interface are thus obtained by combining these two parts of fluxes. The calculation procedure of numerical fluxes on the cell interface is summarized as below:

(a) For each cell interface, use Eq. (5.35) to interpolate the macroscopic flow variables on the left and the right sides of the cell interface, $\mathbf{W}^L$ and $\mathbf{W}^R$;

(b) Use Roe averages in Eqs. (6.58a)-(6.58d) to calculate u_1^+, u_2^+ and c^+ on the cell interface;

(c) Calculate the streaming time step δ_t by Eq. (6.137);

(d) Compute the conservative variables on the discrete points along the circle $\mathbf{W}_m^{cir}$ by Eq. (7.29);
(e) Compute the particle velocities $\xi_{m,1}^{cir}$ and $\xi_{m,2}^{cir}$, the particle potential energy $e_{p,m}^{cir}$ and the equilibrium distribution function g_m^{cir} by Eqs. (7.31)-(7.34);
(f) Calculate the conservative variables on the cell interface $\overline{\mathbf{W}}^{face}$ by Eq. (7.36), and further compute the flux vector $\mathbf{F}^I$ by Eq. (7.37);
(g) Compute $\mathbf{F}^{II}$ by Eq. (7.38);
(h) Calculate τ_0 by Eq. (7.27);
(i) Substitute $\mathbf{F}^I$, $\mathbf{F}^{II}$ and τ_0 into Eq. (7.28) to obtain the full expression of numerical fluxes.

Comparing with the C-GKFS introduced in the last chapter, the formulations of the present C-DGKFS is more concise. One should also be reminded that, since the present C-DGKFS maintains the consistency with the moment relationships in C-GKFS, the Prandtl number of the recovered macroscopic governing equations is also fixed at γ. Modifications are needed to model different Prandtl numbers. The details can be found in the work of Yang *et al.* [2014b] or Eqs. (6.139) and (6.140). In addition, the choice of the streaming time step δ_t in C-DGKFS is also based on the principle of avoiding extrapolations, which is shown in Eq. (6.137).

7.3 Sphere Function-Based DGKFS

In this section, the derivation is extended to the 3D scenario. The D3Q8 discrete velocity model forges the basis here to reconstruct numerical fluxes of the 3D Navier-Stokes equations on the cell interface with the local solution to the Boltzmann equation. The corresponding scheme is termed as the sphere function-based discrete gas kinetic flux solver and abbreviated as S-DGKFS.

For the convenience of derivation, the 3D local coordinate system defined on the cell interface is introduced into the S-DGKFS. In the local coordinate system, x_1-axis is taken as the normal direction pointing outwards of the cell interface of the control volume, x_2-axis and x_3-axis are chosen as two tangential directions of the cell interface, which are mutually orthogonal and form a right-hand coordinate system. For simplicity, x_2-axis can be set parallel to one of the edges of the cell interface and x_3-axis is then determined by using the right-hand rule. The unit vector in the directions 1, 2 and 3 of the local coordinate system are respectively denoted as $\mathbf{n}_1 = (n_{1x}, n_{1y}, n_{1z})$, $\mathbf{n}_2 = (n_{2x}, n_{2y}, n_{2z})$ and $\mathbf{n}_3 = (n_{3x}, n_{3y}, n_{3z})$. By employing discrete quadrature points of D3Q8 model, the connections between the discrete distribution functions f_q and the conservative variables and flux vector in 3D conditions can be expressed as follows:

$$\overline{\mathbf{W}} = (\rho, \rho u_1, \rho u_2, \rho u_3, \rho E)^T = \sum_{q=1}^{Q} \overline{\boldsymbol{\varphi}}_q f_q \Delta S, \tag{7.39}$$

$$\overline{\mathbf{F}} = (F_1, F_2, F_3, F_4, F_5)^T = \sum_{q=1}^{Q} \xi_{q,1} \overline{\boldsymbol{\varphi}}_q f_q \Delta S, \tag{7.40}$$

where $\overline{\boldsymbol{\varphi}}_q$ is the moment vector for the discrete velocity model expressed in the local coordinate system, which is given by

$$\overline{\boldsymbol{\varphi}}_q = \left(1, \xi_{q,1}, \xi_{q,2}, \xi_{q,3}, \frac{1}{2}\left|\boldsymbol{\xi}_q\right|^2 + e_{p,q}\right)^T. \tag{7.41}$$

By applying the coordinate transformation from the local coordinate system to the global Cartesian coordinate system, the following relationships are obtained:

$$\mathbf{W} = \begin{bmatrix} \rho \\ \rho(u_1 n_{1x} + u_2 n_{2x} + u_3 n_{3x}) \\ \rho(u_1 n_{1y} + u_2 n_{2y} + u_3 n_{3y}) \\ \rho(u_1 n_{1z} + u_2 n_{2z} + u_3 n_{3z}) \\ \rho E \end{bmatrix}, \quad \mathbf{F}_n = \begin{bmatrix} F_1 \\ F_2 n_{1x} + F_3 n_{2x} + F_4 n_{3x} \\ F_2 n_{1y} + F_3 n_{2y} + F_4 n_{3y} \\ F_2 n_{1z} + F_3 n_{2z} + F_4 n_{3z} \\ F_5 \end{bmatrix}. \tag{7.42}$$

Detailed expressions of $\mathbf{W}$ and $\mathbf{F}_n$ for the 3D Navier-Stokes equations can be found in **Chapter 1**. Like the C-DGKFS, the conservative variables at cell centers are not given from Eq. (7.39) directly. Instead, they are obtained by marching the Navier-Stokes equations in time. In addition, the numerical fluxes on the cell interface are reconstructed by the local solution to the Boltzmann equation.

The discrete form of the 3D Navier-Stokes equations given by FVM can be written as

$$\frac{\partial \mathbf{W}_i}{\partial t} = -\frac{1}{\Omega_i}\sum_{j=1}^{N_f} \mathbf{F}_{nj} S_j . \tag{7.43}$$

As shown in Eq. (7.42), the calculation of $\mathbf{F}_n$ on the cell interface in the global Cartesian coordinate system is equivalent to evaluating $\overline{\mathbf{F}}$ in the local coordinate system by Eq. (7.40) and the key issue is to obtain the discrete distribution functions f_q and the moments $\overline{\boldsymbol{\varphi}}_q$. Suppose that the cell interface is located at $\mathbf{x} = \mathbf{0}$. Like the C-DGKFS, the distribution function on the cell interface for the S-DGKFS can be expressed as

$$\begin{aligned} f_q(\mathbf{0},t) &= g_q(\mathbf{0},t) + \tau_0\left[g_q\left(-\boldsymbol{\xi}_q\delta_t, t-\delta_t\right) - g_q(\mathbf{0},t)\right] \\ &\quad + O\left(\tau^2, \tau\delta_t, \left|\boldsymbol{\xi}_q\right|^2 \tau\delta_t\right), \end{aligned} \tag{7.44}$$

where $g_q(\mathbf{0},t)$ and $g_q\left(-\boldsymbol{\xi}_q\delta_t, t-\delta_t\right)$ are respectively the equilibrium distribution function on the cell interface and at discrete points along the spherical surface. τ_0 is the dimensionless collision time given by Eq. (7.27). For simplicity, we denote "$(\mathbf{0},t)$" by the superscript "$face$" and "$\left(-\boldsymbol{\xi}_q\delta_t, t-\delta_t\right)$" by the superscript "$sph$" in the following text. Substituting Eq. (7.44) into Eq. (7.40) and neglecting the second-order minor terms, we have

$$\begin{aligned} \overline{\mathbf{F}} &= \sum_{q=1}^{Q} \xi_{q,1}^{face}\overline{\boldsymbol{\varphi}}_q^{face} g_q^{face}\Delta S \\ &\quad + \tau_0\left[\sum_{q=1}^{Q} \xi_{q,1}^{sph}\overline{\boldsymbol{\varphi}}_q^{sph} g_q^{sph}\Delta S - \sum_{q=1}^{Q} \xi_{q,1}^{face}\overline{\boldsymbol{\varphi}}_q^{face} g_q^{face}\Delta S\right] \\ &= \mathbf{F}^I + \tau_0\left[\mathbf{F}^{II} - \mathbf{F}^I\right]. \end{aligned} \tag{7.45}$$

In the above Eq. (7.45), the flux vector attributed to the equilibrium distribution functions and the moments on the cell interface is denoted as $\mathbf{F}^{I}$, and the flux vector attributed to the equilibrium distribution functions and the moments on the spherical surface is referred as $\mathbf{F}^{II}$.

As shown in Eq. (7.45), in order to calculate the flux vectors $\mathbf{F}^{I}$ and $\mathbf{F}^{II}$, $\bar{\boldsymbol{\varphi}}_q^{face}$, g_q^{face}, $\bar{\boldsymbol{\varphi}}_q^{sph}$ and g_q^{sph} should be computed in advance. Furthermore, since the equilibrium distribution functions and the moments are functions of conservative variables, the conservative variables at corresponding physical locations must be determined first. The conservative variables at the discrete points on the spherical surface can be computed by

$$\mathbf{W}_q^{sph} = \begin{cases} \mathbf{W}^L - \nabla\mathbf{W}^L \cdot \boldsymbol{\xi}_q^+ \delta_t, & \text{if } \xi_{q,1}^+ \geq 0, \\ \mathbf{W}^R - \nabla\mathbf{W}^R \cdot \boldsymbol{\xi}_q^+ \delta_t, & \text{if } \xi_{q,1}^+ < 0. \end{cases} \tag{7.46}$$

where $\boldsymbol{\xi}_q^+$ is the predicted particle velocity on the cell interface expressed in the local coordinate system, which is given by

$$\begin{aligned} \boldsymbol{\xi}_q^+ = \big(& u_1^+ + c^+ \sin(\varphi_n)\cos(\theta_m), \\ & u_2^+ + c^+ \sin(\varphi_n)\sin(\theta_m), u_3^+ + c^+ \cos(\varphi_n)\big). \end{aligned} \tag{7.47}$$

Furthermore, the predicted mean flow velocity components u_1^+, u_2^+ and u_3^+ and the effective peculiar velocity c^+ can be calculated by the Roe averages as follows [Roe, 1981]:

$$u_1^+ = \frac{u_1^L \sqrt{\rho^L} + u_1^R \sqrt{\rho^R}}{\sqrt{\rho^L} + \sqrt{\rho^R}}, \tag{7.48a}$$

$$u_2^+ = \frac{u_2^L \sqrt{\rho^L} + u_2^R \sqrt{\rho^R}}{\sqrt{\rho^L} + \sqrt{\rho^R}}, \tag{7.48b}$$

$$u_3^+ = \frac{u_3^L \sqrt{\rho^L} + u_3^R \sqrt{\rho^R}}{\sqrt{\rho^L} + \sqrt{\rho^R}}, \tag{7.48c}$$

$$H^+ = \frac{H^L\sqrt{\rho^L} + H^R\sqrt{\rho^R}}{\sqrt{\rho^L} + \sqrt{\rho^R}}, \tag{7.48d}$$

$$c^+ = \sqrt{D\left(H^+ - \frac{1}{2}\left(\left(u_1^+\right)^2 + \left(u_2^+\right)^2 + \left(u_3^+\right)^2\right)\right)\frac{(\gamma-1)}{\gamma}}, \tag{7.48e}$$

where H is the total enthalpy defined by $H = \left(u_1^2 + u_2^2 + u_3^2\right)/2 + \gamma e$.

Once the conservative variables on the discrete points along the spherical surface are obtained, the corresponding particle velocities, particle potential energy and equilibrium distribution function can be calculated by

$$\xi_{q,1}^{sph} = u_{q,1}^{sph} + c^+ \sin(\varphi_n)\cos(\theta_m), \tag{7.49}$$

$$\xi_{q,2}^{sph} = u_{q,2}^{sph} + c^+ \sin(\varphi_n)\sin(\theta_m), \tag{7.50}$$

$$\xi_{q,3}^{sph} = u_{q,3}^{sph} + c^+ \cos(\varphi_n), \tag{7.51}$$

$$e_{p,q}^{sph} = \left[1 - \frac{D}{2}(\gamma-1)\right]e_q^{sph}, \tag{7.52}$$

$$g_q^{sph} = \frac{\rho_q^{sph}}{4\pi}, \tag{7.53}$$

where, ρ_q^{sph}, e_q^{sph}, $u_{q,1}^{sph}$, $u_{q,2}^{sph}$ and $u_{q,3}^{sph}$ are the density, the potential energy, the normal velocity component and 2 tangential velocity components of the mean flow at the discrete points. According to the compatibility condition, the conservative variables on the cell interface can be computed by

$$\overline{\mathbf{W}}^{face} = \sum_{q=1}^{Q} \overline{\boldsymbol{\varphi}}_q^{sph} g_q^{sph} \Delta S. \tag{7.54}$$

Substituting Eqs. (7.49)-(7.53) into Eq. (7.54) yields

$$\overline{\mathbf{W}}^{face} = \begin{bmatrix} \sum_{q=1}^{Q} g_q^{sph} \Delta S \\ \sum_{q=1}^{Q} \xi_{q,1}^{sph} g_q^{sph} \Delta S \\ \sum_{q=1}^{Q} \xi_{q,2}^{sph} g_q^{sph} \Delta S \\ \sum_{q=1}^{Q} \xi_{q,3}^{sph} g_q^{sph} \Delta S \\ \sum_{q=1}^{Q} \left(\frac{1}{2} \left| \boldsymbol{\xi}_q^{sph} \right|^2 + e_{p,q}^{sph} \right) g_q^{sph} \Delta S \end{bmatrix}. \tag{7.55}$$

Furthermore, the moments $\overline{\boldsymbol{\varphi}}_q^{face}$ and the equilibrium distribution function g_q^{face} on the cell interface can be calculated in a straightforward way via Eqs. (7.20a) and (7.20b).

Like the C-DGKFS, $\mathbf{F}^I$ can be calculated directly by substituting the conservative variables $\overline{\mathbf{W}}^{face}$ into the expression of inviscid fluxes for the 3D case, i.e.,

$$\mathbf{F}^I = \begin{bmatrix} \rho u_1 \\ \rho u_1 u_1 + p \\ \rho u_1 u_2 \\ \rho u_1 u_3 \\ (\rho E + p) u_1 \end{bmatrix}^{face}. \tag{7.56}$$

For the flux vector $\mathbf{F}^{II}$, it can be computed by substituting $\overline{\boldsymbol{\varphi}}_q^{sph}$ and g_q^{sph} into the following equation:

$$\mathbf{F}^{II} = \sum_{q=1}^{Q} \xi_{q,1}^{sph} \overline{\boldsymbol{\varphi}}_q^{sph} g_q^{sph} \Delta S . \tag{7.57}$$

Detailed expression of Eq. (7.57) is as follow:

$$\mathbf{F}^{II} = \begin{bmatrix} \sum_{q=1}^{Q} \xi_{q,1}^{sph} g_q^{sph} \Delta S \\ \sum_{q=1}^{Q} \xi_{q,1}^{sph} \xi_{q,1}^{sph} g_q^{sph} \Delta S \\ \sum_{q=1}^{Q} \xi_{q,1}^{sph} \xi_{q,2}^{sph} g_q^{sph} \Delta S \\ \sum_{q=1}^{Q} \xi_{q,1}^{sph} \xi_{q,3}^{sph} g_q^{sph} \Delta S \\ \sum_{q=1}^{Q} \xi_{q,1}^{sph} \left(\frac{1}{2} \left| \xi_q^{sph} \right|^2 + e_{p,q}^{sph} \right) g_q^{sph} \Delta S \end{bmatrix}. \tag{7.58}$$

Substituting Eqs. (7.56) and (7.58) into Eq. (7.45), we can obtain the full expression of numerical fluxes for the 3D Navier-Stokes equations.

The last undetermined variable in Eq. (7.45) is the streaming time step δ_t, which is used to reconstruct the local solution to the Boltzmann equation on the cell interface and is related to the dimensionless collision time τ_0. For the 3D case, δ_t can be determined by

$$\delta_t = \frac{0.2 \times \min\{\Delta l, \Delta r\}}{\left(\max\{u_1^+, u_2^+, u_3^+\} + c^+ \right)}, \tag{7.59}$$

where Δl and Δr are the lengths of the shortest edges of the left and the right cells around the cell interface, respectively. In addition, like the C-DGKFS, the Prandtl number of the macroscopic governing equations recovered by the sphere function is also fixed at γ. To consider different Prandtl numbers, modifications of the energy flux, which has been shown in Eqs. (6.139) and (6.140), should be applied to the S-DGKFS.

7.4 Sample Applications

In this section, three sample applications of C-DGKFS and S-DGKFS will be presented to shed light on their performance in simulating incompressible and compressible viscous flows. In the simulations, the conservative variables on two sides of the cell interface are interpolated from known values defined at cell centers, and the Venkatakrishnan's limiter [Venkatakrishnan, 1995] is used. For temporal discretization of Eqs. (7.25) and (7.43), the three-stages Runge-Kutta scheme is applied. Unless otherwise stated, in all numerical examples reported here, the CFL number is set as 1, the specific heat ratio is chosen as 1.4 and the Prandtl number is taken as 0.72.

7.4.1 *Two-dimensional viscous incompressible flow*

The lid-driven cavity flow is firstly presented to validate the performance of C-DGKFS in simulating viscous incompressible flows. In the simulation, the Mach number is set as 0.15 to fulfill the incompressible

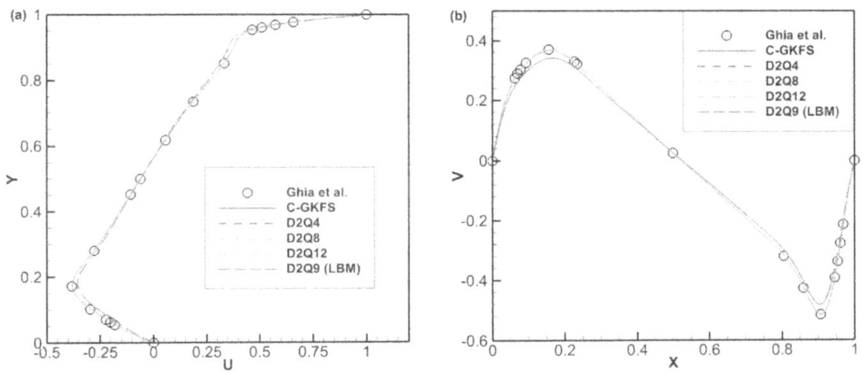

Figure 7.4. Comparison of velocity profiles for lid-driven cavity flow at *Re*=1000. (a) *u*-velocity profiles along the vertical central line. (b) *v*-velocity profiles along the horizontal central line.

Table 7.1. Comparison of computational effort for different discrete velocity models.

Discrete velocity models	Computational effort
D2Q4	1
D2Q8	1.539
D2Q12	1.820

limit. The Reynolds number is defined as $Re = \rho_0 u_0 L/\mu = 1000$, where L=1 is the length of the cavity, ρ_0 and u_0 are respectively the density at the reference state and the velocity of the top lid. The computational domain is discretized uniformly by 125×125 cells.

Numerical tests are carried out by the C-DGKFS with D2Q4, D2Q8 and D2Q12 models, and the computed velocity distributions along the centerlines are presented in Figure 7.4 and compared with the benchmark solutions given by Ghia *et al.* [1982], the simulation results of C-GKFS [Yang *et al.*, 2014b] and the results obtained by the standard LBM with the D2Q9 model [Qian *et al.*, 1992]. It is noted that the standard LBM with the D2Q9 model presents poorer accuracy, while the results of C-DGKFS with D2Q4, D2Q8 and D2Q12 models and C-GKFS are very close and all agree well with the benchmark data. It is thus suggested that 4 discrete velocity points can fulfill the accuracy requirement in 2D simulations. In terms of efficiency, it is expected that models with more discrete velocities should consume more computational efforts. Such expectation is validated in the quantitative comparisons shown in Table 7.1. In this table, the computational efforts of D2Q8 and D2Q12 models have been normalized by that of D2Q4 model. Given the fact that the three discrete velocity models have comparable accuracy, the D2Q4 model seems to be the optimal 2D model.

7.4.2 *Two-dimensional viscous compressible flow*

To investigate the performance of C-DGKFS in simulating 2D viscous compressible flows with complex geometries, the test cases of transonic flow around a NACA0012 airfoil and around a staggered-biplane configuration are carried out. In this simulation, the free-stream Mach

Table 7.2. Comparison of lift and drag coefficients for NACA0012 airfoil.

References	$C_{d,p}$	$C_{d,f}$	$C_{d,total}$	$C_{l,total}$
GAMM [Bristeau, 2013]	--	--	0.243-0.2868	0.4145-0.517
Jawahar & Kamath [2000]	0.15287	0.12439	0.27726	0.50231
Roe	0.14905	0.12668	0.27573	0.44468
C-DGKFS (D2Q4)	0.15477	0.13110	0.28587	0.47449

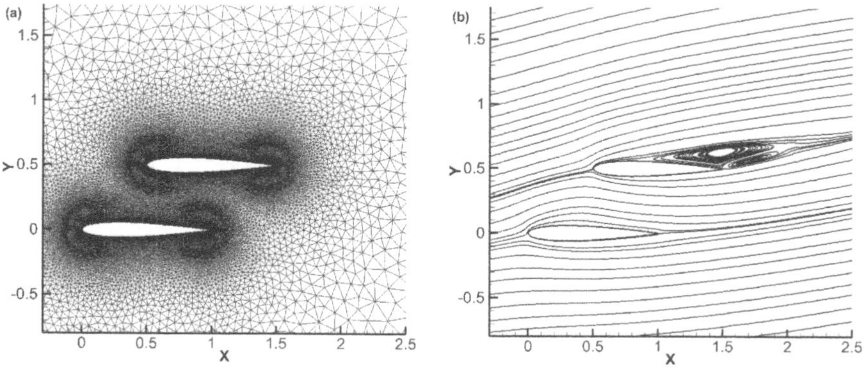

Figure 7.5. (a) Partial view of computational mesh for biplane configuration. (b) Streamline pattern for transonic flow around a biplane configuration.

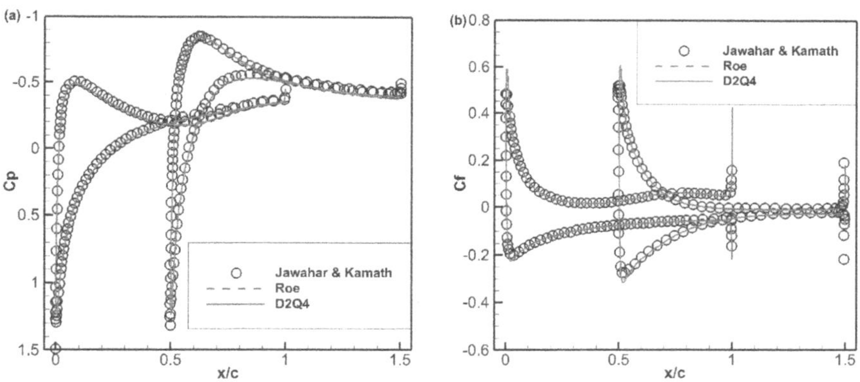

Figure 7.6. (a) Comparison of pressure coefficient distribution for biplane configuration. (b) Comparison of skin friction coefficient distribution for biplane configuration.

Table 7.3. Comparison of computational effort for different numerical schemes.

Schemes	One NACA0012 airfoil	Biplane configuration
Roe	1	1
C-GKFS	1.071	1.013
C-DGKFS (D2Q4)	1.105	1.029

number is taken as 0.8, the chord length-based Reynolds number is chosen as 500 and the angle of attack is set as 10 degrees. For the test case of one NACA0012 airfoil, the O-type grid with 160 points on the airfoil surface and 70 points in the radial direction is utilized, and the computational domain is about 10 times of the chord length. The lift and drag coefficients obtained by the C-DGKFS with the D2Q4 model and the improved Roe scheme [Kim *et al.*, 2002] are compared in Table 7.2 with the reference data of GAMM [Bristeau, 2013] and Jawahar and Kamath [2000]. In this table, $C_{d,p}$, $C_{d,f}$, $C_{d,total}$ and $C_{l,total}$ are the pressure drag coefficient, the friction drag coefficient, the total drag coefficient and the total lift coefficient, respectively. The results of C-DGKFS with the D2Q4 model are within the range reported in the GAMM workshop and agree well with other reference data.

For the test case of biplane configuration, the unstructured grid with 320 points on each airfoil surface and 54,387 triangular cells in the computational domain is utilized, and its partial view is shown in Figure 7.5a. This test case comprises two NACA0012 airfoils, which are staggered by half a chord length in the pitchwise as well as chordwise directions. The streamline pattern obtained by the C-DGKFS with the D2Q4 model is shown in Figure 7.5b. It can be found that the separation region on the upper surface of the top airfoil reveals two vortices, which is in line with the results reported in the work of Jawahar and Kamath [2000]. Figures 7.6a and 7.6b respectively show the comparison of pressure coefficient and skin friction coefficient distributions on the airfoil surface. The results obtained by the C-DGKFS with D2Q4 model and the improved Roe scheme are close to each other, and they agree well with the results of Jawahar and Kamath [2000]. In addition, as reported in Table 7.3, the computational time consumed by the C-DGKFS with the D2Q4 model is almost the same as that of the improved Roe scheme for this test case. Note that, in this table, the computational efforts of C-GKFS and C-DGKFS with the D2Q4 model have been normalized by that of the improved Roe scheme. To facilitate the readers' implementation, a Fortran code for this test example is attached in **Code D**. The computational mesh can be generated by the commercial software POINTWISE with COBALT format and then transformed into a mesh input file by **Code B**.

7.4.3 *Three-dimensional viscous compressible flow*

In this case, the transonic flow around a DPW-W1 wing [Vassberg *et al.*, 2007] is simulated to validate the performance of S-DGKFS with the D3Q8 model for simulation of 3D viscous compressible flows with complex geometries. The geometric configuration and the computational mesh of this test case are extracted from the summary report of the third AIAA CFD drag prediction workshop (DPW-III) and NASA website*.

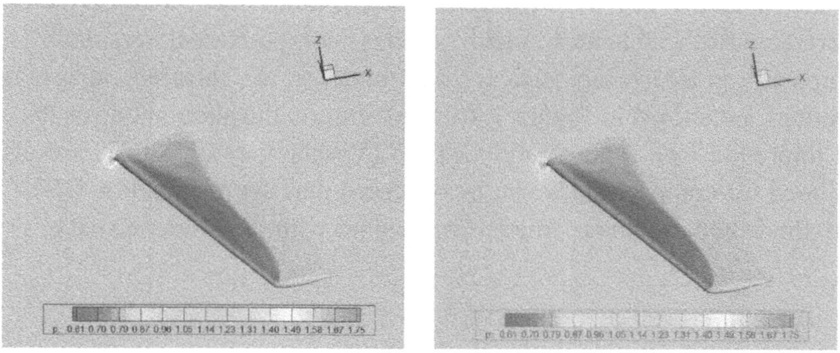

Figure 7.7. Pressure contours of DPW-W1 wing obtained by the S-DGKFS with the D3Q8 model (left) and the improved Roe scheme (right).

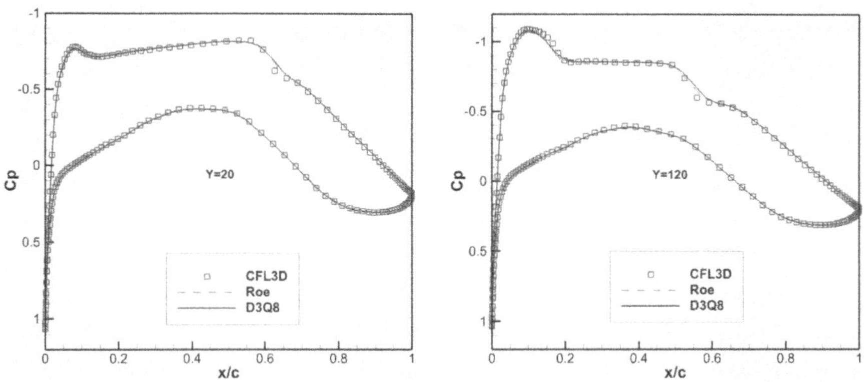

Figure 7.8. Comparison of pressure coefficient distributions at selected sections on DPW-W1 wing.

*https://dpw.larc.nasa.gov/DPW3/

The computational mesh contains 11 blocks and 1,602,651 grid points. In the simulation, the free-stream Mach number is taken as 0.76, the mean-chord based Reynolds number is chosen as 5×10^6 and the angle of attack is set as 0.5 degrees. The fully turbulent flow is modeled by the Spalart-Allmaras turbulence model [Spalart and Allmaras, 1992]. Figure 7.7 compares the pressure contours obtained by the S-DGKFS with the D3Q8 model and the improved Roe scheme [Kim *et al.*, 2002]. In the simulation, the settings of the S-DGKFS and the improved Roe scheme are the same except for the reconstruction of numerical fluxes on the cell interface. Both schemes yield similar computational results. The computed pressure coefficient distributions at selected span-wise locations are shown in Figure 7.8. In this figure, the simulation results of the improved Roe scheme and CFL3D [Vassberg *et al.*, 2007] are also included for comparison. It can be observed that the results of S-DGKFS with the D3Q8 model are in good agreement with the reference data.

7.5 Summary

In this chapter, the C-DGKFS and S-DGKFS are presented as accurate and efficient solvers for simulation of incompressible and compressible viscous flows. Different from the previous C-GKFS and S-GKFS, which are based on the continuous velocity models in the particle velocity space, the C-DGKFS and S-DGKFS are established on the discrete velocity models and thus expect much simpler implementation. In the discrete models, the analytical integrals in the original circular function and sphere function are replaced by the quadrature algorithm by distributing discrete points along the circle or on the spherical surface in the particle velocity space. The basic requirement is to preserve the moment relationships of the circular function or the sphere function so that the recovered macroscopic governing equations maintain consistency with the Navier-Stokes equations. Specifically, the D2Q4, D2Q8 and D2Q12 models are presented for the 2D case, while for the 3D case, the D3Q8 model is derived from the sphere function. By applying these discrete velocity models, the complicated integral formulations of conservative variables and numerical fluxes on the cell interface can be replaced by a simpler solution process.

The effectiveness of C-DGKFS and S-DGKFS in simulating viscous incompressible and compressible flows has been validated by various test cases. Good agreements are achieved between the simulation results and reference data, which demonstrates the accuracy of these solvers. In addition, it is found that the number of discrete velocities in the particle velocity space has limited effect on the accuracy of numerical results, while the discrete velocity model with fewer points is more efficient. In practice, the computational efficiency of C-DGKFS with the D2Q4 model is comparable to the conventional Navier-Stokes solver. At the same time, since the moment relationships for recovering the Navier-Stokes equations are accurately satisfied by the discrete models, the inherent advantages of GKFS over the conventional Navier-Stokes solver are well kept in C-DGKFS and S-DGKFS.

Chapter 8

Gas Kinetic Flux Solvers for Incompressible Flows

The flux solvers presented in Chapters 5-7 aim to resolve general fluid flow problems in a wide range of Mach numbers. The introduced methods include Maxwellian function-based gas kinetic scheme (M-GKS), Maxwellian function-based gas kinetic flux solver (M-GKFS) and circular/sphere function-based gas kinetic flux solver (C-GKFS/S-GKFS). The M-GKS and M-GKFS are constructed from the continuous Boltzmann equation with Maxwellian distribution function, while the C-GKFS and S-GKFS are derived from the simplified distribution functions, namely the circular function and the sphere function. To interpret the abrupt changes of properties in compressible flows, the flow variables and their derivatives on two sides of the cell interface are independently evaluated, based on which the numerical fluxes on the cell interface are reconstructed by various flux solvers. For the M-GKS and the M-GKFS, this process brings in a large number of coefficients related to the flow variables and their derivatives on every cell interface and at each time step, which is tedious and costly in virtual memory [Tang, 2012]. As for the C-GKFS and the S-GKFS, the integral domain is usually asymmetric about the cell interface, which makes the formulations of C-GKFS for viscous flows more complicated than the conventional Navier-Stokes solvers [Van Leer, 1979; Roe, 1981; Toro *et al.*, 1994].

However, the above situations may alter if the flows are incompressible. The incompressible limit assumes smooth change of flow variables in the computational domain, and that the characteristic velocity is much smaller than the sound speed c. The M-GKS can benefit from these assumptions. By calculating the flow variables and their derivatives on the cell interface from the smooth function approximation, several Maxwellian function-based incompressible gas kinetic schemes (M-IGKSs) have been reported in the literature [Su *et al.*, 1999; Xu and Liu, 1999; Xu and He, 2003; Guo *et al.*, 2008]. The C-GKFS and the S-GKFS can also be simplified by applying the incompressible limit. Since the flow velocity (u, v, w) is much smaller than the sound speed c, the circle/sphere on the cell interface can be approximately considered symmetric about the interface, which yields simple and explicit expressions for the conservative variables and numerical fluxes on the cell interface. The resultant schemes are named as the circular/sphere function-based incompressible gas kinetic flux solvers (C-IGKFS/S-IGKFS) [Yang *et al*, 2017b; Yang *et al*, 2017c; Yang *et al*, 2018c]. In this chapter, the M-IGKS, C-IGKFS and S-IGKFS will be introduced.

8.1 Governing Equations and FVM Discretization

The M-IGKS, C-IGKFS and S-IGKFS are all based on the finite volume discretization of macroscopic governing equations. Instead of employing strict incompressibility, the governing equations here allow weak compressibility under the incompressible limit, which reads

$$\frac{\partial \rho}{\partial t} + \nabla \cdot (\rho \mathbf{u}) = 0, \tag{8.1}$$

$$\frac{\partial \rho \mathbf{u}}{\partial t} + \nabla \cdot (\rho \mathbf{u}\mathbf{u} + p\mathbf{I}) = \nabla \cdot \left[\mu \left(\nabla \mathbf{u} + (\nabla \mathbf{u})^T \right) \right], \tag{8.2}$$

where ρ, $\mathbf{u}$, p and μ are respectively the density, the velocity vector, the pressure and the dynamic viscosity of the mean flow. $\mathbf{I}$ is the unit tensor. $\mathbf{u} = (u, v)$ is for the two-dimensional (2D) case, while $\mathbf{u} = (u, v, w)$ is for the three-dimensional (3D) case, where u, v and w are the velocity components in x-, y- and z- directions, respectively.

With the finite volume method (FVM), the discrete form of Eqs. (8.1) and (8.2) can be written as

$$\frac{d\mathbf{W}_i}{dt} = -\frac{1}{\Omega_i}\sum_{j=1}^{N_f}\mathbf{F}_{nj}S_j \,, \tag{8.3}$$

where i is the index of the control volume, Ω_i and N_f represent the volume and the number of faces of the control volume i, respectively. S_j denotes the area of the jth interface of the control volume. The conservative variable vector $\mathbf{W}$ and the flux vector $\mathbf{F}_n$ for the 2D case are given by

$$\mathbf{W} = \left(\rho, \rho u, \rho v\right)^T, \tag{8.4}$$

$$\mathbf{F}_n = \left(F_\rho, F_{\rho u}, F_{\rho v}\right)^T, \tag{8.5}$$

and for the 3D case, they are expressed as

$$\mathbf{W} = \left(\rho, \rho u, \rho v, \rho w\right)^T, \tag{8.6}$$

$$\mathbf{F}_n = \left(F_\rho, F_{\rho u}, F_{\rho v}, F_{\rho w}\right)^T. \tag{8.7}$$

Eq. (8.3) is a set of ordinary differential equations (ODEs). Its solution updates the conservative variable vector $\mathbf{W}$ defined at the cell center. The key issue is to evaluate the flux vector $\mathbf{F}_n$ on the cell interface from the conservative variables $\mathbf{W}$ at cell centers.

For the convenience of derivation and application, the local coordinate system defined on the cell interface is introduced. In the 2D case, x_1-axis points to the normal direction and x_2-axis is aligned with the tangential direction of the cell interface; and in the 3D case, x_1-axis is taken as the normal direction pointing outwards of the cell interface, while x_2-axis and x_3-axis are chosen as two tangential directions of the cell interface, which are mutually orthogonal and form a right-hand coordinate system. Under the local coordinate system, the conservative variable vector and the flux vector can be expressed as

$$\overline{\mathbf{W}} = \left(\rho, \rho u_1, \rho u_2\right)^T, \tag{8.8}$$

$$\overline{\mathbf{F}} = \left(F_{\rho}, F_{\rho u_1}, F_{\rho u_2}\right)^T, \tag{8.9}$$

for the 2D case, and

$$\overline{\mathbf{W}} = \left(\rho, \rho u_1, \rho u_2, \rho u_3\right)^T, \tag{8.10}$$

$$\overline{\mathbf{F}} = \left(F_{\rho}, F_{\rho u_1}, F_{\rho u_2}, F_{\rho u_3}\right)^T, \tag{8.11}$$

for the 3D case. u_1, u_2 and u_3 are the velocity components expressed in the local coordinate system. The coordinate transformation can be performed to correlate the local coordinate system with the global Cartesian coordinate system. In the 2D condition, we have

$$\begin{cases} u = n_x u_1 - n_y u_2, \\ v = n_x u_2 + n_y u_1, \end{cases} \tag{8.12}$$

$$\begin{cases} F_{\rho u} = n_x F_{\rho u_1} - n_y F_{\rho u_2}, \\ F_{\rho v} = n_x F_{\rho u_2} + n_y F_{\rho u_1}, \end{cases} \tag{8.13}$$

where $\mathbf{n} = \left(n_x, n_y\right)$ denotes the unit normal vector of the cell interface in the global Cartesian coordinate system. And for the 3D case, we have

$$\begin{cases} u = n_{1x} u_1 + n_{2x} u_2 + n_{3x} u_3, \\ v = n_{1y} u_1 + n_{2y} u_2 + n_{3y} u_3, \\ w = n_{1z} u_1 + n_{2z} u_2 + n_{3z} u_3, \end{cases} \tag{8.14}$$

$$\begin{cases} F_{\rho u} = n_{1x} F_{\rho u_1} + n_{2x} F_{\rho u_2} + n_{3x} F_{\rho u_3}, \\ F_{\rho v} = n_{1y} F_{\rho u_1} + n_{2y} F_{\rho u_2} + n_{3y} F_{\rho u_3}, \\ F_{\rho w} = n_{1z} F_{\rho u_1} + n_{2z} F_{\rho u_2} + n_{3z} F_{\rho u_3}, \end{cases} \tag{8.15}$$

where $\mathbf{n}_1 = \left(n_{1x}, n_{1y}, n_{1z}\right)$, $\mathbf{n}_2 = \left(n_{2x}, n_{2y}, n_{2z}\right)$ and $\mathbf{n}_3 = \left(n_{3x}, n_{3y}, n_{3z}\right)$ represent the unit vector of x_1-axis, x_2-axis and x_3-axis, respectively. Eqs. (8.13) and (8.15) suggest that the calculation of $\mathbf{F}_{\underline{n}}$ in the global Cartesian coordinate system is equivalent to evaluating $\overline{\mathbf{F}}$ in the local coordinate system, which will be utilized in constructing different solvers in the following sections.

8.2 Maxwellian Function-Based IGKS

The M-GKS utilizes the local integral solution to the Boltzmann equation to reconstruct numerical fluxes on the cell interface. Once the distribution function on the cell interface is obtained, the numerical fluxes can be calculated by the integral for moments. By assuming that the flow variables and their derivatives over the cell interface are smooth, simplifications of the conventional M-GKS were made by Su *et al.* [1999] for incompressible isothermal flows and by Xu and Lui [1999] for incompressible thermal flows. These methods can be viewed as the reduced form of the conventional M-GKS with the reduced local integral solution to the continuous Boltzmann equation on the cell interface. Subsequently, Xu and He [2003] and Guo *et al.* [2008] discarded the energy equation and applied the isothermal distribution function to further simplify the conventional M-GKS for simulation of incompressible isothermal flows. In this section, the incompressible version of M-GKS (i.e., M-IGKS) developed by Xu and He [2003] will be introduced in detail.

8.2.1 *Moment relationships of Maxwellian function*

The continuous Boltzmann equation with the Bhatnagar-Gross-Krook (BGK) collision model can be written as

$$\frac{\partial f}{\partial t} + \boldsymbol{\xi} \cdot \nabla f = \frac{g - f}{\tau}, \tag{8.16}$$

where f is the gas distribution function that varies with the physical space $\mathbf{x} = (x_1, x_2, \ldots, x_D)$, particle velocity space $\boldsymbol{\xi} = (\xi_1, \xi_2, \ldots, \xi_D)$, and time t. g is the equilibrium state approached by f through particle collisions within a collision time scale τ. For incompressible flows, the equilibrium distribution function is defined by

$$g=\rho\left(\frac{\lambda}{\pi}\right)^{\frac{D}{2}}e^{-\lambda\left[\sum_{i=1}^{D}\left(\xi_i-u_i\right)^2\right]}. \tag{8.17}$$

Here, u_i is the macroscopic flow velocity in the ith direction of D-dimensional space; and $\lambda=m/(2kT)=1/(2RT)$, where m, k, R and T are the molecular mass, the Boltzmann constant, the gas constant and the temperature, respectively. Note that the above equilibrium distribution function may deviate from the physical reality. For the case of $D=3$, the above g is the equilibrium state for a monatomic gas. However, for the case of $D=2$, the gas described by the above equilibrium state is limited to move only in a plane without any random motion in the third direction, which does not correspond to any real gas distribution function. Therefore, the above g only serves as a mathematical model for the 2D incompressible flows.

Utilizing Eq. (8.17), the following 4 moment relationships (from the zeroth order to the third order) can be obtained:

$$\int g d\Xi=\rho, \tag{8.18a}$$

$$\int g\xi_\alpha d\Xi=\rho u_\alpha, \tag{8.18b}$$

$$\int g\xi_\alpha\xi_\beta d\Xi=\rho u_\alpha u_\beta+p\delta_{\alpha\beta}, \tag{8.18c}$$

$$\int g\xi_\alpha\xi_\beta\xi_\chi d\Xi=p\left(u_\alpha\delta_{\beta\chi}+u_\beta\delta_{\chi\alpha}+u_\chi\delta_{\alpha\beta}\right)+\rho u_\alpha u_\beta u_\chi, \tag{8.18d}$$

where ξ_α, ξ_β, ξ_χ and u_α, u_β, u_χ are the components of particle velocity and macroscopic flow velocity in the α -, β - and χ -directions, respectively. $d\Xi=d\xi_\alpha d\xi_\beta d\xi_\chi$ is the volume element in the particle velocity space. In addition, $p=\rho RT$ is the pressure of the mean flow. For isothermal flows, the temperature T is a constant, which is noted as T_{ref} in the following text.

To apply the local solution to the Boltzmann equation for reconstructing numerical fluxes of the macroscopic governing equations, the connection between Eqs. (8.1)-(8.2) and Eq. (8.16) has to be established in advance. This can be done by the Chapman-Enskog expansion analysis. The first-order expansion to Eq. (8.16) in terms of the collision time scale can be written as

$$f = f^{eq} - \tau Df = f^{eq} - \tau Df^{eq} + O(\tau^2), \tag{8.19}$$

where $D = (\partial_t + \boldsymbol{\xi} \cdot \nabla)$ is the operator of substantial derivative and $f^{eq} = g$ is the equilibrium state. Note that the expression of $f = f^{eq} - \tau Df$ has been substituted into Df to derive the above expression. To simplify the notation, the integration of an arbitrary function ϕ over the particle velocity space is defined as

$$\langle \phi \rangle = \int_{-\infty}^{+\infty} \phi d\Xi . \tag{8.20}$$

To fulfill the compatibility condition, the non-equilibrium part of the distribution function should not contribute to the calculation of conservative flow variables, i.e.,

$$\left\langle \boldsymbol{\varphi}_a \left(f - f^{eq} \right) \right\rangle = 0, \tag{8.21}$$

where $\boldsymbol{\varphi}_\alpha$ stands for the moment vector and is given by

$$\boldsymbol{\varphi}_\alpha = (1, \boldsymbol{\xi})^T . \tag{8.22}$$

To derive the macroscopic governing equations, let us apply the operator D to Eq. (8.19),

$$Df = Df^{eq} - D\left(\tau Df^{eq}\right) + O\left(\tau^2\right). \tag{8.23}$$

From Eq. (8.16), we have $Df = \left(f^{eq} - f\right)/\tau$. Furthermore, multiplying $\boldsymbol{\varphi}_a$ on both sides of Eq. (8.23) and performing integration over the particle velocity space for the resultant equation yield

$$\begin{aligned} &\partial_t \left\langle \boldsymbol{\varphi}_a f^{eq} \right\rangle + \nabla \cdot \left\langle \boldsymbol{\xi} \boldsymbol{\varphi}_a f^{eq} \right\rangle \\ &= \partial_t \left\langle \boldsymbol{\varphi}_a \left(\tau Df^{eq}\right) \right\rangle + \nabla \cdot \left\langle \boldsymbol{\xi} \boldsymbol{\varphi}_a \left(\tau Df^{eq}\right) \right\rangle + O\left(\tau^2\right). \end{aligned} \tag{8.24}$$

Note that the compatibility condition has been used in deriving the above equation. By substituting Eq. (8.19) into the first term on the right-hand side of Eq. (8.24) and utilizing the compatibility condition, we can obtain:

$$\partial_t \left\langle \boldsymbol{\varphi}_a f^{eq} \right\rangle + \nabla \cdot \left\langle \boldsymbol{\xi} \boldsymbol{\varphi}_a f^{eq} \right\rangle = \nabla \cdot \left\langle \boldsymbol{\xi} \boldsymbol{\varphi}_a \left(\tau D f^{eq} \right) \right\rangle + O\left(\tau^2\right). \quad (8.25)$$

The first component of $\boldsymbol{\varphi}_a$ is 1, which gives the following relationship when being substituted into Eq. (8.25):

$$\partial_t \left\langle f^{eq} \right\rangle + \nabla \cdot \left\langle \boldsymbol{\xi} f^{eq} \right\rangle = \nabla \cdot \left\langle \boldsymbol{\xi} \left(\tau D f^{eq} \right) \right\rangle + O\left(\tau^2\right). \quad (8.26)$$

Utilizing Eq. (8.19) and the compatibility condition (8.21) allows further simplification of the above formula into

$$\partial_t \left\langle f^{eq} \right\rangle + \nabla \cdot \left\langle \boldsymbol{\xi} f^{eq} \right\rangle = O\left(\tau^2\right). \quad (8.27)$$

Applying the conservation relationships in Eqs. (8.18a) and (8.18b), the above equation can be rewritten as

$$\frac{\partial \rho}{\partial t} + \nabla \cdot \left(\rho \mathbf{u} \right) = O\left(\tau^2\right), \quad (8.28)$$

which recovers the continuity equation in the order of $O\left(\tau^2\right)$.

Considering the second component of $\boldsymbol{\varphi}_a$ in Eq. (8.25) gives

$$\begin{aligned} &\partial_t \left\langle \boldsymbol{\xi} f^{eq} \right\rangle + \nabla \cdot \left\langle \boldsymbol{\xi}\boldsymbol{\xi} f^{eq} \right\rangle \\ &= \nabla \cdot \left[\tau \left(\partial_t \left\langle \boldsymbol{\xi}\boldsymbol{\xi} f^{eq} \right\rangle + \nabla \cdot \left\langle \boldsymbol{\xi}\boldsymbol{\xi}\boldsymbol{\xi} f^{eq} \right\rangle \right) \right] + O\left(\tau^2\right). \end{aligned} \quad (8.29)$$

Substituting the conservation relationships in Eqs. (8.18b) and (8.18c) into the left-hand side of Eq. (8.29), we have

$$\partial_t \left(\rho \mathbf{u} \right) + \nabla \cdot \left(\rho \mathbf{u}\mathbf{u} + p \mathbf{I} \right) = O\left(\tau\right). \quad (8.30)$$

Further substitution of the conservative relationships (8.18c) and (8.18d) into the right-hand side of Eq. (8.29) yields

$$\begin{aligned}
\partial_t \left\langle \boldsymbol{\xi\xi} f^{eq} \right\rangle &= \partial_t \left(\rho \mathbf{uu} + p\mathbf{I} \right) \\
&= \partial_t \left(\rho \mathbf{u} \right) \mathbf{u} + \mathbf{u} \partial_t \left(\rho \mathbf{u} \right) - \mathbf{uu} \partial_t \rho + RT_{ref} \partial_t \rho \mathbf{I} \\
&= \left[-\nabla \cdot \left(\rho \mathbf{uu} + p\mathbf{I} \right) \right] \mathbf{u} + \mathbf{u} \left[-\nabla \cdot \left(\rho \mathbf{uu} + p\mathbf{I} \right) \right] \\
&\quad - \left(\mathbf{uu} - RT_{ref} \mathbf{I} \right) \left[-\nabla \cdot \left(\rho \mathbf{u} \right) \right] + O(\tau) \\
&= -\nabla \cdot \left(\rho \mathbf{uuu} \right) - \left[\left(\nabla p \right) \mathbf{u} + \mathbf{u} \nabla p \right] - \nabla \cdot \left(p \mathbf{u} \right) \mathbf{I} + O(\tau),
\end{aligned} \tag{8.31}$$

$$\begin{aligned}
\nabla \cdot \left\langle \boldsymbol{\xi\xi\xi} f^{eq} \right\rangle &= \partial_\chi \left\langle \xi_\alpha \xi_\beta \xi_\chi f^{eq} \right\rangle \\
&= \partial_\chi \left[p(u_\alpha \delta_{\beta\chi} + u_\beta \delta_{\chi\alpha} + u_\chi \delta_{\alpha\beta}) + \rho u_\alpha u_\beta u_\chi \right] \\
&= \nabla \cdot \left(\rho \mathbf{uuu} \right) + p \left(\partial_\beta u_\alpha + \partial_\alpha u_\beta + \delta_{\alpha\beta} \partial_\chi u_\chi \right) \\
&\quad + \left(u_\alpha \partial_\beta p + u_\beta \partial_\alpha p + \delta_{\alpha\beta} u_\chi \partial_\chi p \right) \\
&= \nabla \cdot \left(\rho \mathbf{uuu} \right) + p \left[\nabla \mathbf{u} + \left(\nabla \mathbf{u} \right)^T \right] \\
&\quad + \left[\left(\nabla p \right) \mathbf{u} + \mathbf{u} \nabla p \right] + \nabla \cdot \left(p \mathbf{u} \right) \mathbf{I}.
\end{aligned} \tag{8.32}$$

Note that Eq. (8.30) has been used for the derivation of Eq. (8.31). Although the truncation error of Eq. (8.31) is $O(\tau)$, it will be multiplied by τ when being substituted into Eq. (8.29). Thus, the approximation in Eq. (8.31) will not affect the overall accuracy of Eq. (8.29), which yields

$$\frac{\partial \rho \mathbf{u}}{\partial t} + \nabla \cdot \left(\rho \mathbf{uu} + p\mathbf{I} \right) = \nabla \cdot \mathbf{\Pi} + O\left(\tau^2 \right), \tag{8.33}$$

where

$$\mathbf{\Pi} = \mu \left[\left(\nabla \mathbf{u} + \left(\nabla \mathbf{u} \right)^T \right) \right],$$

$$\mu = \tau p.$$

Eq. (8.33) proves that the momentum equation can be recovered from the Boltzmann equation in the order of $O\left(\tau^2\right)$.

The above Chapman-Enskog expansion analysis reveals that the weakly compressible isothermal Navier-Stokes equations (8.1) and (8.2) can be accurately recovered from the Boltzmann equation as long as the conservation relationships in Eqs. (8.18a)-(8.18d) hold. It also suggests that the local solution to the Boltzmann equation can be used to reconstruct the numerical fluxes of Eqs. (8.1) and (8.2) on the cell interface. This is the core idea of M-IGKS.

8.2.2 *Evaluation of fluxes on the cell interface by M-IGKS*

To evaluate the numerical fluxes of macroscopic governing equations, let us multiply $\boldsymbol{\varphi}_\alpha$ on both sides of Eq. (8.16) and perform integration over the particle velocity space for the resultant equation, which yield

$$\frac{\partial}{\partial t}\langle \boldsymbol{\varphi}_\alpha f\rangle + \nabla\cdot\langle \boldsymbol{\xi}\boldsymbol{\varphi}_\alpha f\rangle = \frac{1}{\tau}\langle \boldsymbol{\varphi}_\alpha (g-f)\rangle . \tag{8.34}$$

According to the compatibility condition (8.21) and the relationships (8.18a)-(8.18b), the above equation can be rewritten as

$$\frac{\partial \mathbf{W}}{\partial t} + \nabla\cdot\langle \boldsymbol{\xi}\boldsymbol{\varphi}_\alpha f\rangle = 0 . \tag{8.35}$$

By comparing Eq. (8.35) with Eqs. (8.1) and (8.2), it is found that the numerical fluxes of macroscopic governing equations can be calculated by the moment relationships of the distribution function. In the local coordinate system, the numerical fluxes of macroscopic governing equations can be expressed as

$$\overline{\mathbf{F}} = \langle \xi_1 \overline{\boldsymbol{\varphi}}_\alpha f\rangle \tag{8.36}$$

where ξ_1 is the particle velocity in the normal direction of the cell interface and $\overline{\boldsymbol{\varphi}}_\alpha$ is the moment vector expressed in the local coordinate system. Eq. (8.36) shows that the calculation of $\overline{\mathbf{F}}$ is equivalent to evaluating the distribution function f on the cell interface. In M-IGKS, the local integral solution to the Boltzmann equation is used to determine the distribution function f . Suppose that the midpoint of the cell interface is located at $\mathbf{x}=\mathbf{0}$, the integral solution to the Boltzmann equation with the BGK collision model (8.16) yields

$$f(\mathbf{0},\xi,t)=\frac{1}{\tau}\int_0^t g\left(-\xi(t-t'),\xi,t'\right)e^{-(t-t')/\tau}dt' + e^{-t/\tau}f_0\left(-\xi t,\xi,0\right), \tag{8.37}$$

where $f_0\left(-\xi t,\xi,0\right)$ is the initial distribution function around the cell interface. In the above equation, free transport and collision are coupled at any point in space and time.

In the general flow conditions, the initial distribution function f_0 could be discontinuous across the cell interface. Consideration of such discontinuity complicates the solution in Eq. (8.37), as shown in **Chapter 5**. However, as per the comments of Xu and He [2003], f_0 can be simplified in incompressible flow scenario which assumes smooth change of flow variables and derivatives across the cell interface. According to the Chapman-Enskog expansion analysis, with the assumption of smooth flow field, the initial distribution function around the cell interface can be expanded as

$$f_0\left(\mathbf{x},\xi,0\right)=g+\mathbf{x}\cdot\nabla g-\tau Dg\,, \tag{8.38}$$

where $g=g\left(\mathbf{0},\xi,0\right)$ is the equilibrium distribution function on the cell interface. In addition, the equilibrium distribution function at time t around the cell interface can be obtained from the expansion of its equilibrium state,

$$g\left(\mathbf{x},\xi,t\right)=g+\mathbf{x}\cdot\nabla g+t\,\partial g/\partial t\,. \tag{8.39}$$

Substituting Eqs. (8.38) and (8.39) into Eq. (8.37), we have

$$f\left(\mathbf{0},\xi,t\right)=g-\tau Dg+t\,\partial g/\partial t\,. \tag{8.40}$$

In fact, Eq. (8.40) is the local integral solution to the Boltzmann equation with the assumption of smooth change of flow variables and their derivatives across the cell interface.

To reconstruct the numerical fluxes of macroscopic governing equations by $f\left(\mathbf{0},\xi,t\right)$, the derivatives of g in space and time have to be determined in advance. For the 2D case, Eq. (8.40) can be rewritten in the following form:

$$f(\mathbf{0},\xi,t)=g\left[1-\tau\left(a\xi_1+b\xi_2+A\right)+tA\right], \tag{8.41}$$

where $\partial g/\partial x_1 = ag$, $\partial g/\partial x_2 = bg$, $\partial g/\partial t = Ag$ and

$$a = a_0 + a_1\xi_1 + a_2\xi_2,$$

$$b = b_0 + b_1\xi_1 + b_2\xi_2,$$

$$A = A_0 + A_1\xi_1 + A_2\xi_2.$$

The coefficients a_0, a_1, a_2, b_0, b_1 and b_2 can be determined from the following relationships:

$$\langle a\bar{\boldsymbol{\varphi}}_\alpha g\rangle = \left(\partial\rho/\partial x_1, \partial\rho u_1/\partial x_1, \partial\rho u_2/\partial x_1\right)^T, \tag{8.42}$$

$$\langle b\bar{\boldsymbol{\varphi}}_\alpha g\rangle = \left(\partial\rho/\partial x_2, \partial\rho u_1/\partial x_2, \partial\rho u_2/\partial x_2\right)^T, \tag{8.43}$$

where ξ_1 and ξ_2 are the particle velocity components expressed in the local coordinate system. The detailed expressions of a_0, a_1, a_2, b_0, b_1 and b_2 can be found in **Appendix C**. In addition, the coefficients A_0, A_1 and A_2 can be calculated by the compatibility condition, i.e.,

$$\langle (a\xi_1 + b\xi_2 + A)\bar{\boldsymbol{\varphi}}_\alpha g\rangle = 0. \tag{8.44}$$

The above equation gives

$$\langle A\bar{\boldsymbol{\varphi}}_\alpha g\rangle = -\langle (a\xi_1 + b\xi_2)\bar{\boldsymbol{\varphi}}_\alpha g\rangle. \tag{8.45}$$

Since a_0, a_1, a_2, b_0, b_1 and b_2 are known from Eqs. (8.42) and (8.43), the values of A_0, A_1 and A_2 can also be calculated in the same way as that shown in **Appendix C**. In M-IGKS, the flow variables on the cell interface are evaluated by the second-order/third-order interpolation, and their derivatives are computed by the central difference scheme.

Once the local integral solution to the Boltzmann equation is fully determined, the numerical fluxes of macroscopic governing equations on the cell interface can be computed by substituting Eq. (8.41) into Eq. (8.36), which yields

$$\begin{aligned}\overline{\mathbf{F}} &= \frac{1}{\Delta t}\int_0^{\Delta t}\left\langle \xi_1 \overline{\boldsymbol{\varphi}}_\alpha g\left[1-\tau\left(a\xi_1+b\xi_2+A\right)+tA\right]\right\rangle dt \\ &= \left\langle \xi_1 \overline{\boldsymbol{\varphi}}_\alpha g\left[1-\tau\left(a\xi_1+b\xi_2+A\right)+\frac{\Delta t}{2}A\right]\right\rangle,\end{aligned} \tag{8.46}$$

where Δt is the time step size for solving the macroscopic governing equations, which is determined by the Courant-Friedrichs-Lewy (CFL) condition. The detailed expressions of Eq. (8.46) can be obtained by using the recurrence relationships shown in **Appendix A**. In addition, for isothermal flows, the temperature is a constant, which can be linked to the sound speed and Mach number as

$$c_s^2 = RT_{ref} = \frac{u_{ref}^2}{Ma^2}. \tag{8.47}$$

To satisfy the incompressible limit, in the simulation, we can set the reference velocity as $u_{ref} = 1$ and the Mach number as $Ma < 0.3$. Alternatively, like the lattice Boltzmann method (LBM), we can choose $u_{ref} = 0.1$ and set the sound speed as $c_s = 1/\sqrt{3}$, which corresponds to the Mach number of $Ma \approx 0.173$.

Comparing with the original M-GKS for general flow problems, the M-IGKS for incompressible flows simplifies the solution process by employing the assumption of smooth change of flow variables and their derivatives over the cell interface. However, it should be pointed out that the full expressions of numerical fluxes given by the M-IGKS are still complicated than the conventional Navier-Stokes solvers. Moreover, Chen *et al.* [2011] recently found that the assumption of smooth flow field may deteriorate the stability of M-IGKS at high Reynolds numbers. They commented that the discontinuous derivatives of flow variables across the cell interface should be retained in order to improve the stability of M-IGKS. By doing so, they successfully simulated the lid-driven cavity flow at high Reynolds numbers and the flow past a stationary sphere. Nevertheless, such modification makes the implementation of M-IGKS almost equally complicated as the compressible counterpart.

8.3 Circular Function-Based IGKFS

For general fluid flow problems, the C-GKFS has been developed by replacing the Maxwellian distribution function in M-GKS with the circular function [Yang *et al.*, 2014b]. It brings in two advantages. The first one is to simplify the original integration over the infinite velocity space into the line integral over the circle. The other one is that the non-equilibrium distribution function is approximated by the difference of equilibrium distribution functions on the cell interface and its surrounding points, which yields an explicit algebraic expression of the local solution to the Boltzmann equation. These simplifications make the computation of C-GKFS more compact and efficient than M-GKS. By restricting the application within incompressible flows, the C-GKFS can be further simplified. Since the flow velocity is much smaller than the sound speed under the incompressible limit, the circle established in the local reconstruction can be considered as approximately symmetric about the interface, which forms the C-IGKFS [Yang *et al*, 2017b; Yang *et al*, 2018a]. Such symmetricity could greatly simplify the expressions of conservative variables and numerical fluxes on the cell interface. At the same time, the C-IGKFS can still tackle cases with discontinuous conservative variables and their derivatives across the cell interface, which is beneficial to improving the numerical stability at high Reynolds numbers as suggested by Chen *et al.* [2011]. In this section, the details of C-IGKFS will be presented.

8.3.1 *Moment relationships of circular function*

As shown in **Chapter 6**, in 2D conditions, the Maxwellian distribution function can be simplified to the following circular function:

$$g_C = \begin{cases} \dfrac{\rho}{2\pi} & \text{if } (\xi_1 - u_1)^2 + (\xi_2 - u_2)^2 = c^2, \\ 0 & \text{otherwise.} \end{cases} \tag{8.48}$$

where the square of radius c represents the mean kinetic energy of particles and can be expressed as

$$c^2 = DRT. \tag{8.49}$$

For isothermal incompressible flows considered here, the temperature T can be set a constant and noted as T_{ref} in the following text. The value of RT_{ref} can be determined by Eq. (8.47), and D is the number of dimensions ($D = 2$ for two dimensions).

From Eq. (8.48), the velocity components of the particle on the circle can be written as follows in the local coordinate system:

$$\xi_1 = u_1 + c\cos(\theta), \tag{8.50a}$$

$$\xi_2 = u_2 + c\sin(\theta), \tag{8.50b}$$

where θ is the angle between the radial direction of the circle and the normal direction of the cell interface, as shown in Figure 6.1. With notations $c_1 = c\cos(\theta)$, $c_2 = c\sin(\theta)$ and $\xi_\alpha = u_\alpha + c_\alpha$, the following 4 moment relationships (from the zeroth order to the third order) can be derived:

$$\int_0^{2\pi} g_C d\theta = \rho, \tag{8.51a}$$

$$\int_0^{2\pi} g_C \xi_\alpha d\theta = \rho u_\alpha, \tag{8.51b}$$

$$\int_0^{2\pi} g_C \xi_\alpha \xi_\beta d\theta = \rho u_\alpha u_\beta + p\delta_{\alpha\beta}, \tag{8.51c}$$

$$\int_0^{2\pi} g_C \xi_\alpha \xi_\beta \xi_\chi d\theta = p\left(u_\alpha \delta_{\beta\chi} + u_\beta \delta_{\chi\alpha} + u_\chi \delta_{\alpha\beta}\right) + \rho u_\alpha u_\beta u_\chi. \tag{8.51d}$$

By comparing Eqs. (8.51a)-(8.51d) with Eqs. (8.18a)-(8.18d), one could find that the circular function and the Maxwellian distribution function satisfy the same moment relationships. Thus, the weakly compressible Navier-Stokes equations (8.1) and (8.2) can also be recovered from the Boltzmann equation with the circular function. The derivation is basically the same as the Chapman-Enskog expansion analysis shown in **Section 8.2.1**.

8.3.2 *Evaluation of fluxes on the cell interface by C-IGKFS*

It is indicated in the Chapman-Enskog expansion analysis (see Eq. (8.19)) that, to recover Navier-Stokes equations, the first-order expansion to the distribution function in terms of the collision time scale should be retained. Suppose that the midpoint of the cell interface is located at $\mathbf{x}=\mathbf{0}$, the distribution function on the cell interface can be expanded as

$$\begin{aligned} f(\mathbf{0},t) &= f^{eq}(\mathbf{0},t) - \tau D f^{eq}(\mathbf{0},t) + O(\tau^2) \\ &= g_C(\mathbf{0},t) + \tau_0\left[g_C(-\boldsymbol{\xi}\delta_t, t-\delta_t) - g_C(\mathbf{0},t)\right] \\ &\quad + O\left(\tau^2, \tau\delta_t, |\boldsymbol{\xi}|^2 \tau\delta_t\right), \end{aligned} \tag{8.52}$$

where $g_C(\mathbf{0},t) = f^{eq}(\mathbf{0},t)$ is the equilibrium distribution function on the cell interface, and $g_C(-\boldsymbol{\xi}\delta_t, t-\delta_t)$ is the equilibrium distribution function on the circle. $\tau_0 = \tau/\delta_t$ is the dimensionless collision time and δ_t is the streaming time step. To simplify the notation, "$(\mathbf{0},t)$" is denoted by the superscript "*face*" and "$(-\boldsymbol{\xi}\delta_t, t-\delta_t)$" is noted by the superscript "*cir*" in the following text. Substituting Eq. (8.52) into Eq. (8.36) and neglecting the second-order error terms, we can get

$$\begin{aligned} \overline{\mathbf{F}} &= \int \xi_1^{face} \overline{\boldsymbol{\varphi}}_\alpha^{face} g_C^{face} d\theta \\ &\quad + \tau_0\left[\int \xi_1^{cir} \overline{\boldsymbol{\varphi}}_\alpha^{cir} g_C^{cir} d\theta - \int \xi_1^{face} \overline{\boldsymbol{\varphi}}_\alpha^{face} g_C^{face} d\theta\right] \\ &= \mathbf{F}^I + \tau_0\left[\mathbf{F}^{II} - \mathbf{F}^I\right]. \end{aligned} \tag{8.53}$$

Here, $\overline{\boldsymbol{\varphi}}_\alpha = (1, \xi_1, \xi_2)^T$ is the moment vector expressed in the local coordinate system. From Eq. (8.53), it can be observed that the numerical fluxes on the cell interface consist of two parts. One is the flux attributed to the equilibrium distribution function and the moments on the cell interface, which is denoted as $\mathbf{F}^I$. The other part is the flux attributed to the equilibrium distribution function and the moments on the circle, which is referred as $\mathbf{F}^{II}$.

In order to calculate the fluxes $\mathbf{F}^I$ and $\mathbf{F}^{II}$, $\overline{\boldsymbol{\varphi}}_\alpha^{face}$, g_C^{face}, $\overline{\boldsymbol{\varphi}}_\alpha^{cir}$ and g_C^{cir} should be computed in advance. For any variable ϕ, its value along the circle can be obtained by the linear interpolation as follows:

$$\phi^{cir}=\begin{cases}\phi^L-\nabla\phi^L\cdot\boldsymbol{\xi}^+\delta_t, & \text{if } \xi_1^+\geq 0,\\ \phi^R-\nabla\phi^R\cdot\boldsymbol{\xi}^+\delta_t, & \text{if } \xi_1^+<0.\end{cases}\tag{8.54}$$

where ϕ^L and ϕ^R are the values of ϕ on the left and the right sides of the cell interface, respectively. $\nabla\phi^L$ and $\nabla\phi^R$ are the first-order derivatives of ϕ at the left and the right cells around the cell interface. As can be seen from Eq. (8.54), the possibility of discontinuous conservative variables and their derivatives across the cell interface has been considered in C-IGKFS. $\boldsymbol{\xi}^+$ is the predicted particle velocity on the cell interface given by

$$\boldsymbol{\xi}^+=\left(\xi_1^+,\xi_2^+\right)=\left(u_1^+ +c^+\cos(\theta),u_2^+ +c^+\sin(\theta)\right).\tag{8.55}$$

Like the C-GKFS shown in **Chapter 6**, the predicted mean flow velocities u_1^+ and u_2^+ can be calculated by the Roe averages, and the effective peculiar velocity $c^+=c$ is a constant for incompressible flows (see Eq. (8.49)). Take $\xi_1^+\geq 0$ as an example, the particle velocities and the equilibrium distribution function along the circle can be expressed as

$$\begin{aligned}\xi_1^{cir,L}=u_1^L&-\frac{\partial u_1^L}{\partial x_1}\left(u_1^+ +c^+\cos(\theta)\right)\delta_t\\&-\frac{\partial u_1^L}{\partial x_2}\left(u_2^+ +c^+\sin(\theta)\right)\delta_t+c^+\cos(\theta),\end{aligned}\tag{8.56}$$

$$\begin{aligned}\xi_2^{cir,L}=u_2^L&-\frac{\partial u_2^L}{\partial x_1}\left(u_1^+ +c^+\cos(\theta)\right)\delta_t\\&-\frac{\partial u_2^L}{\partial x_2}\left(u_2^+ +c^+\sin(\theta)\right)\delta_t+c^+\sin(\theta),\end{aligned}\tag{8.57}$$

$$g_C^{cir,L}=g_C^L-\frac{\partial g_C^L}{\partial x_1}\left(u_1^+ +c^+\cos(\theta)\right)\delta_t-\frac{\partial g_C^L}{\partial x_2}\left(u_2^+ +c^+\sin(\theta)\right)\delta_t.\tag{8.58}$$

To facilitate the derivation of Eq. (8.53), we can rearrange Eqs. (8.56)-(8.58) into the following forms:

$$\xi_1^{cir,L} = a_0^L + a_1^L r + a_2^L s, \tag{8.59}$$

$$\xi_2^{cir,L} = b_0^L + b_1^L r + b_2^L s, \tag{8.60}$$

$$g_C^{cir,L} = g_0^L + g_1^L r + g_2^L s, \tag{8.61}$$

where

$$r = \cos(\theta), \ s = \sin(\theta).$$

$$a_0^L = u_1^L - \frac{\partial u_1^L}{\partial x_1} u_1^+ \delta_t - \frac{\partial u_1^L}{\partial x_2} u_2^+ \delta_t, \ a_1^L = c^+ - \frac{\partial u_1^L}{\partial x_1} c^+ \delta_t, \ a_2^L = -\frac{\partial u_1^L}{\partial x_2} c^+ \delta_t.$$

$$b_0^L = u_2^L - \frac{\partial u_2^L}{\partial x_1} u_1^+ \delta_t - \frac{\partial u_2^L}{\partial x_2} u_2^+ \delta_t, \ b_1^L = -\frac{\partial u_2^L}{\partial x_1} c^+ \delta_t, \ b_2^L = c^+ - \frac{\partial u_2^L}{\partial x_2} c^+ \delta_t.$$

$$g_0^L = g_C^L - \frac{\partial g_C^L}{\partial x_1} u_1^+ \delta_t - \frac{\partial g_C^L}{\partial x_2} u_2^+ \delta_t, \ g_1^L = -\frac{\partial g_C^L}{\partial x_1} c^+ \delta_t, \ g_2^L = -\frac{\partial g_C^L}{\partial x_2} c^+ \delta_t.$$

Note that all the above coefficients are local constants except for r and s. For the case of $\xi_1^+ < 0$, the above variables can be easily obtained by replacing the superscript "L" in Eqs. (8.56)-(8.61) with "R".

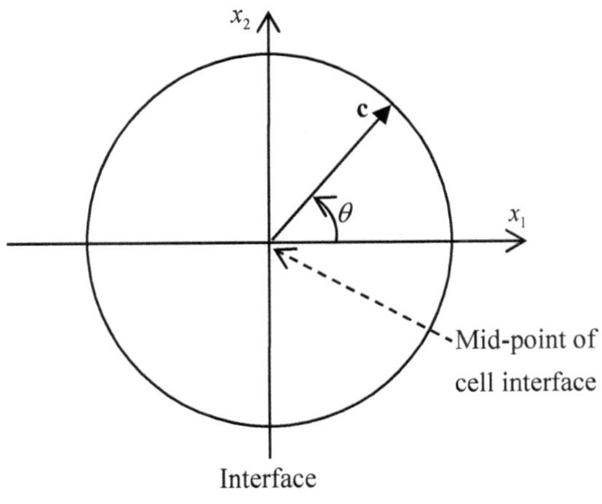

Figure 8.1. Distribution of circular function in the particle velocity space on a cell interface for incompressible flows.

According to the compatibility condition, the non-equilibrium term $\tau Df^{eq}(\mathbf{0},t)$ has no contribution to the calculation of conservative variables. Thus, the conservative variables on the cell interface can be computed by

$$\begin{aligned}\overline{\mathbf{W}}^{face} &= \int \overline{\boldsymbol{\varphi}}_{\alpha}^{face} g_C^{face} d\theta \\ &= \int \overline{\boldsymbol{\varphi}}_{\alpha}^{cir} g_C^{cir} d\theta \\ &= \int_{\xi_1^+ \geq 0} \overline{\boldsymbol{\varphi}}_{\alpha}^{cir,L} g_C^{cir,L} d\theta + \int_{\xi_1^+ < 0} \overline{\boldsymbol{\varphi}}_{\alpha}^{cir,R} g_C^{cir,R} d\theta.\end{aligned} \tag{8.62}$$

In Eq. (8.62), the integral domain is decided by ξ_1^+. As shown in **Chapter 6**, for general cases, Eq. (8.62) can be rewritten as

$$\overline{\mathbf{W}}^{face} = \int_{-\theta_0}^{\theta_0} \overline{\boldsymbol{\varphi}}_{\alpha}^{cir,L} g_C^{cir,L} d\theta + \int_{\theta_0}^{2\pi-\theta_0} \overline{\boldsymbol{\varphi}}_{\alpha}^{cir,R} g_C^{cir,R} d\theta, \tag{8.63}$$

where $\theta_0 = \arccos\left(-u_1^+/c^+\right)$. For incompressible flows, the value of c^+ is much larger than u_1^+ due to the incompressible limit (Mach number is very small). Thus, we have

$$\theta_0 = \arccos\left(-u_1^+/c^+\right) \approx \frac{\pi}{2}. \tag{8.64}$$

With that, Eq. (8.63) can be simplified as

$$\overline{\mathbf{W}}^{face} = \int_{-\frac{\pi}{2}}^{\frac{\pi}{2}} \overline{\boldsymbol{\varphi}}_{\alpha}^{cir,L} g_C^{cir,L} d\theta + \int_{\frac{\pi}{2}}^{\frac{3\pi}{2}} \overline{\boldsymbol{\varphi}}_{\alpha}^{cir,R} g_C^{cir,R} d\theta. \tag{8.65}$$

Eq. (8.65) implies that, for incompressible flows, the circle can be approximately considered to be symmetric about the cell interface. The configuration of the circle on a cell interface is shown in Figure 8.1. Substituting Eqs. (8.59)-(8.61) into Eq. (8.65) and considering the contributions from the " R " side in the same way, we have

$$\overline{\mathbf{W}}^{face}(1) = \left[\pi g_0 + 2g_1\right]^L + \left[\pi g_0 - 2g_1\right]^R, \tag{8.66}$$

$$\overline{\mathbf{W}}^{face}(2)=\left[\pi a_0 g_0+2\left(a_1 g_0+a_0 g_1\right)+\frac{\pi}{2}\left(a_1 g_1+a_2 g_2\right)\right]^L + \left[\pi a_0 g_0-2\left(a_1 g_0+a_0 g_1\right)+\frac{\pi}{2}\left(a_1 g_1+a_2 g_2\right)\right]^R, \tag{8.67}$$

$$\overline{\mathbf{W}}^{face}(3)=\left[\pi b_0 g_0+2\left(b_1 g_0+b_0 g_1\right)+\frac{\pi}{2}\left(b_1 g_1+b_2 g_2\right)\right]^L + \left[\pi b_0 g_0-2\left(b_1 g_0+b_0 g_1\right)+\frac{\pi}{2}\left(b_1 g_1+b_2 g_2\right)\right]^R. \tag{8.68}$$

Here, $\overline{\mathbf{W}}^{face}(i)$ means the ith component of $\overline{\mathbf{W}}^{face}$. Once the conservative variables on the cell interface $\overline{\mathbf{W}}^{face}$ are obtained, $\overline{\boldsymbol{\varphi}}_\alpha^{face}$ and g_C^{face} can be determined in a straightforward way by using Eqs. (8.48) and (8.50), and the flux vector $\mathbf{F}^I$ can be evaluated by substituting $\overline{\boldsymbol{\varphi}}_\alpha^{face}$ and g_C^{face} into Eq. (8.53). An alternative but simpler way to compute $\mathbf{F}^I$ is to substitute the conservative variables $\overline{\mathbf{W}}^{face}$ directly into the expression of inviscid flux vector, i.e.,

$$\mathbf{F}^I=\begin{bmatrix}\rho u_1\\ \rho u_1 u_1+\rho c^+ c^+/D\\ \rho u_1 u_2\end{bmatrix}^{face}. \tag{8.69}$$

Like the derivation of Eq. (8.65), $\mathbf{F}^{II}$ can be computed by

$$\mathbf{F}^{II}=\int_{-\frac{\pi}{2}}^{\frac{\pi}{2}}\xi_1^{cir,L}\overline{\boldsymbol{\varphi}}_\alpha^{cir,L} g_C^{cir,L}\,d\theta+\int_{\frac{\pi}{2}}^{\frac{3\pi}{2}}\xi_1^{cir,R}\overline{\boldsymbol{\varphi}}_\alpha^{cir,R} g_C^{cir,R}\,d\theta. \tag{8.70}$$

Substituting Eqs. (8.59)-(8.61) and similar formulations from the "R" side into Eq. (8.70), we have

$$\mathbf{F}^{II}(1)=\left[\pi a_0 g_0+2\left(a_1 g_0+a_0 g_1\right)+\frac{\pi}{2}\left(a_1 g_1+a_2 g_2\right)\right]^L + \left[\pi a_0 g_0-2\left(a_1 g_0+a_0 g_1\right)+\frac{\pi}{2}\left(a_1 g_1+a_2 g_2\right)\right]^R, \tag{8.71}$$

$$\mathbf{F}^{II}(2)=\begin{bmatrix} a_0 g_0\left(\pi a_0+2a_1\right)+\left(2a_0+\dfrac{\pi a_1}{2}\right)\left(a_0 g_1+a_1 g_0\right) \\ +\dfrac{2}{3}a_2 a_2 g_1+\dfrac{\pi a_2}{2}\left(a_0 g_2+a_2 g_0\right) \\ +\left(\dfrac{\pi a_0}{2}+\dfrac{4a_1}{3}\right)\left(a_1 g_1+a_2 g_2\right) \end{bmatrix}^L + \begin{bmatrix} a_0 g_0\left(\pi a_0-2a_1\right)-\left(2a_0-\dfrac{\pi a_1}{2}\right)\left(a_0 g_1+a_1 g_0\right) \\ -\dfrac{2}{3}a_2 a_2 g_1+\dfrac{\pi a_2}{2}\left(a_0 g_2+a_2 g_0\right) \\ +\left(\dfrac{\pi a_0}{2}-\dfrac{4a_1}{3}\right)\left(a_1 g_1+a_2 g_2\right) \end{bmatrix}^R, \tag{8.72}$$

$$\mathbf{F}^{II}(3)=\begin{bmatrix} b_0 g_0\left(\pi a_0+2a_1\right)+\left(2a_0+\dfrac{\pi a_1}{2}\right)\left(b_0 g_1+b_1 g_0\right) \\ +\dfrac{\pi a_2}{2}\left(b_0 g_2+b_2 g_0\right)+\left(\dfrac{\pi a_0}{2}+\dfrac{2a_1}{3}\right)\left(b_1 g_1+b_2 g_2\right) \\ +\dfrac{2}{3}a_2\left(b_1 g_2+b_2 g_1\right)+\dfrac{2}{3}a_1 b_1 g_1 \end{bmatrix}^L + \begin{bmatrix} b_0 g_0\left(\pi a_0-2a_1\right)-\left(2a_0-\dfrac{\pi a_1}{2}\right)\left(b_0 g_1+b_1 g_0\right) \\ +\dfrac{\pi a_2}{2}\left(b_0 g_2+b_2 g_0\right)+\left(\dfrac{\pi a_0}{2}-\dfrac{2a_1}{3}\right)\left(b_1 g_1+b_2 g_2\right) \\ -\dfrac{2}{3}a_2\left(b_1 g_2+b_2 g_1\right)-\dfrac{2}{3}a_1 b_1 g_1 \end{bmatrix}^R, \tag{8.73}$$

where $\mathbf{F}^{II}(i)$ is the ith component of $\mathbf{F}^{II}$. Furthermore, by substituting Eq. (8.69) and Eqs. (8.71)-(8.73) into Eq. (8.53), we can obtain the full expressions of fluxes on the cell interface, which are expressed explicitly as the functions of conservative variables and their derivatives at cell centers, for the 2D weakly compressible Navier-Stokes equations.

The last undetermined variable in Eq. (8.53) is the streaming time step δ_t, which is used to reconstruct the local solution to the Boltzmann equation on the cell interface and is related to the dimensionless collision time τ_0. The principle of choosing δ_t is that the circle in the physical space must be within the cells sharing the interface in order to avoid extrapolation. In C-IGKFS, δ_t can be determined by

$$\delta_t = \frac{0.2 \times \min\{\Delta l, \Delta r\}}{\left(\max\{u_1^+, u_2^+\} + c^+\right)} \approx \frac{0.2 \times \min\{\Delta l, \Delta r\}}{c^+}, \tag{8.74}$$

where Δl and Δr are the lengths of the shortest edges of the left and the right cells around the cell interface, respectively.

For the convenience of application, the calculation procedure of numerical fluxes on the cell interface is summarized as below:

(a) For each cell interface, use Eq. (5.35) to interpolate the macroscopic flow variables on the left and the right sides of the cell interface, $\mathbf{W}^L$ and $\mathbf{W}^R$;
(b) Use Roe averages (Eqs. (6.58a) and (6.58b)) to calculate u_1^+ and u_2^+ on the cell interface, c^+ is determined by Eq. (8.49);
(c) Calculate the streaming time step δ_t by Eq. (8.74);
(d) Calculate all coefficients defined in Eqs. (8.59)-(8.61);
(e) Compute the conservative variables on the cell interface $\overline{\mathbf{W}}^{face}$ by Eqs. (8.66)-(8.68), and further calculate the flux vector $\mathbf{F}^I$ by Eq. (8.69);
(f) Compute $\mathbf{F}^{II}$ by Eqs. (8.71)-(8.73);
(g) Calculate τ_0 by Eq. (7.27);
(h) Substitute $\mathbf{F}^I$, $\mathbf{F}^{II}$ and τ_0 into Eq. (8.53) to obtain the full expressions of numerical fluxes.

8.4 Sphere Function-Based IGKFS

By using the sphere function as the equilibrium distribution, the S-GKFS has been presented for simulation of inviscid compressible flows in **Chapter 6**. In S-GKFS, the surface integrals along the spherical surface are performed to calculate the inviscid fluxes on the cell interface.

Despite its simpler form as compared with the M-GKFS, extension of S-GKFS to 3D viscous flows is challenging. The reason is that the formulations of numerical fluxes given by the S-GKFS are determined by the integral domain on the spherical surface, which is centered at (*u*, *v*, *w*) with the radius of *c*, where (*u*, *v*, *w*) is the macroscopic flow velocity vector and *c* is proportional to the sound speed. For general flow problems, the sphere on the cell interface is usually not symmetric about the cell interface. This makes the calculation of numerical fluxes in S-GKFS more complicated than the conventional Navier-Stokes solvers [Van Leer, 1979; Roe, 1981; Toro *et al.*, 1994]. However, for incompressible flows in which the flow velocity is much smaller than the sound speed, the sphere on the cell interface can be approximately considered to be symmetric. And the simulation of incompressible isothermal flows can further omit the resolution of the energy equation. Based on these facts, the S-IGKFS for simulation of 3D incompressible flows was developed by Shu and his coworkers [Yang *et al.*, 2017d; Yang *et al.*, 2018c]. In the method, the formulations of numerical fluxes on the cell interface can be expressed concisely and explicitly. In addition, like the original S-GKFS, the discontinuities of conservative variables and their derivatives across the cell interface can be considered in S-IGKFS, which contributes to its numerical stability at high Reynolds numbers. The details of S-IGKFS will be shown in this section.

8.4.1 *Moment relationships of sphere function*

For the 3D case, the Maxwellian distribution function can be simplified into the following sphere function:

$$g_S = \begin{cases} \dfrac{\rho}{4\pi} & \text{if } (\xi_1 - u_1)^2 + (\xi_2 - u_2)^2 + (\xi_3 - u_3)^2 = c^2, \\ 0 & \text{otherwise.} \end{cases} \tag{8.75}$$

where $\boldsymbol{\xi} = (\xi_1, \xi_2, \xi_3)$ and $\mathbf{u} = (u_1, u_2, u_3)$ are the particle velocity vector and the macroscopic flow velocity vector, respectively. The definition of c is the same as in Eq. (8.49), while the dimension should be set as $D = 3$. From Eq. (8.75), the velocity components of the particle on the spherical surface can be written as follows in the local coordinate system:

$$\xi_1 = u_1 + c\sin(\varphi)\cos(\theta), \tag{8.76a}$$

$$\xi_2 = u_2 + c\sin(\varphi)\sin(\theta), \tag{8.76b}$$

$$\xi_3 = u_3 + c\cos(\varphi). \tag{8.76c}$$

The definitions of angles θ and φ can refer to Figure 6.2. With notations of $c_1 = c\sin(\varphi)\cos(\theta)$, $c_2 = c\sin(\varphi)\sin(\theta)$, $c_3 = c\cos(\varphi)$ and $\xi_\alpha = u_\alpha + c_\alpha$, the following 4 moment relationships of the sphere function can be obtained:

$$\int_0^{2\pi}\int_0^{\pi} g_S \sin(\varphi)\, d\varphi d\theta = \rho, \tag{8.77a}$$

$$\int_0^{2\pi}\int_0^{\pi} g_S \xi_\alpha \sin(\varphi)\, d\varphi d\theta = \rho u_\alpha, \tag{8.77b}$$

$$\int_0^{2\pi}\int_0^{\pi} g_S \xi_\alpha \xi_\beta \sin(\varphi)\, d\varphi d\theta = \rho u_\alpha u_\beta + p\delta_{\alpha\beta}, \tag{8.77c}$$

$$\begin{aligned}&\int_0^{2\pi}\int_0^{\pi} g_S \xi_\alpha \xi_\beta \xi_\chi \sin(\varphi)\, d\varphi d\theta \\ &= p\left(u_\alpha \delta_{\beta\chi} + u_\beta \delta_{\chi\alpha} + u_\chi \delta_{\alpha\beta}\right) + \rho u_\alpha u_\beta u_\chi.\end{aligned} \tag{8.77d}$$

It is easy to find that the moment relationships of the sphere function are the same as those of the Maxwellian distribution function and the circular function. Thus, the weakly compressible Navier-Stokes equations (8.1) and (8.2) can be recovered from the Boltzmann equation with the sphere function. Eqs. (8.77a)-(8.77d) play the fundamental role in the development of S-IGKFS.

8.4.2 *Evaluation of fluxes on the cell interface by S-IGKFS*

Like the C-IGKFS, the distribution function on the cell interface for the S-IGKFS can be expressed as

$$
\begin{aligned}
f(\mathbf{0},t) &= f^{eq}(\mathbf{0},t) - \tau D f^{eq}(\mathbf{0},t) + O(\tau^2) \\
&= g_S(\mathbf{0},t) + \tau_0\left[g_S(-\boldsymbol{\xi}\delta_t, t-\delta_t) - g_S(\mathbf{0},t)\right] \\
&\quad + O\left(\tau^2, \tau\delta_t, |\boldsymbol{\xi}|^2\tau\delta_t\right),
\end{aligned}
\tag{8.78}
$$

where $g_S(\mathbf{0},t)$ and $g_S(-\boldsymbol{\xi}\delta_t, t-\delta_t)$ are respectively the equilibrium distribution functions on the cell interface and on the spherical surface. $\tau_0 = \tau/\delta_t$ is the dimensionless collision time and δ_t is the streaming time step. For simplicity, we denote "$(\mathbf{0},t)$" by the superscript "*face*" and "$(-\boldsymbol{\xi}\delta_t, t-\delta_t)$" by the superscript "*sph*" in the following text. Substituting Eq. (8.78) into Eq. (8.36) and neglecting the second-order error terms, we have

$$
\begin{aligned}
\overline{\mathbf{F}} &= \iint \xi_1^{face}\overline{\boldsymbol{\varphi}}_\alpha^{face} g_S^{face}\sin(\varphi)\,d\varphi d\theta \\
&\quad + \tau_0\left[\iint \xi_1^{sph}\overline{\boldsymbol{\varphi}}_\alpha^{sph} g_S^{sph}\sin(\varphi)\,d\varphi d\theta\right. \\
&\quad \left. -\iint \xi_1^{face}\overline{\boldsymbol{\varphi}}_\alpha^{face} g_S^{face}\sin(\varphi)\,d\varphi d\theta\right] \\
&= \mathbf{F}^I + \tau_0\left[\mathbf{F}^{II} - \mathbf{F}^I\right].
\end{aligned}
\tag{8.79}
$$

Here, $\overline{\boldsymbol{\varphi}}_\alpha = (1, \xi_1, \xi_2, \xi_3)^T$ is the moment vector expressed in the local coordinate system. In Eq. (8.79), $\mathbf{F}^I$ denotes the flux vector attributed to the equilibrium distribution function and the moments on the cell interface and $\mathbf{F}^{II}$ represents the flux vector attributed to the equilibrium distribution function and the moments on the spherical surface.

In S-IGKFS, the equilibrium distribution function and the moments are the functions of conservative variables. The conservative variables on the spherical surface can be obtained by interpolation from those at cell centers. And the conservative variables on the cell interface can be computed by the compatibility condition, which gives

$$
\begin{aligned}
\overline{\mathbf{W}}^{face} &= \iint \overline{\boldsymbol{\varphi}}_\alpha^{face} g_S^{face}\sin(\varphi)\,d\varphi d\theta \\
&= \iint \overline{\boldsymbol{\varphi}}_\alpha^{sph} g_S^{sph}\sin(\varphi)\,d\varphi d\theta.
\end{aligned}
\tag{8.80}
$$

It can be seen from Eqs. (8.79)-(8.80) that the calculation of the flux vector $\overline{\mathbf{F}}$ is now reduced to the determination of the equilibrium distribution function and the moments on the spherical surface. For any variable ϕ, its value on the spherical surface can be computed by

$$\phi^{sph} = \begin{cases} \phi^L - \nabla\phi^L \cdot \boldsymbol{\xi}^+ \delta_t, & \text{if } \xi_1^+ \geq 0, \\ \phi^R - \nabla\phi^R \cdot \boldsymbol{\xi}^+ \delta_t, & \text{if } \xi_1^+ < 0. \end{cases} \tag{8.81}$$

where ϕ^L and ϕ^R are the values of ϕ on the left and the right sides of the cell interface, respectively. $\nabla\phi^L$ and $\nabla\phi^R$ are respectively the first-order derivatives of ϕ at the left and the right adjacent cells of the cell interface. $\boldsymbol{\xi}^+$ is the predicted particle velocity on the cell interface given by

$$\begin{aligned} \boldsymbol{\xi}^+ &= \left(\xi_1^+, \xi_2^+, \xi_3^+\right) \\ &= \left(u_1^+ + c^+ \sin(\varphi)\cos(\theta), u_2^+ + c^+ \sin(\varphi)\sin(\theta), \right. \\ &\quad \left. u_3^+ + c^+ \cos(\varphi)\right). \end{aligned} \tag{8.82}$$

Furthermore, the predicted mean flow velocities u_1^+, u_2^+ and u_3^+ can be calculated by the Roe averages, and the effective peculiar velocity $c^+ = c$ is a constant for incompressible isothermal flows, which is given by Eq. (8.49).

Once the flow variables on the spherical surface are obtained, the particle velocities and the equilibrium distribution function for the case of $\xi_1^+ \geq 0$ can be calculated by

$$\begin{aligned} \xi_1^{sph,L} = u_1^L &- \frac{\partial u_1^L}{\partial x_1}\left(u_1^+ + c^+ \sin(\varphi)\cos(\theta)\right)\delta_t \\ &- \frac{\partial u_1^L}{\partial x_2}\left(u_2^+ + c^+ \sin(\varphi)\sin(\theta)\right)\delta_t \\ &- \frac{\partial u_1^L}{\partial x_3}\left(u_3^+ + c^+ \cos(\varphi)\right)\delta_t \\ &+ c^+ \sin(\varphi)\cos(\theta), \end{aligned} \tag{8.83}$$

$$\begin{aligned}\xi_2^{sph,L} = u_2^L &- \frac{\partial u_2^L}{\partial x_1}\left(u_1^+ + c^+ \sin(\varphi)\cos(\theta)\right)\delta_t \\ &- \frac{\partial u_2^L}{\partial x_2}\left(u_2^+ + c^+ \sin(\varphi)\sin(\theta)\right)\delta_t \\ &- \frac{\partial u_2^L}{\partial x_3}\left(u_3^+ + c^+ \cos(\varphi)\right)\delta_t \\ &+ c^+ \sin(\varphi)\sin(\theta),\end{aligned} \tag{8.84}$$

$$\begin{aligned}\xi_3^{sph,L} = u_3^L &- \frac{\partial u_3^L}{\partial x_1}\left(u_1^+ + c^+ \sin(\varphi)\cos(\theta)\right)\delta_t \\ &- \frac{\partial u_3^L}{\partial x_2}\left(u_2^+ + c^+ \sin(\varphi)\sin(\theta)\right)\delta_t \\ &- \frac{\partial u_3^L}{\partial x_3}\left(u_3^+ + c^+ \cos(\varphi)\right)\delta_t \\ &+ c^+ \cos(\varphi),\end{aligned} \tag{8.85}$$

$$\begin{aligned}g_S^{sph,L} = g_S^L &- \frac{\partial g_S^L}{\partial x_1}\left(u_1^+ + c^+ \sin(\varphi)\cos(\theta)\right)\delta_t \\ &- \frac{\partial g_S^L}{\partial x_2}\left(u_2^+ + c^+ \sin(\varphi)\sin(\theta)\right)\delta_t \\ &- \frac{\partial g_S^L}{\partial x_3}\left(u_3^+ + c^+ \cos(\varphi)\right)\delta_t .\end{aligned} \tag{8.86}$$

To facilitate the derivation of Eqs. (8.79) and (8.80), Eqs. (8.83)-(8.86) can be rearranged as follows:

$$\xi_1^{sph,L} = a_0^L + a_1^L s_1 + a_2^L s_2 + a_3^L s_3 , \tag{8.87}$$

$$\xi_2^{sph,L} = b_0^L + b_1^L s_1 + b_2^L s_2 + b_3^L s_3 , \tag{8.88}$$

$$\xi_3^{sph,L} = c_0^L + c_1^L s_1 + c_2^L s_2 + c_3^L s_3 , \tag{8.89}$$

$$g_S^{sph,L} = g_0^L + g_1^L s_1 + g_2^L s_2 + g_3^L s_3, \tag{8.90}$$

where

$$s_1 = \sin(\varphi)\cos(\theta),\ s_2 = \sin(\varphi)\sin(\theta),\ s_3 = \cos(\varphi).$$

$$a_0^L = u_1^L - \frac{\partial u_1^L}{\partial x_1} u_1^+ \delta_t - \frac{\partial u_1^L}{\partial x_2} u_2^+ \delta_t - \frac{\partial u_1^L}{\partial x_3} u_3^+ \delta_t,$$

$$a_1^L = c^+ - \frac{\partial u_1^L}{\partial x_1} c^+ \delta_t,\ a_2^L = -\frac{\partial u_1^L}{\partial x_2} c^+ \delta_t,\ a_3^L = -\frac{\partial u_1^L}{\partial x_3} c^+ \delta_t.$$

$$b_0^L = u_2^L - \frac{\partial u_2^L}{\partial x_1} u_1^+ \delta_t - \frac{\partial u_2^L}{\partial x_2} u_2^+ \delta_t - \frac{\partial u_2^L}{\partial x_3} u_3^+ \delta_t,$$

$$b_1^L = -\frac{\partial u_2^L}{\partial x_1} c^+ \delta_t,\ b_2^L = c^+ - \frac{\partial u_2^L}{\partial x_2} c^+ \delta_t,\ b_3^L = -\frac{\partial u_2^L}{\partial x_3} c^+ \delta_t.$$

$$c_0^L = u_3^L - \frac{\partial u_3^L}{\partial x_1} u_1^+ \delta_t - \frac{\partial u_3^L}{\partial x_2} u_2^+ \delta_t - \frac{\partial u_3^L}{\partial x_3} u_3^+ \delta_t,$$

$$c_1^L = -\frac{\partial u_3^L}{\partial x_1} c^+ \delta_t,\ c_2^L = -\frac{\partial u_3^L}{\partial x_2} c^+ \delta_t,\ c_3^L = c^+ - \frac{\partial u_3^L}{\partial x_3} c^+ \delta_t.$$

$$g_0^L = g_S^L - \frac{\partial g_S^L}{\partial x_1} u_1^+ \delta_t - \frac{\partial g_S^L}{\partial x_2} u_2^+ \delta_t - \frac{\partial g_S^L}{\partial x_3} u_3^+ \delta_t,$$

$$g_1^L = -\frac{\partial g_S^L}{\partial x_1} c^+ \delta_t,\ g_2^L = -\frac{\partial g_S^L}{\partial x_2} c^+ \delta_t,\ g_3^L = -\frac{\partial g_S^L}{\partial x_3} c^+ \delta_t.$$

Note that all the above coefficients are local constants except for s_1, s_2 and s_3. For the case of $\xi_1^+ < 0$, the above variables can be easily obtained by replacing the superscript "L" with "R".

Given the fact that the flow variables on the left and the right sides of the cell interface could be different as shown in Eq. (8.81), Eq. (8.80) can be further rewritten as

$$\begin{aligned}\overline{\mathbf{W}}^{face} &= \int_{-\theta_0}^{\theta_0}\int_0^{\pi} \overline{\boldsymbol{\varphi}}_{\alpha}^{sph,L} g_S^{sph,L} \sin(\varphi)\, d\varphi d\theta \\ &+ \int_{\theta_0}^{2\pi-\theta_0}\int_0^{\pi} \overline{\boldsymbol{\varphi}}_{\alpha}^{sph,R} g_S^{sph,R} \sin(\varphi)\, d\varphi d\theta,\end{aligned} \tag{8.91}$$

where $\theta_0 = \arccos(-u_1^+ / c^+)$. For incompressible flows, the value of c^+ is much larger than u_1^+ due to the incompressible limit (Mach number is very small). Thus, we have

$$\theta_0 = \arccos\left(-u_1^+/c^+\right) \approx \pi/2\,. \tag{8.92}$$

As a result, the sphere on the cell interface can be approximately considered to be symmetric about the cell interface as shown in Figure 8.2. Correspondingly, Eq. (8.91) can be simplified as

$$\begin{aligned}\overline{\mathbf{W}}^{face} &= \int_{-\frac{\pi}{2}}^{\frac{\pi}{2}}\int_0^{\pi} \overline{\boldsymbol{\varphi}}_{\alpha}^{sph,L} g_S^{sph,L} \sin(\varphi)\, d\varphi d\theta \\ &+ \int_{\frac{\pi}{2}}^{\frac{3\pi}{2}}\int_0^{\pi} \overline{\boldsymbol{\varphi}}_{\alpha}^{sph,R} g_S^{sph,R} \sin(\varphi)\, d\varphi d\theta.\end{aligned} \tag{8.93}$$

Substituting Eqs. (8.87)-(8.90) into Eq. (8.93) and considering the contributions from the " R " side in the same way, we have

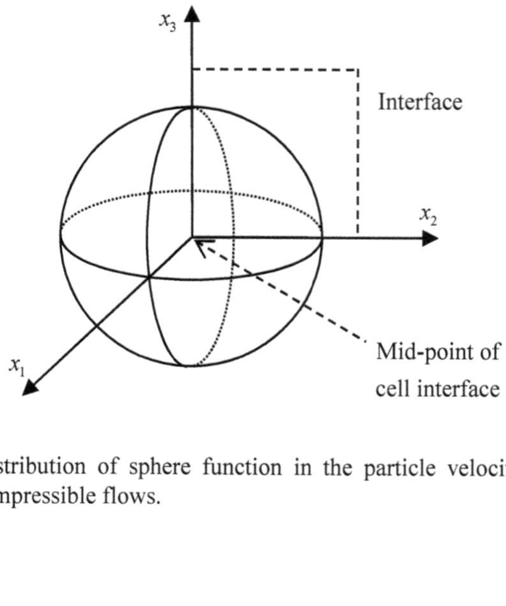

Figure 8.2. Distribution of sphere function in the particle velocity space on a cell interface for incompressible flows.

$$\overline{\mathbf{W}}^{face}(1) = \left[2\pi g_0 + \pi g_1\right]^L + \left[2\pi g_0 - \pi g_1\right]^R, \tag{8.94}$$

$$\begin{aligned}\overline{\mathbf{W}}^{face}(2) = &\left[2\pi a_0 g_0 + \pi\left(a_0 g_1 + a_1 g_0\right) + \chi_a\right]^L \\ &+ \left[2\pi a_0 g_0 - \pi\left(a_0 g_1 + a_1 g_0\right) + \chi_a\right]^R,\end{aligned} \tag{8.95}$$

$$\begin{aligned}\overline{\mathbf{W}}^{face}(3) = &\left[2\pi b_0 g_0 + \pi\left(b_0 g_1 + b_1 g_0\right) + \chi_b\right]^L \\ &+ \left[2\pi b_0 g_0 - \pi\left(b_0 g_1 + b_1 g_0\right) + \chi_b\right]^R,\end{aligned} \tag{8.96}$$

$$\begin{aligned}\overline{\mathbf{W}}^{face}(4) = &\left[2\pi c_0 g_0 + \pi\left(c_0 g_1 + c_1 g_0\right) + \chi_c\right]^L \\ &+ \left[2\pi c_0 g_0 - \pi\left(c_0 g_1 + c_1 g_0\right) + \chi_c\right]^R.\end{aligned} \tag{8.97}$$

where

$$\chi_a = \frac{2\pi}{3}\left(a_1 g_1 + a_2 g_2 + a_3 g_3\right),$$

$$\chi_b = \frac{2\pi}{3}\left(b_1 g_1 + b_2 g_2 + b_3 g_3\right),$$

$$\chi_c = \frac{2\pi}{3}\left(c_1 g_1 + c_2 g_2 + c_3 g_3\right).$$

Here, $\overline{\mathbf{W}}^{face}(i)$ means the ith component of $\overline{\mathbf{W}}^{face}$. Once the conservative variables on the cell interface are obtained, the flux vector $\mathbf{F}^I$ can be calculated straightforwardly by substituting $\overline{\mathbf{W}}^{face}$ into the expression of inviscid flux vector for the 3D weakly compressible Navier-Stokes equations, i.e.,

$$\mathbf{F}^I = \begin{bmatrix} \rho u_1 \\ \rho u_1 u_1 + \rho c^+ c^+ / D \\ \rho u_1 u_2 \\ \rho u_1 u_3 \end{bmatrix}^{face}. \tag{8.98}$$

Like the derivation of Eq. (8.93), the flux vector $\mathbf{F}^{II}$ can be computed by

$$\begin{aligned}\mathbf{F}^{II} &= \int_{-\frac{\pi}{2}}^{\frac{\pi}{2}} \int_0^{\pi} \xi_1^{sph,L} \overline{\boldsymbol{\varphi}}_\alpha^{sph,L} g_S^{sph,L} \sin(\varphi) d\varphi d\theta \\ &+ \int_{\frac{\pi}{2}}^{\frac{3\pi}{2}} \int_0^{\pi} \xi_1^{sph,R} \overline{\boldsymbol{\varphi}}_\alpha^{sph,R} g_S^{sph,R} \sin(\varphi) d\varphi d\theta.\end{aligned} \tag{8.99}$$

Substituting Eqs. (8.87)-(8.90) and similar formulations from the " R " side into Eq. (8.99), we have

$$\mathbf{F}^{II}(1) = \overline{\mathbf{W}}^{face}(2), \tag{8.100}$$

$$\begin{aligned}\mathbf{F}^{II}(2) &= \left[\hbar_1 + \lambda\!\!\!^-_1 + 2\pi a_0 g_0 (a_0 + a_1)\right]^L \\ &+ \left[\hbar_1 - \lambda\!\!\!^-_1 + 2\pi a_0 g_0 (a_0 - a_1)\right]^R,\end{aligned} \tag{8.101}$$

$$\mathbf{F}^{II}(3) = \left[\hbar_2 + \lambda\!\!\!^-_2\right]^L + \left[\hbar_2 - \lambda\!\!\!^-_2\right]^R, \tag{8.102}$$

$$\mathbf{F}^{II}(4) = \left[\hbar_3 + \lambda\!\!\!^-_3\right]^L + \left[\hbar_3 - \lambda\!\!\!^-_3\right]^R, \tag{8.103}$$

where

$$\hbar_1 = \frac{2\pi g_0}{3}\left(a_1^2 + a_2^2 + a_3^2\right) + \frac{4\pi a_0}{3}\left(a_1 g_1 + a_2 g_2 + a_3 g_3\right),$$

$$\lambda\!\!\!^-_1 = \frac{\pi g_1}{4}\left(4a_0^2 + a_2^2 + a_3^2\right) + \frac{\pi a_1}{2}\left(a_1 g_1 + a_2 g_2 + a_3 g_3\right),$$

$$\begin{aligned}\hbar_2 &= \frac{2\pi g_0}{3}\left(3a_0 b_0 + a_1 b_1 + a_2 b_2 + a_3 b_3\right) \\ &+ \frac{2\pi a_0}{3}\left(b_1 g_1 + b_2 g_2 + b_3 g_3\right) + \frac{2\pi b_0}{3}\left(a_1 g_1 + a_2 g_2 + a_3 g_3\right),\end{aligned}$$

$$\bar{\lambda}_2 = \pi g_0 \left(a_0 b_1 + a_1 b_0\right) + \frac{\pi g_1}{4}\left(4a_0 b_0 + 2a_1 b_1 + a_2 b_2 + a_3 b_3\right)$$
$$+ \frac{\pi g_2}{4}\left(a_1 b_2 + a_2 b_1\right) + \frac{\pi g_3}{4}\left(a_1 b_3 + a_3 b_1\right),$$

$$\hbar_3 = \frac{2\pi g_0}{3}\left(3a_0 c_0 + a_1 c_1 + a_2 c_2 + a_3 c_3\right)$$
$$+ \frac{2\pi a_0}{3}\left(c_1 g_1 + c_2 g_2 + c_3 g_3\right) + \frac{2\pi c_0}{3}\left(a_1 g_1 + a_2 g_2 + a_3 g_3\right),$$

$$\bar{\lambda}_3 = \pi g_0 \left(a_0 c_1 + a_1 c_0\right) + \frac{\pi g_1}{4}\left(4a_0 c_0 + 2a_1 c_1 + a_2 c_2 + a_3 c_3\right)$$
$$+ \frac{\pi g_2}{4}\left(a_1 c_2 + a_2 c_1\right) + \frac{\pi g_3}{4}\left(a_1 c_3 + a_3 c_1\right).$$

Here, $\mathbf{F}^{II}(i)$ is the ith component of $\mathbf{F}^{II}$. Finally, by substituting Eq. (8.98) and Eqs. (8.100)-(8.103) into Eq. (8.79), we can obtain the full expressions of numerical fluxes on the cell interface for the 3D weakly compressible Navier-Stokes equations. It should be indicated that the formulations of S-IGKFS are expressed explicitly as the functions of conservative variables and their derivatives at the centers of the left and the right cells. Hence, the present scheme can be easily embedded into the existing codes under the FVM framework for solving the weakly compressible Navier-Stokes equations.

8.5 Sample Applications

In this section, four sample applications will be presented for validating the performance of C-IGKFS and S-IGKFS. The C-IGKFS and S-IGKFS provide numerical strategies of reconstructing fluxes on the cell interface. And the temporal derivative in Eq. (8.3) is approximated by the three-stages Runge-Kutta scheme for the 2D case and the implicit Lower-Upper Symmetric Gauss-Seidel (LU-SGS) method [Yoon and Jameson, 1988; Yoon and Kwak, 1991] for the 3D case. All the computations are done on a PC with 3.2 GHz CPU.

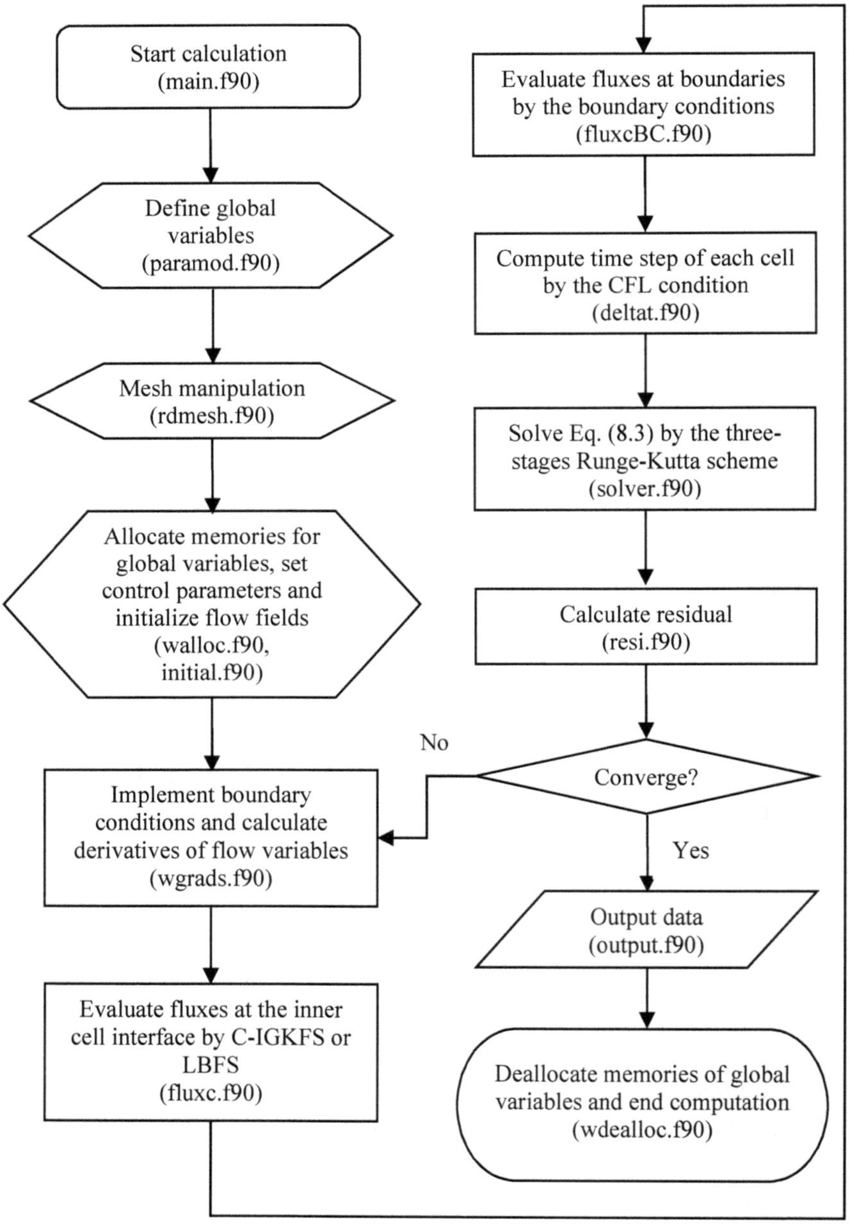

Figure 8.3. Flowchart of C-IGKFS and LBFS for simulation of incompressible flows.

For the convenience of application, the flowchart of C-IGKFS and lattice Boltzmann flux solver (LBFS) for simulation of incompressible flows on the unstructured meshes is shown in Figure 8.3 and the sample code for solving flow over a circular cylinder is provided in **Code E**. The computational mesh can be generated by the commercial software POINTWISE with COBALT format and then transformed into a mesh input file by **Code B**.

8.5.1 *Two-dimensional decaying vortex flow*

The numerical accuracy of C-IGKFS is first examined by simulating the decaying vortex flow, which has analytical solutions as follows:

$$\rho(x,y,t)=\rho_0-\frac{\rho_0 u_0^2 D}{4c^2}\left[\cos(2\pi x/L)+\cos(2\pi y/L)\right]e^{-4\pi^2 u_0 t/(ReL)}, \tag{8.104}$$

$$u(x,y,t)=-u_0\cos(\pi x/L)\sin(\pi y/L)e^{-2\pi^2 u_0 t/(ReL)}, \tag{8.105}$$

$$v(x,y,t)=+u_0\sin(\pi x/L)\cos(\pi y/L)e^{-2\pi^2 u_0 t/(ReL)}, \tag{8.106}$$

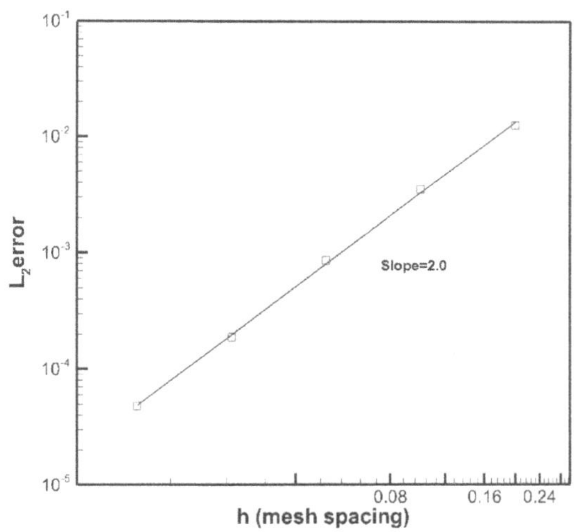

Figure 8.4. L_2 norm of relative error of u-velocity versus mesh spacing h for decaying vortex flow.

where ρ_0 and u_0 are the reference density and velocity. The Reynolds number is set as $Re = \rho_0 u_0 L/\mu = 10$ and the dimensionless collision time is fixed at $\tau_0 = 0.3$. The computational domain of $[-L, L]\times[-L, L]$ is discretized by five different uniform grids with 10×10, 20×20, 40×40, 80×80 and 160×160 cells. The solutions at $t = L/u_0$ are selected and the relative error of u-velocity is measured by the L_2 norms defined as

$$L_2(u)_{retative} = \sqrt{\frac{1}{N\times N}\sum_{N\times N}\left(\frac{u^{numerical} - u^{exact}}{u_0}\right)^2}, \tag{8.107}$$

where $u^{numerical}$ and u^{exact} are respectively the numerical and the exact results. Figure 8.4 shows the L_2 norm of relative error of u-velocity versus the mesh spacing in the log scale. As can be seen, the accuracy of C-IGKFS is validated to be the second-order in space.

8.5.2 *Two-dimensional lid-driven cavity flow*

To compare the numerical accuracy and computational efficiency of C-IGKFS with those of the original C-GKFS [Yang *et al.*, 2014b] and the LBFS [Shu, 2014], the lid-driven cavity flows at $Re = 1000$, 3200 and 5000 are simulated. In the simulation, non-uniform grids with 100×100 cells are used. The smallest mesh spacings are 0.005, 0.002 and 0.001 for $Re = 1000$, 3200 and 5000, respectively. The convergence criterion for this test case is set to be

$$\frac{\sum_{Ncells}\sqrt{\left(u^{n+1} - u^{n}\right)^2 + \left(v^{n+1} - v^{n}\right)^2}}{\sum_{Ncells}\sqrt{\left(u^{n+1}\right)^2 + \left(v^{n+1}\right)^2}} < 10^{-7}. \tag{8.108}$$

The computed u-velocity along the vertical central line and v-velocity along the horizontal central line at $Re = 1000$ and 5000 are shown in Figure 8.5. Clearly, the results given by the C-IGKFS are in good agreement with those of Ghia *et al.* [1982], which demonstrates the accuracy of C-IGKFS at high Reynolds numbers and its flexibility on the non-uniform grid. The comparison of computational time required by the

C-GKFS, LBFS and C-IGKFS is made in Table 8.1. It can be seen that the computational efficiency of C-IGKFS is slightly superior to that of LBFS and is much higher than C-GKFS (it takes about 60% of the computational time cost by the C-GKFS).

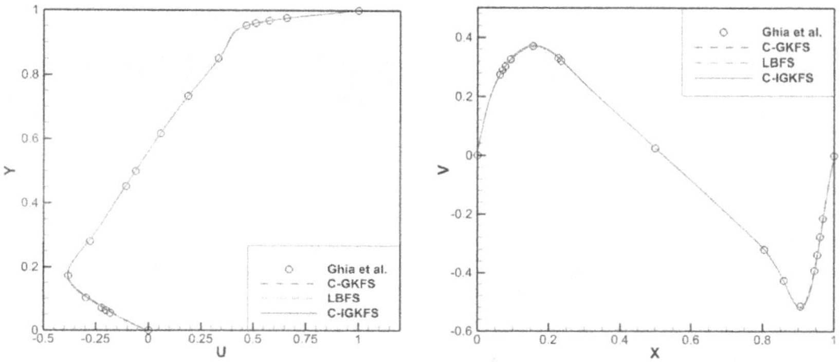

(a) *Re*=1000

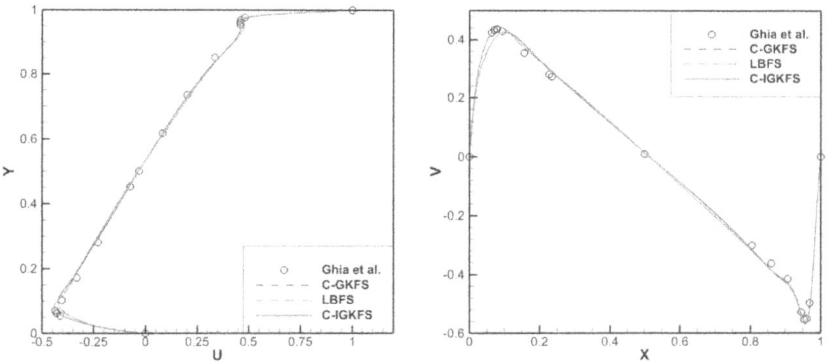

(b) *Re*=5000

Figure 8.5. Comparison of u-velocity (left) and v-velocity (right) profiles along vertical and horizontal central lines at Re=1000 and 5000.

Table 8.1. Comparison of computational time (hours) consumed by the C-GKFS, LBFS and C-IGKFS for lid-driven cavity flow at different Reynolds numbers.

Schemes	*Re*=1000	*Re*=3200	*Re*=5000
C-GKFS	0.6757	1.2666	1.5226
LBFS	0.4313	0.8088	1.0000
C-IGKFS	0.3922	0.7439	0.9216

8.5.3 *Two-dimensional flow over a circular cylinder*

To further illustrate the capability of C-IGKFS for solving flow problems with curved boundary, the flow past a stationary circular cylinder with various Reynolds numbers ($Re = 20$ and 40) is simulated. The Reynolds number of this problem is defined as $Re = \rho_0 u_0 L / \mu$, where ρ_0 is the

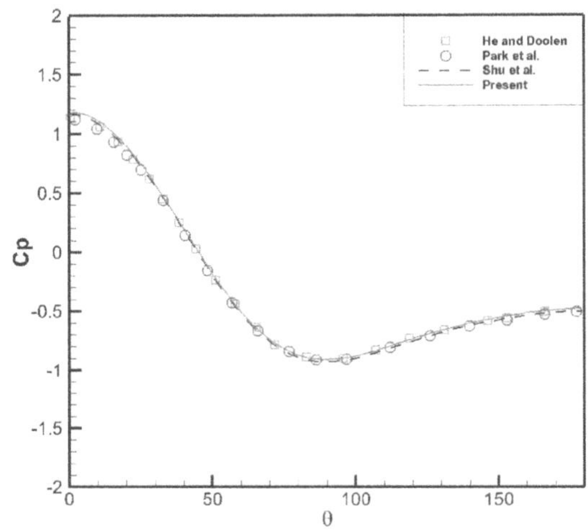

Figure 8.6. Comparison of pressure coefficient distribution around the cylinder surface at *Re*=40.

Table 8.2. Comparison of drag coefficient, recirculation length and separation angle for flow past a stationary circular cylinder at *Re*=20 and 40.

Re	References	C_d	L_s / L	θ_s
20	Dennis and Chang [1970]	2.05	0.94	43.7
	Shukla *et al.* [2007]	2.07	0.92	43.3
	Ding *et al.* [2007a]	2.14	0.94	43.8
	Shu *et al.* [2014]	2.062	0.935	42.94
	Present	2.059	0.921	43.73
40	Dennis and Chang [1970]	1.52	2.35	53.8
	Shukla *et al.* [2007]	1.55	2.34	52.7
	Ding *et al.* [2007a]	1.58	2.32	52.8
	Shu *et al.* [2014]	1.53	2.240	52.69
	Present	1.540	2.229	53.71

reference density, u_0 is the free stream velocity and L is the diameter of the cylinder. In the simulation, the O-type grid with the size of 201×151 is adopted and the far-field boundary is put at $25.5L$ away from the center of the cylinder. The distance from the wall to the nearest grid point in the computational domain is taken as $0.01L$. Besides that, the free stream density ρ_0 and the velocity u_0 are respectively set as 1.0 and 0.1, and the sound speed is chosen as $c_s = 1/\sqrt{3}$. Initially, the flow properties are taken the same as the free stream values. The pressure coefficient C_p, the lift coefficient C_l and the drag coefficient C_d are used to quantify the simulation results, which are defined as

$$C_p = \frac{p_w - p_0}{\rho_0 u_0^2/2}, \tag{8.109}$$

$$C_l = \frac{F_l}{\rho_0 u_0^2 L/2}, \tag{8.110}$$

$$C_d = \frac{F_d}{\rho_0 u_0^2 L/2}, \tag{8.111}$$

where p_0 is the free stream pressure, p_w is the pressure on the cylinder surface, F_l is the lift force and F_d is the drag force. In addition, the geometrical quantities of the eddy, such as the recirculation length L_s and the separation angle θ_s, are also measured.

Since steady state solutions can be obtained for the Reynolds numbers considered here, the convergence criterion of Eq. (8.108) is applied in the simulation. The comparison of the drag coefficient, the recirculation length and the separation angle is made in Table 8.2, which presents good agreement between the results of C-IGKFS and the reference data. To further validate the C-IGKFS, the computed pressure coefficient distribution around the cylinder surface at $Re = 40$ is depicted in Figure 8.6 and compared with previous results in the literature [He and Doolen, 1997; Park *et al.*, 1998; Shu *et al.*, 2014]. Note that, the results of Shu *et al.* [2014] are obtained by the LBFS. Once again, the results of C-IGKFS match very well with the published data.

8.5.4 *Three-dimensional flow over a backward-facing step*

To investigate the performance of S-IGKFS, a 3D numerical test of the laminar channel flow over a backward-facing step is carried out. This example has been widely studied both experimentally [Armaly *et al.*, 1983; Lee, 2000; Lee and Wu, 2008] and numerically [Williams and Baker, 1997; Chiang and Sheu, 1999]. The geometry of the channel is shown in Figure 8.7. In the simulation, the length of the inlet channel is taken as $L_u = 0$; the height of the inlet channel and the step are respectively chosen as $h = 1$ and $S = 0.9423$; the width and the length of the main channel are respectively set as $B = 35h$ and $L_d = 55h$. The flow characteristic of this test case is governed by the Reynolds number, which is defined as $Re = \rho_0 u_0 (2h)/\mu$, where ρ_0 is the reference density and u_0 is the average inflow velocity. At the inlet boundary, the inflow velocity profile was prescribed as a fully developed laminar flow in the straight square duct as suggested by White [1991]

$$\mathbf{u}_{inlet} = (u(y,z), 0, 0), \tag{8.112}$$

where

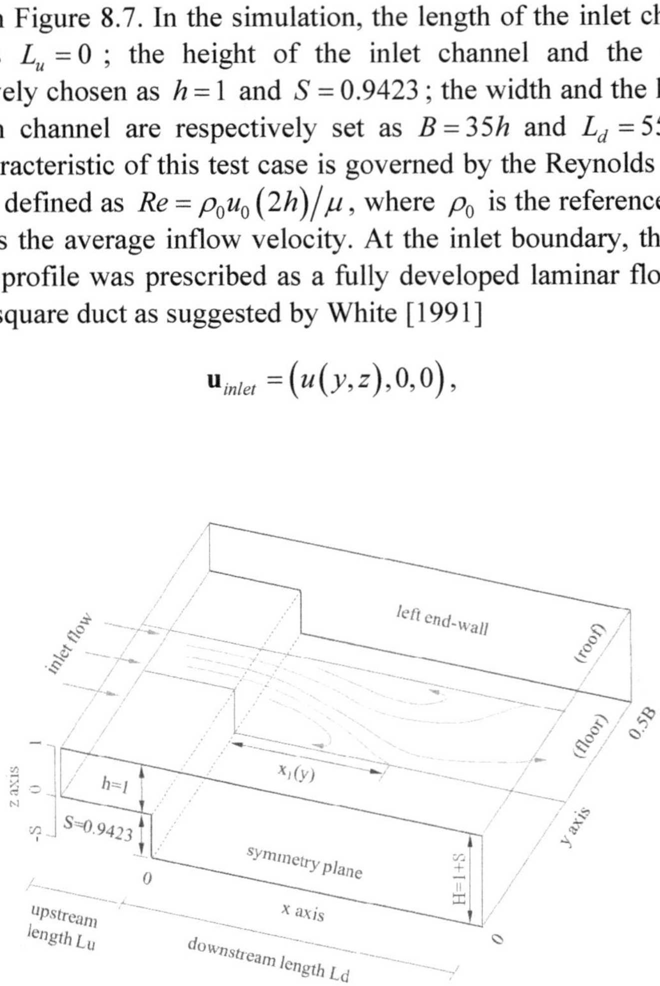

Figure 8.7. Illustration of flow over a backward-facing step.

$$u(y,z)=\frac{48u_0}{\pi^3}\frac{\sum_{i=1,3,5,\ldots}^{\infty}\left[(-1)^{\frac{i-1}{2}}\left(1-\frac{\cosh((2z-1)\xi)}{\cosh(\xi)}\right)\frac{\cos(2y\xi)}{i^3}\right]}{1-\frac{192B}{\pi^5}\sum_{i=1,3,5,\ldots}^{\infty}\frac{\tanh(\xi)}{i^5}},$$

$$\xi=\frac{i\pi}{2B}.$$

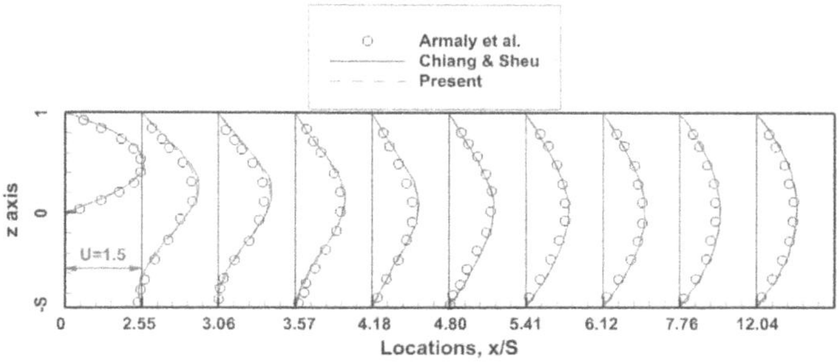

(a) *Re*=100

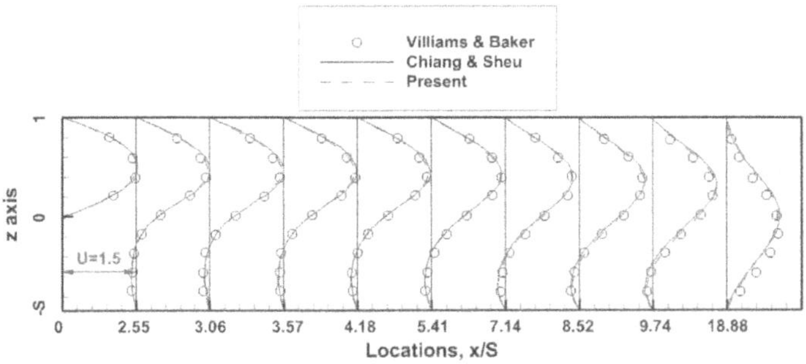

(b) *Re*=800

Figure 8.8. Comparison of streamwise velocity profiles at different streamwise locations on the symmetry plane.

Table 8.3. Comparison of reattachment length on the symmetry plane for flow over a backward-facing step.

References	*Re*=100	*Re*=389	*Re*=800
Chiang and Sheu [1999]	3.26	8.89	16.19
Present	3.30	8.95	15.87

At the outlet boundary, the natural boundary condition is imposed. The symmetric boundary condition is used on the midplane, while the no-slip boundary condition is implemented on the roof and the floor of the channel. The computational domain is discretized by the non-uniform mesh with $Nx \times Ny \times Nz = 130 \times 85 \times 45$ cells.

As indicated by Armaly *et al.* [1983], the present channel flow remains laminar for $Re < 1200$. To avoid the complexities that result from turbulence, three Reynolds numbers of $Re = 100$, 389 and 800 are considered in the present simulations. Figure 8.8 shows the streamwise velocity profiles at different streamwise locations on the symmetric plane obtained by the S-IGKFS for the cases of $Re = 100$ and 800. Also displayed in this figure are the experimental data of Armaly *et al.* [1983] and the numerical results of Williams and Baker [1997] and Chiang and Sheu [1999]. Basically, the results of S-IGKFS compare well with the reference data. Moreover, as can be seen from Table 8.3, the computed reattachment length x_1/S on the symmetric plane shows good agreement with the results of Chiang and Sheu [1999], which further demonstrates the accuracy of S-IGKFS.

8.6 Summary

In this chapter, three gas kinetic flux solvers for simulation of incompressible flows are introduced. These solvers are simplified from the compressible counterparts by implementing the incompressible limit. With the assumption of smooth flow field, the M-IGKS is developed by several researchers [Su *et al.*, 1999; Xu and Liu, 1999; Xu and He, 2003; Guo *et al.*, 2008]. In M-IGKS, the discontinuities of conservative variables and their derivatives on the cell interface are ignored in reconstructing the initial distribution function. This simplification makes the M-IGKS more compact than the original M-GKS which was designed for general fluid problems. However, it was later found by Chen *et al.* [2011] that this strategy may trigger numerical instability at high Reynolds numbers. They commented that the discontinuity assumption should be maintained for the derivatives of flow variables across the cell interface, and thus the modified incompressible flux solver shows little advantage to the initial compressible one in terms of the implementation and computational efficiency. To further simply the M-IGKS, Shu and his coworkers [Yang *et al*, 2017b; Yang *et al*, 2017c; Yang *et al*, 2018c] proposed the C-IGKFS and S-IGKFS by replacing the Maxwellian distribution function with the circular function and the sphere function, respectively. Under the incompressible limit, the circle or the spherical surface established across the cell interface can be approximately considered to be symmetric about the cell interface. By adopting this approximation, the formulations for computing the conservative variables and numerical fluxes on the cell interface are explicit and more concise than the compressible solvers. At the same time, the discontinuous derivatives at the cell interface are kept, which enhances the numerical stability.

Sample applications of C-IGKFS and S-IGKFS to the 2D and 3D incompressible flows have been presented. The good agreements between the computational results and the reference data demonstrate the accuracy and robustness of these solvers. Comparisons also indicate that the C-IGKFS and the S-IGKFS are much more efficient than the compressible solvers and slightly faster than the LBFS.

Appendix A

Recurrence Relationships for Moment Integration

The moment relationships of the Maxwellian distribution function with bounded and unbounded integration limits are widely used in the gas kinetic scheme (GKS) and the gas kinetic flux solver (GKFS). For convenience, the recurrence relationships for moment integration are summarized in this appendix.

For the 1D case, the Maxwellian distribution function Eq. (2.28) can be written as

$$f^{eq} = \rho\left(\frac{\lambda}{\pi}\right)^{\frac{1+K}{2}} e^{-\lambda\left[\left(\xi_1 - u_1\right)^2 + \varsigma^2\right]}. \tag{A.1}$$

Here, $\varsigma^2 = \sum_{j=1}^{K} \varsigma_j^2$. With the following notation:

$$\rho\langle\phi\rangle = \int_{-\infty}^{+\infty} \phi f^{eq} d\Xi, \tag{A.2}$$

where ϕ is an arbitrary function and $d\Xi$ is the volume element in the particle velocity space. Note that

$$\left\langle \xi_1^n \varsigma^l \right\rangle = \left\langle \xi_1^n \right\rangle \left\langle \varsigma^l \right\rangle, \tag{A.3}$$

where n is an integer and l is an even integer (owing to the symmetrical property of ς).

The recovery of the Navier-Stokes equations only utilizes the moments of $\left\langle \varsigma^l \right\rangle$ with l=2 and 4, which are given by

$$\langle \zeta^2 \rangle = \frac{K}{2\lambda}, \tag{A.4}$$

$$\langle \zeta^4 \rangle = \frac{3K}{4\lambda^2} + \frac{K(K-1)}{4\lambda^2}. \tag{A.5}$$

The values of $\langle \xi_1^n \rangle$ depend on the integration limits. If the integration is from $-\infty$ to $+\infty$, we have

$$\langle \xi_1^0 \rangle = 1, \tag{A.6}$$

$$\langle \xi_1^1 \rangle = u_1, \tag{A.7}$$

$$\langle \xi_1^{n+2} \rangle = u_1 \langle \xi_1^{n+1} \rangle + \frac{n+1}{2\lambda} \langle \xi_1^n \rangle. \tag{A.8}$$

If the integration is from 0 to $+\infty$, by introducing the following notation

$$\rho \langle \xi_1^n \rangle_{>0} = \int_{\xi_1 > 0} \int_{-\infty}^{+\infty} \xi_1^n f^{eq} d\Xi, \tag{A.9}$$

we have

$$\langle \xi_1^0 \rangle_{>0} = \frac{1}{2} erfc\left(-\sqrt{\lambda} u_1\right), \tag{A.10}$$

$$\langle \xi_1^1 \rangle_{>0} = u_1 \langle \xi_1^0 \rangle_{>0} + \frac{1}{2} \frac{e^{-\lambda u_1^2}}{\sqrt{\pi \lambda}}, \tag{A.11}$$

$$\langle \xi_1^{n+2} \rangle_{>0} = u_1 \langle \xi_1^{n+1} \rangle_{>0} + \frac{n+1}{2\lambda} \langle \xi_1^n \rangle_{>0}, \tag{A.12}$$

where $erfc(\bullet)$ is the complementary error function. Similarly, by introducing the following notation

$$\rho \langle \xi_1^n \rangle_{<0} = \int_{\xi_1 < 0} \int_{-\infty}^{+\infty} \xi_1^n f^{eq} d\Xi, \tag{A.13}$$

the moment integration from $-\infty$ to 0 of ξ_1 yields

$$\left\langle \xi_1^0 \right\rangle_{<0} = \frac{1}{2} erfc\left(\sqrt{\lambda} u_1\right), \tag{A.14}$$

$$\left\langle \xi_1^1 \right\rangle_{<0} = u_1 \left\langle \xi_1^0 \right\rangle_{<0} - \frac{1}{2} \frac{e^{-\lambda u_1^2}}{\sqrt{\pi \lambda}}, \tag{A.15}$$

$$\left\langle \xi_1^{n+2} \right\rangle_{<0} = u_1 \left\langle \xi_1^{n+1} \right\rangle_{<0} + \frac{n+1}{2\lambda} \left\langle \xi_1^n \right\rangle_{<0}. \tag{A.16}$$

For 2D and 3D cases, the equilibrium state can be decomposed into the form of the 1D case, and the above integration formula can still be used. For example, in 3D conditions, we have

$$f^{eq} = \rho \left(\frac{\lambda}{\pi} \right)^{\frac{3+K}{2}} e^{-\lambda\left[\left(\xi_1 - u_1\right)^2 + \left(\xi_2 - u_2\right)^2 + \left(\xi_3 - u_3\right)^2 + \zeta^2\right]}, \tag{A.17}$$

and

$$\left\langle \xi_1^n \xi_2^m \xi_3^p \zeta^l \right\rangle = \left\langle \xi_1^n \right\rangle \left\langle \xi_2^m \right\rangle \left\langle \xi_3^p \right\rangle \left\langle \zeta^l \right\rangle, \tag{A.18}$$

where the integration limits can be either the full or the half of the velocity space.

Appendix B

Coefficients for the M-GKS

The coefficients of a^L, b^L, a^R and b^R in Eqs. (5.37a) and (5.37b) can all be calculated in a similar procedure. Taking a^L as an example, Eq. (5.37a) can be expressed as

$$\left\langle \left(a_0 + a_1\xi_1 + a_2\xi_2 + a_3\frac{1}{2}\left(\sum_{i=1}^{D}\xi_i^2 + \sum_{j=1}^{K}\varsigma_j^2 \right) \right) \overline{\boldsymbol{\varphi}}_a g \right\rangle = \begin{pmatrix} \partial\rho/\partial x_1 \\ \partial\rho u_1/\partial x_1 \\ \partial\rho u_2/\partial x_1 \\ \partial\rho E/\partial x_1 \end{pmatrix}, \quad \text{(B.1)}$$

where

$$\overline{\boldsymbol{\varphi}}_a = \left(1, \xi_1, \xi_2, \frac{1}{2}\left(\sum_{i=1}^{D}\xi_i^2 + \sum_{j=1}^{K}\varsigma_j^2 \right) \right)^T .$$

Note that the superscript "L" and the subscript "c" have been omitted for simplicity in the above equation. The term on the right-hand side of Eq. (B.1) is known since the derivatives of conservative variables have been determined after the initial reconstruction. If we denote the vector on the right-hand side of Eq. (B.1) as

$$\hat{\mathbf{H}} = \left(\hat{h}_0, \hat{h}_1, \hat{h}_2, \hat{h}_3 \right)^T , \quad \text{(B.2)}$$

and introduce

$$\mathbf{H} = \hat{\mathbf{H}}/\rho = \left(h_0, h_1, h_2, h_3 \right)^T , \quad \text{(B.3)}$$

Eq. (B.1) can be rewritten as

$$\mathbf{M}\begin{pmatrix} a_0 \\ a_1 \\ a_2 \\ a_3 \end{pmatrix} = \begin{pmatrix} h_0 \\ h_1 \\ h_2 \\ h_3 \end{pmatrix}. \tag{B.4}$$

The matrix **M** can be determined by using the recurrence relationships shown in **Appendix A**. After some manipulations, **M** has the following form:

$$\mathbf{M} = \begin{pmatrix} 1 & u_1 & u_2 & M_1 \\ u_1 & u_1^2 + \dfrac{1}{2\lambda} & u_1 u_2 & M_2 \\ u_2 & u_1 u_2 & u_2^2 + \dfrac{1}{2\lambda} & M_3 \\ M_1 & M_2 & M_3 & M_4 \end{pmatrix}, \tag{B.5}$$

where

$$M_1 = \frac{1}{2}\left(u_1^2 + u_2^2 + \frac{K+2}{2\lambda} \right),$$

$$M_2 = \frac{1}{2}\left(u_1^3 + u_1 u_2^2 + \frac{K+4}{2\lambda} u_1 \right),$$

$$M_3 = \frac{1}{2}\left(u_2^3 + u_2 u_1^2 + \frac{K+4}{2\lambda} u_2 \right),$$

$$M_4 = \frac{1}{4}\left[\left(u_1^2 + u_2^2 \right)^2 + \frac{K+4}{\lambda}\left(u_1^2 + u_2^2 \right) + \frac{K^2 + 6K + 8}{4\lambda^2} \right].$$

In the calculation of M_1, M_2, M_3 and M_4, the setting of $D = 2$ has been used as the 2D case is considered here. Defining

$$r_0 = u_1^2 + u_2^2 + \frac{K+2}{2\lambda}, \tag{B.6}$$

$$r_1 = h_1 - u_1 h_0, \tag{B.7}$$

$$r_2 = h_2 - u_2 h_0, \tag{B.8}$$

$$r_3 = 2h_3 - r_0 h_0, \tag{B.9}$$

the solutions of Eq. (B.4) yield

$$a_3 = \frac{4\lambda^2}{K+2}\left(r_3 - 2u_1 r_1 - 2u_2 r_2\right), \tag{B.10}$$

$$a_2 = 2\lambda r_2 - u_2 a_3, \tag{B.11}$$

$$a_1 = 2\lambda r_1 - u_1 a_3, \tag{B.12}$$

$$a_0 = h_0 - u_1 a_1 - u_2 a_2 - \frac{1}{2} a_3 r_0 . \tag{B.13}$$

The above solutions can also be used to evaluate A^L, A^R, $\overline{a}^L$, $\overline{a}^R$, $\overline{b}$ and $\overline{A}$, because the matrix $\mathbf{M}$ is identical in all these cases.

Appendix C

Coefficients for the M-IGKS

The coefficients of a , b and A can all be calculated in a similar procedure. Taking a as an example, Eq. (8.42) can be expressed as

$$\left\langle (a_0 + a_1\xi_1 + a_2\xi_2)\begin{pmatrix} 1 \\ \xi_1 \\ \xi_2 \end{pmatrix} g \right\rangle = \begin{pmatrix} \partial\rho/\partial x_1 \\ \partial\rho u_1/\partial x_1 \\ \partial\rho u_2/\partial x_1 \end{pmatrix}, \tag{C.1}$$

where the right-hand side of Eq. (C.1) is known. If we denote the vector on the right-hand side as

$$\hat{\mathbf{H}} = \left(\hat{h}_0, \hat{h}_1, \hat{h}_2\right)^T, \tag{C.2}$$

and introduce

$$\mathbf{H} = \hat{\mathbf{H}}/\rho = (h_0, h_1, h_2)^T, \tag{C.3}$$

Eq. (C.1) can be rewritten as

$$\mathbf{M}\begin{pmatrix} a_0 \\ a_1 \\ a_2 \end{pmatrix} = \begin{pmatrix} h_0 \\ h_1 \\ h_2 \end{pmatrix}. \tag{C.4}$$

The matrix **M** can be determined by using the recurrence relationships shown in **Appendix A**. After some manipulations, **M** has the following form:

$$\mathbf{M} = \begin{pmatrix} 1 & u_1 & u_2 \\ u_1 & u_1^2 + \dfrac{1}{2\lambda} & u_1 u_2 \\ u_2 & u_1 u_2 & u_2^2 + \dfrac{1}{2\lambda} \end{pmatrix}, \tag{C.5}$$

Defining

$$r_2 = h_2 - u_2 h_0, \tag{C.6}$$

$$r_1 = h_1 - u_1 h_0, \tag{C.7}$$

the solutions of Eq. (C.4) yield

$$a_2 = 2\lambda r_2, \tag{C.8}$$

$$a_1 = 2\lambda r_1, \tag{C.9}$$

$$a_0 = h_0 - u_1 a_1 - u_2 a_2. \tag{C.10}$$

The above solutions can also be used to evaluate both b and A in Eqs. (8.43) and (8.45), because the matrix $\mathbf{M}$ is identical in all these cases.

Appendix D

Description of Attached Computer Codes

In this book, five source code packages are provided to help readers for the application of the introduced methods. They can be compiled using the FORTRAN compiler. All the codes can be downloaded from https://www.worldscientific.com/worldscibooks/10.1142/11949#t=suppl. The descriptions of each package are shown below.

(1) Code A: Simulation of 2D Lid-Driven Cavity Flow by LBFS

This package provides all source codes of LBFS for simulation of two-dimensional (2D) lid-driven cavity flow described in **Chapter 3**. It contains 6 subroutines, which are edited by FORTRAN 77. The implementation procedures of LBFS can refer to page 89 and the flowchat of this code can refer to Figure 3.3. The functions of each subroutine are summarized as below:

(a) *main.for*: This is the main subroutine, which controls the overall solution procedures and prints some key information for monitoring the computation. The basic procedures include the mesh manipulation, the initialization of flow fields and the evolution of Navier-Stokes equations.

(b) *head.for*: This subroutine defines the flow variables, fluxes, mesh variables and other auxiliary variables used for computation.

(c) *grid.for*: This module calls the mesh generation subroutine "gridgen" to generate a non-uniform structured mesh and then calculates the volume of each cell and the length, streaming distance, relaxation parameter of each interface.

(d) *slope.for*: This subroutine implements the boundary conditions by applying dummy cells and calculates the derivatives of flow variables by the finite difference scheme. Details of the calculation of derivatives can be found in Eqs. (3.28)-(3.30).

(e) *fluxff.for*: This subroutine calculates the fluxes by the LBFS. The details of LBFS can be found in **Section 3.2**.

(f) *feq.for*: In this module, the equilibrium density distribution function is calculated by using the subroutine "feq" for all directions and the subroutine "feqi" for one specific direction.

(2) Code B: Mesh Transformation

This code transforms the mesh generated by the commercial software POINTWISE with COBALT format to the computational mesh input file of the in-house code including **Codes C**, **D** and **E**. It is applicable for both the triangular cell and the quadrilateral cell. The input file is "mesh.grd" and the output files are "mesh.plt", "mesh.my" and "dcell.my". This code is edited by FORTRAN 90.

When generating a mesh by POINTWISE with COBALT format, the boundary ID should be set according to the following rules. For the solid wall boundary, the ID takes a value of 1; for the far-field boundary, the ID should be set as 2; for the inlet boundary, the ID should be taken as 3; for the outlet boundary, the ID should be set as 4; and for the user-specified boundary such as the moving wall, the ID should be set as 5.

(3) Code C: Simulation of Transonic Inviscid Flows around Airfoil by C-GKFS and LBFS with D1Q4 Model

This package provides all source codes of C-GKFS and LBFS with the D1Q4 model for simulation of transonic inviscid flows around RAE2822 airfoil described in **Chapter 6**. It is applicable for the unstructured grid with the triangular cell and the quadrilateral cell. This package contains 14 subroutines, which are edited by FORTRAN 90. The implementation procedures of C-GKFS can refer to page 212 and the flowchat of this code can refer to Figure 6.5. The functions of each subroutine are summarized as below:

(a) *main.f90*: This subroutine controls the overall solution procedures and prints some key information for monitoring the computation. The basic procedures include the mesh manipulation, the dynamic memory allocation, the initialization of flow fields, the evolution of macroscopic governing equations and the dynamic memory deallocation.
(b) *paramod.f90*: This subroutine defines the global variables used in all subroutines. They include the mesh variables, the flow field variables, the initial conservative variables, the gradients, the limiter functions and some constants.
(c) *rdmesh.f90*: This subroutine reads the mesh file and calculates the coordinates of cell centers. The input file is "mesh.my".
(d) *walloc.f90*: This subroutine allocates memories for global variables defined in "paramod.f90".
(e) *initial.f90*: This subroutine reads the configuration file and initializes the flow fields. The input file is "config.dat".
(f) *solver.f90*: In this subroutine, the three-stage Runge-Kutta method is applied as the temporal discretization strategy for evolution of conservative variables at cell centers. Details of the temporal discretization can be found in Eqs. (6.145a)-(6.145c).
(g) *wgrads.f90*: This subroutine calculates the gradients of conservative variables by the Green-Gauss approach, which is shown in Eq. (6.142).
(h) *limiter.f90*: In this subroutine, the values of the limiter function are calculated by the Venkatakrishnan's limiter, which is shown in Eq. (6.143).
(i) *fluxc.f90*: This subroutine calculates the numerical fluxes by the C-GKFS or LBFS with the D1Q4 model for inviscid flows. When changing the C-GKFS to LBFS, the texts in "fluxc.f90" should be replaced by those of "fluxc_lbfs.f90". The details of C-GKFS and LBFS with the D1Q4 model can be found in **Section 6.2** and **Section 6.4**, respectively.

(j) *fluxcBC.f90*: In this subroutine, the numerical fluxes at the boundary are calculated according to the boundary conditions. For the solid wall boundary, the no-penetration condition is used, and for the far-field boundary, the characteristics-based boundary condition is adopted.
(k) *deltat.f90*: This subroutine calculates the time step of each cell by the Courant-Friedrichs-Lewy (CFL) condition, as shown in Eqs. (6.146) and (6.147).
(l) *resi.f90*: In this subroutine, the residual of the continuity equation is evaluated for monitoring the convergence history.
(m) *output.f90*: This subroutine calculates the dynamic forces and outputs the flow fields in the TECPLOT format. The output files include "CP.plt", "Dynamic.txt" and "Output.plt".
(n) *wdealloc.f90*: This module deallocates the memories of global variables.

(4) Code D: Simulation of Transonic Viscous Flows around Airfoil by C-DGKFS with D2Q4 Model

This package provides all source codes of C-DGKFS with the D2Q4 model for simulation of transonic viscous flow around NACA0012 airfoil described in **Chapter 7**. It is applicable for the unstructured grid with the triangular cell and the quadrilateral cell. This package contains 14 subroutines, which are edited by FORTRAN 90. The implementation procedures of C-DGKFS can refer to pages 263-264. The functions of each subroutine are summarized as below:
(a) *main.f90*: This subroutine is the same as that of **Code C**.
(b) *paramod.f90*: This subroutine is the same as that of **Code C**.
(c) *rdmesh.f90*: This subroutine is the same as that of **Code C**.
(d) *walloc.f90*: This subroutine is the same as that of **Code C**.
(e) *initial.f90*: This subroutine reads the configuration file and initializes the flow fields. The input files are "config.dat" and "dcell.my".
(f) *solver.f90*: This subroutine is the same as that of **Code C**.
(g) *wgrads.f90*: This subroutine calculates the gradients of conservative variables by the Green-Gauss approach, which is

shown in Eq. (6.142). Since the viscous flows are considered in this code, the no-slip condition is utilized to determine the flow variables on the solid wall boundary.

(h) *limiter.f90*: This subroutine is the same as that of **Code C**.
(i) *fluxc.f90*: This subroutine calculates the numerical fluxes by the C-GKFS with the D2Q4 model. To utilize the D2Q8 and D2Q12 models for simulation, one can change the value of "Ntheta" in this code to 8 and 12, respectively. The details of this flux solver can be found in **Section 7.2**.
(j) *fluxcBC.f90*: In this subroutine, the numerical fluxes at the boundary are calculated according to the boundary conditions. For the solid wall boundary, the no-slip condition is used, and for the far-field boundary, the characteristics-based boundary condition is adopted.
(k) *deltat.f90*: This subroutine is the same as that of **Code C**.
(l) *resi.f90*: This subroutine is the same as that of **Code C**.
(m) *output.f90*: This subroutine calculates the dynamic forces and outputs the flow fields in the TECPLOT format. The output files include "CPCF.plt", "Dynamic.txt" and "Output.plt".
(n) *wdealloc.f90*: This subroutine is the same as that of **Code C**.

(5) Code E: Simulation of Incompressible Flows around Circular Cylinder by C-IGKFS and LBFS

This package provides all source codes of C-IGKFS and LBFS for simulation of incompressible flow over a circular cylinder described in **Chapter 8**. It is applicable for the unstructured grid with the triangular cell and the quadrilateral cell. This package contains 13 subroutines, which are edited by FORTRAN 90. The implementation procedures of C-IGKFS can refer to page 299 and the flowchat of this code can refer to Figure 8.3. The functions of each subroutine are summarized as below:

(a) *main.f90*: This subroutine is the same as that of **Code C**.
(b) *paramod.f90*: This subroutine is the same as that of **Code C**, except that the last line is commented out.
(c) *rdmesh.f90*: This subroutine is the same as that of **Code C**.

(d) *walloc.f90*: This subroutine allocates memories for global variables defined in "paramod.f90". Since the incompressible isothermal flows are considered in this code, the limiter function is not needed and the conservative variable vector contains only 3 components for 2D conditions.
(e) *initial.f90*: This subroutine reads the configuration file and initializes the flow fields. The input files are "config.dat" and "dcell.my".
(f) *solver.f90*: This subroutine is the same as that of **Code C**, except that the limiter function is commented out.
(g) *wgrads.f90*: This subroutine calculates the gradients of conservative variables by the Green-Gauss approach, which is shown in Eq. (6.142). Since the viscous incompressible flows are considered in this code, the no-slip condition is utilized on the solid wall boundary and the natural boundary condition is applied on the far-field boundary.
(h) *fluxc.f90*: This subroutine calculates the numerical fluxes by the C-IGKFS or LBFS. When changing the C-IGKFS to LBFS, the texts in "fluxc.f90" should be replaced by those of "fluxc_lbfs.f90". The details of LBFS and C-IGKFS can be found in **Section 3.2** and **Section 8.3**, respectively.
(i) *fluxcBC.f90*: In this subroutine, the numerical fluxes at the boundary are calculated according to the boundary conditions. For the solid wall boundary, the no-slip condition is used, and for the far-field boundary, the natural boundary condition is adopted.
(j) *deltat.f90*: This subroutine is the same as that of **Code C**.
(k) *resi.f90*: In this subroutine, the residual of the velocity field is evaluated for monitoring the convergence history.
(l) *output.f90*: This subroutine calculates the dynamic forces and outputs the flow fields in the TECPLOT format. The output files include "CP.plt", "Dynamic.txt" and "Output.plt".
(m) *wdealloc.f90*: This subroutine is the same as that of **Code C**.

Appendix E

Supplementary Material

The supplementary material includes:

(1) Code A: Simulation of 2D Lid-Driven Cavity Flow by LBFS

(2) Code B: Mesh Transformation

(3) Code C: Simulation of Transonic Inviscid Flows around Airfoil by C-GKFS and LBFS with D1Q4 Model

(4) Code D: Simulation of Transonic Viscous Flows around Airfoil by C-DGKFS with D2Q4 Model

(5) Code E: Simulation of Incompressible Flows around Circular Cylinder by C-IGKFS and LBFS

Please follow the instructions below to download the files:

1. Register an account/login at https://www.worldscientific.com.
2. Go to https://www.worldscientific.com/r/11949-supp to activate the access.
3. Access the solutions from: https://www.worldscientific.com/worldscibooks/10.1142/11949#t=suppl

For subsequent access, simply log in with the same login details. For enquiries, please email: sales@wspc.com.sg.

Bibliography

Aidun, C. K. and Clausen, J. R. (2010). Lattice-Boltzmann method for complex flows, *Annu. Rev. Fluid Mech.*, 42, pp. 439–472.

Anderson, W. K., Thomas, J. L. and Van Leer, B. (1986). Comparison of finite volume flux vector splittings for the Euler equations, *AIAA J.*, 24, pp. 1453–1460.

Ansumali, S., Karlin, I. V., Frouzakis, C. E. and Boulouchos, K. B. (2006). Entropic lattice Boltzmann method for microflows, *Physica A Stat. Mech. Appl.*, 359, pp. 289–305.

Armaly, B. F., Durst, F., Pereira, J. C. F. and Schönung, B. (1983). Experimental and theoretical investigation of backward-facing step flow, *J. Fluid Mech.*, 127, pp. 473–496.

Ba, Y., Liu, H., Li, Q., Kang, Q. and Sun, J. (2016). Multiple-relaxation-time color-gradient lattice Boltzmann model for simulating two-phase flows with high density ratio, *Phys. Rev. E*, 94, pp. 023310.

Barth, T. and Jespersen, D. (1989). The design and application of upwind schemes on unstructured meshes, *27th Aerosp. Sci. Mtg.*, pp. 366.

Batten, P., Clarke, N., Lambert, C. and Causon, D. M. (1997). On the choice of wavespeeds for the HLLC Riemann solver, *SIAM J. Sci. Comput.*, 18, pp. 1553–1570.

Benson, M. G., Bellamy-Knights, P. G., Gerrard, J. H. and Gladwell, I. (1989). A viscous splitting algorithm applied to low Reynolds number flows round a circular cylinder, *J. Fluids Struct.*, 3, pp. 439–479.

Benzi, R., Succi, S. and Vergassola, M. (1992). The lattice Boltzmann equation: Theory and applications, *Phys. Rep.*, 222, pp. 145–197.

Bhatnagar, P. L., Gross, E. P. and Krook, M. (1954). A model for collision processes in gases. I. Small amplitude processes in charged and neutral one-component systems, *Phys. Rev.*, 94, pp. 511–525.

Blazek, J. (2015). *Computational Fluid Dynamics: Principles and Applications*, 3rd Ed. (Butterworth-Heinemann, UK).

Bouwhuis, W., van der Veen, R. C., Tran, T., Keij, D. L., Winkels, K. G., Peters, I. R., van der Meer, D., Sun, C., Snoeijer, J. H. and Lohse, D. (2012). Maximal air bubble entrainment at liquid-drop impact, *Phys. Rev. Lett.*, 109, pp. 264501.

Braza, M., Chassaing, P. and Minh, H. H. (1986). Numerical study and physical analysis of the pressure and velocity fields in the near wake of a circular cylinder, *J. Fluid Mech.*, 165, pp. 79–130.

Briant, A. J. and Yeomans, J. M. (2004). Lattice Boltzmann simulations of contact line motion. II. Binary fluids, *Phys. Rev. E*, 69, pp. 031603.

Briant, A. J., Wagner, A. J. and Yeomans, J. M. (2004). Lattice Boltzmann simulations of contact line motion. I. Liquid-gas systems, *Phys. Rev. E*, 69, pp. 031602.

Bristeau, M. O. (2013). *Numerical Simulation of Compressible Navier–Stokes Flows: A GAMM Workshop*, (Springer Science & Business Media, Germany).

Bruneau, C. H. and Saad, M. (2006). The 2D lid-driven cavity problem revisited, *Comput. Fluids*, 35, pp. 326–348.

Cao, Y. (2016). Variable property-based lattice Boltzmann flux solver for thermal flows in the low Mach number limit, *Int. J. Heat Mass Tran.*, 103, pp. 254–264.

Cao, Y. and Zhang, Y. (2017). Investigation on the natural convection in horizontal concentric annulus using the variable property-based lattice Boltzmann flux solver, *Int. J. Heat Mass Tran.*, 111, pp. 1260–1271.

Cebeci, T., Shao, J. P., Kafyeke, F. and Laurendeau, E. (2009). *Computational Fluid Dynamics for Engineers: From Panel to Navier–Stokes Methods with Computer Programs*, 2nd Ed. (Springer Berlin Heidelberg, Germany).

Chae, D., Kim, C. and Rho, O. H. (2000). Development of an improved gas-kinetic BGK scheme for inviscid and viscous flows, *J. Comput. Phys.*, 158, pp. 1–27.

Chen, S., Chen, H., Martnez, D. and Matthaeus, W. (1991). Lattice Boltzmann model for simulation of magnetohydrodynamics, *Phys. Rev. Lett.*, 67, pp. 3776–3779.

Chen, S., Martinez, D. and Mei, R. (1996). On boundary conditions in lattice Boltzmann methods, *Phys. Fluids*, 8, pp. 2527–2536.

Chen, S. and Doolen, G. D. (1998). Lattice Boltzmann method for fluid flows, *Annu. Rev. Fluid Mech.*, 30, pp. 329–364.

Chen, S., Jin, C., Li, C. and Cai, Q. (2011). Gas-kinetic scheme with discontinuous derivative for low speed flow computation, *J. Comput. Phys.*, 230, pp. 2045–2059.

Chen, Z., Shu, C. and Tan, D. (2018). High-order simplified thermal lattice Boltzmann method for incompressible thermal flows, *Int. J. Heat Mass Tran.*, 127, pp. 1–16.

Chew, Y. T., Shu, C. and Peng, Y. (2002). On implementation of boundary conditions in the application of finite volume lattice Boltzmann method, *J. Stat. Phys.*, 107, pp. 539–556.

Chiang, T. P. and Sheu, T. W. H. (1999). A numerical revisit of backward-facing step flow problem, *Phys. Fluids*, 11, pp. 862–874.

Chou, S. Y. and Baganoff, D. (1997). Kinetic flux-vector splitting for the Navier–Stokes equations, *J. Comput. Phys.*, 130, pp. 217–230.

Contrino, D., Lallemand, P., Asinari, P. and Luo, L. S. (2014). Lattice-Boltzmann simulations of the thermally driven 2D square cavity at high Rayleigh numbers, *J. Comput. Phys.*, 275, pp. 257–272.

Cossali, G. E., Coghe, A. and Marengo, M. (1997). The impact of a single drop on a wetted solid surface, *Exp. Fluids*, 22, pp. 463–472.

Courant, R., Isaacson, E. and Rees, M. (1952). On the solution of nonlinear hyperbolic differential equations by finite differences, *Commun. Pure Appl. Math.*, 5, pp. 243–255.

Daly, B. J. (1967). Numerical study of two fluid Rayleigh–Taylor instability, *Phys. Fluids*, 10, pp. 297–307.

Davis, S. F. (1988). Simplified second-order Godunov-type methods, *SIAM J. Sci. Statist. Comput.*, 9, pp. 445–473.

De Vahl Davis, G. (1983). Natural convection of air in a square cavity: A bench mark numerical solution, *Int. J. Numer. Methods Fluids*, 3, pp. 249–264.

Dellar, P. J. (2008). Two routes from the Boltzmann equation to compressible flow of polyatomic gases, *Prog. Comput. Fluid Dyn.*, 8, pp. 84–96.

Dennis, S. C. R. and Chang, G. Z. (1970). Numerical solutions for steady flow past a circular cylinder at Reynolds numbers up to 100, *J. Fluid Mech.*, 42, pp. 471–489.

Ding, H., Shu, C., Yeo, K. S. and Xu, D. (2004). Simulation of incompressible viscous flows past a circular cylinder by hybrid FD scheme and meshless least square-based finite difference method, *Comput. Methods Appl. Mech. Eng.*, 193, pp. 727–744.

Ding, H., Shu, C., Yeo, K. S. and Xu, D. (2006). Numerical computation of three-dimensional incompressible viscous flows in the primitive variable form by local multiquadric differential quadrature method, *Comput. Methods Appl. Mech. Eng.*, 195, pp. 516–533.

Ding, H., Shu, C. and Cai, Q. D. (2007a). Applications of stencil-adaptive finite difference method to incompressible viscous flows with curved boundary, *Comput. Fluids*, 36, pp. 786–793.

Ding, H., Spelt, P. D. and Shu, C. (2007b). Diffuse interface model for incompressible two-phase flows with large density ratios, *J. Comput. Phys.*, 226, pp. 2078–2095.

Dixit, H. N. and Babu, V. (2006). Simulation of high Rayleigh number natural convection in a square cavity using the lattice Boltzmann method, *Int. J. Heat Mass Tran.*, 49, pp. 727–739.

D'Humières, D. (1994). Generalized lattice-Boltzmann equations, *Prog. Astronaut. Aeronaut.*, 159, pp. 450–458.

Einfeldt, B. (1988). On Godunov-type methods for gas dynamics, *SIAM J. Numer. Anal.*, 25, pp. 294–318.

Frisch, U., d'Humieres, D., Hasslacher, B., Lallemand, P., Pomeau, Y. and Rivet, J. P. (1987). Lattice gas hydrodynamics in two and three dimensions, *Complex Syst.*, 1, pp. 649–707.

Fusegi, T., Hyun, J. M., Kuwahara, K. and Farouk, B. (1991). A numerical study of three-dimensional natural convection in a differentially heated cubical enclosure, *Int. J. Heat Mass Tran.*, 34, pp. 1543–1557.

Gan, Y., Xu, A., Zhang, G., Yu, X. and Li, Y. (2008). Two-dimensional lattice Boltzmann model for compressible flows with high Mach number, *Physica A Stat. Mech. Appl.*, 387, pp. 1721–1732.

Ghia, U., Ghia, K. N. and Shin, C. T. (1982). High-Re solutions for incompressible flow using the Navier–Stokes equations and a multigrid method, *J. Comput. Phys.*, 48, pp. 387–411.

Godunov, S. K. (1959). A difference method for numerical calculation of discontinuous solutions of the equations of hydrodynamics, *Mat. Sb.*, 89, pp. 271–306.

Gorban, A. N. and Packwood, D. J. (2014). Enhancement of the stability of lattice Boltzmann methods by dissipation control, *Physica A Stat. Mech. Appl.*, 414, pp. 285–299.

Grad, H. (1949). On the kinetic theory of rarefied gases, *Commun. Pur. Appl. Math.*, 2, pp. 331–407.

Grad, H. (1958). *Principles of the Kinetic Theory of Gases*, eds. Flügge S., "Thermodynamik der Gase / Thermodynamics of Gases. Handbuch der Physik / Encyclopedia of Physics," (Springer, Germany) pp. 205–294.

Guermond, J. L. and Quartapelle, L. (2000). A projection FEM for variable density incompressible flows, *J. Comput. Phys.*, 165, pp. 167–188.

Guo, Z. and Zhao, T. S. (2002). Lattice Boltzmann model for incompressible flows through porous media, *Phys. Rev. E*, 66, pp. 036304.

Guo, Z., Shi, B. and Zheng, C. (2002a). A coupled lattice BGK model for the Boussinesq equations, *Int. J. Numer. Methods Fluids*, 39, pp. 325–342.

Guo, Z., Zheng, C. and Shi, B. (2002b). An extrapolation method for boundary conditions in lattice Boltzmann method, *Phys. Fluids*, 14, pp. 2007–2010.

Guo, Z., Zheng, C. and Shi, B. (2002c). Non-equilibrium extrapolation method for velocity and pressure boundary conditions in the lattice Boltzmann method, *Chin. Phys.*, 11, pp. 366–374.

Guo, Z., Liu, H., Luo, L. S. and Xu, K. (2008). A comparative study of the LBE and GKS methods for 2D near incompressible laminar flows, *J. Comput. Phys.*, 227, pp. 4955–4976.

Guo, Z. and Shu, C. (2013). *Lattice Boltzmann Method and Its Applications in Engineering*, (World Scientific, Singapore).

Guo, Z., Xu, K. and Wang, R. (2013). Discrete unified gas kinetic scheme for all Knudsen number flows: Low-speed isothermal case, *Phys. Rev. E*, 88, pp. 033305.

Ha, M. Y. and Jung, M. J. (2000). A numerical study on three-dimensional conjugate heat transfer of natural convection and conduction in a differentially heated cubic enclosure with a heat-generating cubic conducting body, *Int. J. Heat Mass Tran.*, 43, pp. 4229–4248.

Hakkinen, R. J., Greber, I., Trilling, L. and Abarbanel, S. S. (1959). The interaction of an oblique shock wave with a laminar boundary layer, NASA-MEMO-2-18-59W.

Hänel, D., Schwane, R. and Seider, G. (1987). On the accuracy of upwind schemes for the solution of the Navier–Stokes equations, *8th Comput. Fluid Dyn. Conf.*, pp. 1105.

Hänel, D. and Schwane, R. (1989). An implicit flux-vector splitting scheme for the computation of viscous hypersonic flow, *27th Aerosp. Sci. Mtg.*, pp. 274.

Harten, A. and Hyman, J. M. (1983). Self adjusting grid methods for one-dimensional hyperbolic conservation laws, *J. Comput. Phys.*, 50, pp. 235–269.

Harten, A., Lax, P. D. and Van Leer, B. (1983). On upstream differencing and Godunov-type schemes for hyperbolic conservation laws, *SIAM Rev.*, 25, pp. 35–61.

Haselbacher, A. and Blazek, J. (2000). Accurate and efficient discretization of Navier–Stokes equations on mixed grids, *AIAA J.*, 38, pp. 2094–2102.

He, X., Luo, L. S. and Dembo, M. (1996). Some progress in lattice Boltzmann method. Part I. Nonuniform mesh grids, *J. Comput. Phys.*, 129, pp. 357–363.

He, X. and Doolen, G. D. (1997). Lattice Boltzmann method on curvilinear coordinates system: Flow around a circular cylinder, *J. Comput. Phys.*, 134, pp. 306–315.

He, X., Chen, S. and Zhang, R. (1999). A lattice Boltzmann scheme for incompressible multiphase flow and its application in simulation of Rayleigh–Taylor instability, *J. Comput. Phys.*, 152, pp. 642–663.

Holdych, D. J., Rovas, D., Georgiadis, J. G. and Buckius, R. O. (1998). An improved hydrodynamics formulation for multiphase flow lattice-Boltzmann models, *Int. J. Mod. Phys. C*, 9, pp. 1393–1404.

Holway Jr, L. H. (1966). New statistical models for kinetic theory: Methods of construction, *Phys. Fluids*, 9, pp. 1658–1673.

Hu, S. X., Yan, G. W. and Shi, W. P. (1997). A lattice Boltzmann model for compressible perfect gas, *Acta Mech. Sinica-PRC*, 13, pp. 218–226.

Huang, H. B. and Lu, X. Y. (2009). Relative permeabilities and coupling effects in steady-state gas-liquid flow in porous media: A lattice Boltzmann study, *Phys. Fluids*, 21, pp. 092104.

Huang, H. B., Huang, J. J., Lu, X. Y. and Sukop, M. C. (2013). On simulations of high-density ratio flows using color-gradient multiphase lattice Boltzmann models, *Int. J. Mod. Phys. C*, 24, pp. 1350021.

Huser, A. and Biringen, S. (1992). Calculation of two-dimensional shear-driven cavity flows at high Reynolds numbers, *Int. J. Numer. Methods Fluids*, 14, pp. 1087–1109.

Inamuro, T., Yoshino, M. and Ogino, F. (1997). Accuracy of the lattice Boltzmann method for small Knudsen number with finite Reynolds number, *Phys. Fluids*, 9, pp. 3535–3542.

Inamuro, T., Ogata, T., Tajima, S. and Konishi, N. (2004). A lattice Boltzmann method for incompressible two-phase flows with large density differences, *J. Comput. Phys.*, 198, pp. 628–644.

Jameson, A., Schmidt, W. and Turkel, E. (1981). Numerical solution of the Euler equations by finite volume methods using Runge–Kutta time stepping schemes, *14th Fluid Plasma Dyn. Conf.*, pp. 1259.

Jameson, A. (2017). Origins and further development of the Jameson–Schmidt–Turkel scheme, *AIAA J.*, pp. 1487–1510.

Jawahar, P. and Kamath, H. (2000). A high-resolution procedure for Euler and Navier–Stokes computations on unstructured grids, *J. Comput. Phys.*, 164, pp. 165–203.

Jiang, S. and Ni, G. (2004). A γ-model BGK scheme for compressible multifluids, *Int. J. Numer. Methods Fluids*, 46, pp. 163–182.

Jiang, S. and Ni, G. (2007). A second-order γ-model BGK scheme for multimaterial compressible flows, *Appl. Numer. Math.*, 57, pp. 597–608.

Jiang, J. and Qian, Y. (2012). Implicit gas-kinetic BGK scheme with multigrid for 3D stationary transonic high-Reynolds number flows, *Comput. Fluids*, 66, pp. 21–28.

Karlin, I. V., Ferrante, A. and Öttinger, H. C. (1999). Perfect entropy functions of the lattice Boltzmann method, *EPL-Europhys. Lett.*, 47, pp. 182–188.

Kataoka, T. and Tsutahara, M. (2004a). Lattice Boltzmann method for the compressible Euler equations, *Phys. Rev. E*, 69, pp. 056702.

Kataoka, T. and Tsutahara, M. (2004b). Lattice Boltzmann model for the compressible Navier–Stokes equations with flexible specific-heat ratio, *Phys. Rev. E*, 69, pp. 035701.

Kim, S. S., Kim, C., Rho, O. H. and Hong, S. (2002). Cure for shock instability-Development of an improved Roe scheme, *40th AIAA Aerosp. Sci. Mtg. Exh.*, pp. 548.

Krüger, T., Kusumaatmaja, H., Kuzmin, A., Shardt, O., Silva, G. and Viggen, E. M. (2016). *The Lattice Boltzmann Method: Principles and Practice*, (Springer International Publishing, Switzerland).

Ku, H. C., Hirsh, R. S. and Taylor, T. D. (1987). A pseudospectral method for solution of the three-dimensional incompressible Navier–Stokes equations, *J. Comput. Phys.*, 70, pp. 439–462.

Kumar, G., Girimaji, S. S. and Kerimo, J. (2013). WENO-enhanced gas-kinetic scheme for direct simulations of compressible transition and turbulence, *J. Comput. Phys.*, 234, pp. 499–523.

Ladd, A. J. (1994). Numerical simulations of particulate suspensions via a discretized Boltzmann equation. Part 1. Theoretical foundation, *J. Fluid Mech.*, 271, pp. 285–309.

Laflin, K. R., Klausmeyer, S. M., Zickuhr, T., Vassberg, J. C., Wahls, R. A., Morrison, J. H., Brodersen, O. P., Rakowitz, M. E., Tinoco, E. N. and Godard, J. L. (2005). Data summary from second AIAA computational fluid dynamics drag prediction workshop, *J. Aircraft*, 42, pp. 1165–1178.

Latt, J. and Chopard, B. (2006). Lattice Boltzmann method with regularized pre-collision distribution functions, *Math. Comput. Simulat.*, 72, pp. 165–168.

Le Quéré, P. (1991). Accurate solutions to the square thermally driven cavity at high Rayleigh number, *Comput. Fluids*, 20, pp. 29–41.

Lee, C. B. (2000). Possible universal transitional scenario in a flat plate boundary layer: Measurement and visualization, *Phys. Rev. E*, 62, pp. 3659–3670.

Lee, C. B. and Wu, J. Z. (2008). Transition in wall-bounded flows, *Appl. Mech. Rev.*, 61, pp. 030802.

Lee, T. and Lin, C. L. (2005). A stable discretization of the lattice Boltzmann equation for simulation of incompressible two-phase flows at high density ratio, *J. Comput. Phys.*, 206, pp. 16–47.

Lee, T. and Liu, L. (2010). Lattice Boltzmann simulations of micron-scale drop impact on dry surfaces, *J. Comput. Phys.*, 229, pp. 8045–8063.

Li, Q., Fu, S. and Xu, K. (2005a). A compressible Navier–Stokes flow solver with scalar transport, *J. Comput. Phys.*, 204, pp. 692–714.

Li, Q., Fu, S. and Xu, K. (2005b). Application of gas-kinetic scheme with kinetic boundary conditions in hypersonic flow, *AIAA J.*, 43, pp. 2170–2176.

Li, Q., He, Y. L., Wang, Y. and Tao, W. Q. (2007). Coupled double-distribution-function lattice Boltzmann method for the compressible Navier–Stokes equations, *Phys. Rev. E*, 76, pp. 056705.

Li, Z. H. and Zhang, H. X. (2009). Gas-kinetic numerical studies of three-dimensional complex flows on spacecraft re-entry, *J. Comput. Phys.*, 228, pp. 1116–1138.

Li, Q., Xu, K. and Fu, S. (2010). A high-order gas-kinetic Navier–Stokes flow solver, *J. Comput. Phys.*, 229, pp. 6715–6731.

Li, Q. and Fu, S. (2011). Application of gas-kinetic BGK scheme in three-dimensional flow, *49th AIAA Aerosp. Sci. Mtg.*, pp. 386.

Li, J., Li, Q. and Xu, K. (2011). Comparison of the generalized Riemann solver and the gas-kinetic scheme for inviscid compressible flow simulations, *J. Comput. Phys.*, 230, pp. 5080–5099.

Li, Q., Luo, K. H., Gao, Y. J. and He, Y. L. (2012). Additional interfacial force in lattice Boltzmann models for incompressible multiphase flows, *Phys. Rev. E*, 85, pp. 026704.

Li, J., Zhong, C., Wang, Y. and Zhuo, C. (2017). Implementation of dual time-stepping strategy of the gas-kinetic scheme for unsteady flow simulations, *Phys. Rev. E*, 95, pp. 053307.

Li, J., Zhong, C. and Zhuo, C. (2019a). A third order gas-kinetic scheme for unstructured grid, *Comput. Math. Appl.*, 78, pp. 92–109.

Li, J., Zhong, C., Pan, D. and Zhuo, C. (2019b). A gas-kinetic scheme coupled with SST model for turbulent flows, *Comput. Math. Appl.*, 78, pp. 1227–1242.

Lin, H. C. (1995). Dissipation additions to flux-difference splitting, *J. Comput. Phys.*, 117, pp. 20–27.

Liou, M. S. (1993). On a new class of flux splittings, *13th Int. Conf. Numer. Methods Fluid Dyn.*, pp. 115–119.

Liou, M. S. and Steffen Jr, C. J. (1993). A new flux splitting scheme, *J. Comput. Phys.*, 107, pp. 23–39.

Liou, M. S. (1996). A sequel to AUSM: AUSM+, *J. Comput. Phys.*, 129, pp. 364–382.

Liou, M. S. (2001). Ten years in the making–AUSM-family, *15th AIAA Comput. Fluid Dyn. Conf.*, pp. 2521.

Liu, X. D., Osher, S. and Chan, T. (1994). Weighted essentially non-oscillatory schemes, *J. Comput. Phys.*, 115, pp. 200–212.

Liu, H. and Xu, K. (2007). A Runge–Kutta discontinuous Galerkin method for viscous flow equations, *J. Comput. Phys.*, 224, pp. 1223–1242.

Lomax, H., Pulliam, T. H. and Zingg, D. W. (2013). *Fundamentals of Computational Fluid Dynamics*, 2nd Ed. (Springer Science & Business Media, Germany).

Luo, L. S. (2000). Theory of the lattice Boltzmann method: Lattice Boltzmann models for nonideal gases, *Phys. Rev. E*, 62, pp. 4982–4996.

Mandal, J. C. and Deshpande, S. M. (1994). Kinetic flux vector splitting for Euler equations, *Comput. Fluids*, 23, pp. 447–478.

Mani, M., Ladd, J. A., Cain, A. B., Bush, R. H. (1997). An assessment of one-and two-equation turbulence models for internal and external flows, *28th Fluid Dyn. Conf.*, pp. 2010.

Mattila, K. K., Hegele Jr, L. A. and Philippi, P. C. (2015). Investigation of an entropic stabilizer for the lattice-Boltzmann method, *Phys. Rev. E*, 91, pp. 063010.

Mattila, K. K., Philippi, P. C. and Hegele Jr, L. A. (2017). High-order regularization in lattice-Boltzmann equations, *Phys. Fluids*, 29, pp. 046103.

May, G., Srinivasan, B. and Jameson, A. (2007). An improved gas-kinetic BGK finite-volume method for three-dimensional transonic flow, *J. Comput. Phys.*, 220, pp. 856–878.

Mei, R. and Shyy, W. (1998). On the finite difference-based lattice Boltzmann method in curvilinear coordinates, *J. Comput. Phys.*, 143, pp. 426–448.

Meng, Z. X., Yang, L. M., Wang, D. H., Shu, C. and Zhang, W. H. (2018). Circular function-based gas-kinetic scheme for simulation of viscous compressible flows, *Int. Conf. Comput. Sci.*, pp. 37–47.

Mezrhab, A., Bouzidi, M. H. and Lallemand, P. (2004). Hybrid lattice-Boltzmann finite-difference simulation of convective flows, *Comput. Fluids*, 33, pp. 623–641.

Mieussens, L. (2000). Discrete-velocity models and numerical schemes for the Boltzmann-BGK equation in plane and axisymmetric geometries, *J. Comput. Phys.*, 162, pp. 429–466.

Mohamad, A. A. (2019). *Lattice Boltzmann Method: Fundamentals and Engineering Applications with Computer Codes*, 2nd Ed. (Springer, UK).

Montessori, A. and Falcucci, G. (2018). *Lattice Boltzmann Modeling of Complex Flows for Engineering Applications*, (Morgan & Claypool Publishers, UK).

Moukalled, F., Mangani, L. and Darwish, M. (2015). *The Finite Volume Method in Computational Fluid Dynamics: An Advanced Introduction with OpenFOAM® and Matlab*, (Springer International Publishing, Switzerland).

Munz, C. D. (1994). A tracking method for gas flow into vacuum based on the vacuum Riemann problem, *Math. Method Appl. Sci.*, 17, pp. 597–612.

Ni, G., Jiang, S. and Xu, K. (2008). A DGBGK scheme based on WENO limiters for viscous and inviscid flows, *J. Comput. Phys.*, 227, pp. 5799–5815.

Ni, G., Jiang, S. and Xu, X. (2012). A moving mesh BGK scheme for multi-material flows, *Int. J. Numer. Methods Fluids*, 69, pp. 878–896.

Nieuwstadt, F. and Keller, H. B. (1973). Viscous flow past circular cylinders, *Comput. Fluids*, 1, pp. 59–71.

Niu, X. D., Shu, C., Chew, Y. T. and Wang, T. G. (2004). Investigation of stability and hydrodynamics of different lattice Boltzmann models, *J. Stat. Phys.*, 117, pp. 665–680.

Nourgaliev, R. R., Dinh, T. N. and Theofanous, T. G. (2004). A pseudocompressibility method for the numerical simulation of incompressible multifluid flows, *Int. J. Multiph. Flow*, 30, pp. 901–937.

Pan, L., Zhao, G., Tian, B. and Wang, S. (2012). A gas kinetic scheme for the Baer–Nunziato two-phase flow model, *J. Comput. Phys.*, 231, pp. 7518–7536.

Pan, L. and Xu, K. (2015). A third-order gas-kinetic scheme for three-dimensional inviscid and viscous flow computations, *Comput. Fluids*, 119, pp. 250–260.

Pan, D., Zhong, C., Li, J. and Zhuo, C. (2016a). A gas-kinetic scheme for the simulation of turbulent flows on unstructured meshes, *Int. J. Numer. Methods Fluids*, 82, pp. 748–769.

Pan, L., Xu, K., Li, Q. and Li, J. (2016b). An efficient and accurate two-stage fourth-order gas-kinetic scheme for the Euler and Navier–Stokes equations, *J. Comput. Phys.*, 326, pp. 197–221.

Pan, D., Zhong, C. and Zhuo, C. (2018). An implicit gas-kinetic scheme for turbulent flow on unstructured hybrid mesh, *Comput. Math. Appl.*, 75, pp. 3825–3848.

Park, J., Kwon, K. and Choi, H. (1998). Numerical solutions of flow past a circular cylinder at Reynolds numbers up to 160, *KSME Int. J.*, 12, pp. 1200–1205.

Patankar, S. (2018). *Numerical Heat Transfer and Fluid Flow*, (Taylor & Francis, UK).

Patil, D. V. and Lakshmisha, K. N. (2009). Finite volume TVD formulation of lattice Boltzmann simulation on unstructured mesh, *J. Comput. Phys.*, 228, pp. 5262–5279.

Peery, K. and Imlay, S. (1988). Blunt-body flow simulations, *24th Joint Propul. Conf.*, pp. 2904.

Pellerin, N., Leclaire, S. and Reggio, M. (2017). Solving incompressible fluid flows on unstructured meshes with the lattice Boltzmann flux solver, *Eng. Appl. Comput. Fluid Mech.*, 11, pp. 310–327.

Peng, G., Xi, H., Duncan, C. and Chou, S. H. (1999). Finite volume scheme for the lattice Boltzmann method on unstructured meshes, *Phys. Rev. E*, 9, pp. 4675–4682.

Peng, Y., Shu, C. and Chew, Y. T. (2003). Simplified thermal lattice Boltzmann model for incompressible thermal flows, *Phys. Rev. E*, 68, pp. 026701.

Peng, Y., Shu, C. and Chew, Y. T. (2004). A 3D incompressible thermal lattice Boltzmann model and its application to simulate natural convection in a cubic cavity, *J. Comput. Phys.*, 193, pp. 260–274.

Peyret, R. and Taylor, T. D. (2002). *Computational Methods for Fluid Flow*, 3rd Ed. (Springer Science & Business Media, Germany).

Prendergast, K. H. and Xu, K. (1993). Numerical hydrodynamics from gas-kinetic theory, *J. Comput. Phys.*, 109, pp. 53–66.

Pullin, D. I. (1980). Direct simulation methods for compressible inviscid ideal-gas flow, *J. Comput. Phys.*, 34, pp. 231–244.

Qian, Y. H., d'Humières, D. and Lallemand, P. (1992). Lattice BGK models for Navier–Stokes equation, *EPL-Europhys. Lett.*, 17, pp. 479–484.

Qu, K., Shu, C. and Chew, Y. T. (2007a). Alternative method to construct equilibrium distribution functions in lattice-Boltzmann method simulation of inviscid compressible flows at high Mach number, *Phys. Rev. E*, 75, pp. 036706.

Qu, K., Shu, C. and Chew, Y. T. (2007b). Simulation of shock-wave propagation with finite volume lattice Boltzmann method, *Int. J. Mod. Phys. C*, 18, pp. 447–454.

Qu, K. (2009). *Development of Lattice Boltzmann Method for Compressible Flows*, Ph.D. Thesis, Department of Mech. Eng., Univ. of Singapore, 148p.

Radespiel, R. and Kroll, N. (1995). Accurate flux vector splitting for shocks and shear layers, *J. Comput. Phys.*, 121, pp. 66–78.

Rieber, M. and Frohn, A. (1999). A numerical study on the mechanism of splashing, *Int. J. Heat Fluid Flow*, 20, pp. 455–461.

Roe, P. L. (1981). Approximate Riemann solvers, parameter vectors, and difference schemes, *J. Comput. Phys.*, 43, pp. 357–372.

Roe, P. L. and Pike, J. (1985). Efficient construction and utilisation of approximate Riemann solutions, *Proc. 6th Int. Symp. Comput. Methods in Appl. Sci. Eng.*, pp. 499–518.

Ruan, Y. and Jameson, A. (2002). Gas-kinetic BGK method for three-dimensional compressible flows, *40th AIAA Aerosp. Sci. Mtg. Exh.*, pp. 550.

Schmitt, V. and Charpin, F. (1979). "Pressure distributions on the ONERA-M6-Wing at transonic Mach numbers," experimental data base for computer program assessment, *Report of the Fluid Dynamics Panel Working Group 04*, AGARD AR 138.

Shakhov, E. M. (1968). Generalization of the Krook kinetic relaxation equation, *Fluid Dynam+*, 3, pp. 95–96.

Shan, X. and Chen, H. (1993). Lattice Boltzmann model for simulating flows with multiple phases and components, *Phys. Rev. E*, 47, pp. 1815–1819.

Shan, X. and Chen, H. (1994). Simulation of nonideal gases and liquid-gas phase transitions by the lattice Boltzmann equation, *Phys. Rev. E*, 49, pp. 2941–2948.

Shao, J. Y., Shu, C., Huang, H. B. and Chew, Y. T. (2014). Free-energy-based lattice Boltzmann model for the simulation of multiphase flows with density contrast, *Phys. Rev. E*, 89, pp. 033309.

Sharp, D. H. (1984). An overview of Rayleigh–Taylor instability, *Physica D*, 12, pp. 3–10.

Shu, C. W. and Osher, S. (1988). Efficient implementation of essentially non-oscillatory shock-capturing schemes, *J. Comput. Phys.*, 77, pp. 439–471.

Shu, C. and Xue, H. (1998). Comparison of two approaches for implementing stream function boundary conditions in DQ simulation of natural convection in a square cavity, *Int. J. Heat Fluid Flow*, 19, pp. 59–68.

Shu, C., Niu, X. D. and Chew, Y. T. (2002). Taylor-series expansion and least-squares-based lattice Boltzmann method: Two-dimensional formulation and its applications, *Phys. Rev. E*, 65, pp. 036708.

Shu, C., Niu, X. D. and Chew, Y. T. (2005). A lattice Boltzmann kinetic model for microflow and heat transfer, *J. Stat. Phys.*, 121, pp. 239–255.

Shu, C., Wang, Y., Teo, C. J. and Wu, J. (2014). Development of lattice Boltzmann flux solver for simulation of incompressible flows, *Adv. Appl. Math. Mech.*, 6, pp. 436–460.

Shukla, R. K., Tatineni, M. and Zhong, X. (2007). Very high-order compact finite difference schemes on non-uniform grids for incompressible Navier–Stokes equations, *J. Comput. Phys.*, 224, pp. 1064–1094.

Spalart, P. R. and Allmaras, S. R. (1992). A one-equation turbulence model for aerodynamic flows, *30th Aerosp. Sci. Mtg. Exh.*, pp. 439.

Sterling, J. D. and Chen, S. (1996). Stability analysis of lattice Boltzmann methods, *J. Comput. Phys.*, 123, pp. 196–206.

Stiebler, M., Tölke, J. and Krafczyk, M. (2006). An upwind discretization scheme for the finite volume lattice Boltzmann method, *Comput. Fluids*, 35, pp. 814–819.

Struchtrup, H. (2005). *Macroscopic Transport Equations for Rarefied Gas Flows: Approximate Methods in Kinetic Theory*, (Springer Science & Business Media, Germany).

Su, M., Xu, K. and Ghidaoui, M. S. (1999). Low-speed flow simulation by the gas-kinetic scheme, *J. Comput. Phys.*, 150, pp. 17–39.

Succi, S. (2001). *The Lattice Boltzmann Equation: For Fluid Dynamics and Beyond*, (Oxford University Press, UK).

Sun, C. (1998). Lattice-Boltzmann models for high speed flows, *Phys. Rev. E*, 58, pp. 7283–7287.

Sun, Y., Shu, C., Teo, C. J., Wang, Y. and Yang, L. M. (2015). Explicit formulations of gas-kinetic flux solver for simulation of incompressible and compressible viscous flows, *J. Comput. Phys.*, 300, pp. 492–519.

Sun, Y., Shu, C., Yang, L. M. and Teo, C. J. (2016). A switch function-based gas-kinetic scheme for simulation of inviscid and viscous compressible flows, *Adv. Appl. Math. Mech.*, 8, pp. 703–721.

Sun, Y., Shu, C., Wang, Y., Teo, C. J. and Chen, Z. (2017). An immersed boundary-gas kinetic flux solver for simulation of incompressible flows, *Comput. Fluids*, 142, pp. 45–56.

Sutherland, W. (1893). The viscosity of gases and molecular force, *London, Edinburgh, and Dublin Philos. Mag. J. Sci.*, 36, pp. 507–531.

Swift, M. R., Orlandini, E., Osborn, W. R. and Yeomans, J. M. (1996). Lattice Boltzmann simulations of liquid-gas and binary fluid systems, *Phys. Rev. E*, 54, pp. 5041–5052.

Tang, H. Z. and Xu, K. (2000). A high-order gas-kinetic method for multidimensional ideal magnetohydrodynamics, *J. Comput. Phys.*, 165, pp. 69–88.

Tang, H. Z. (2001). Gas-kinetic schemes for compressible flow of real gases, *Comput. Math. Appl.*, 41, pp. 723–734.

Tang, H. Z., Tang, T. and Xu, K. (2004). A gas-kinetic scheme for shallow-water equations with source terms, *Z. Angew. Math. Phys.*, 55, pp. 365–382.

Tang, L. (2012). Progress in gas-kinetic upwind schemes for the solution of Euler/Navier–Stokes equations–I: Overview, *Comput. Fluids*, 56, pp. 39–48.

Tian, C. T., Xu, K., Chan, K. L. and Deng, L. C. (2007). A three-dimensional multidimensional gas-kinetic scheme for the Navier–Stokes equations under gravitational fields, *J. Comput. Phys.*, 226, pp. 2003–2027.

Toro, E. F., Spruce, M. and Speares, W. (1994). Restoration of the contact surface in the HLL-Riemann solver, *Shock Waves*, 4, pp. 25–34.

Toro, E. F. and Titarev, V. A. (2006). Derivative Riemann solvers for systems of conservation laws and ADER methods, *J. Comput. Phys.*, 212, pp. 150–165.

Toro, E. F. (2009). *Riemann Solvers and Numerical Methods for Fluid Dynamics: A Practical Introduction*, 3rd Ed. (Springer Science & Business Media, Germany).

Tran, T., de Maleprade, H., Sun, C. and Lohse, D. (2013). Air entrainment during impact of droplets on liquid surfaces, *J. Fluid Mech.*, 726, R3.

Tryggvason, G. (1988). Numerical simulations of the Rayleigh–Taylor instability, *J. Comput. Phys.*, 75, pp. 253–282.

Valero-Lara, P. (2018). *Analysis and Applications of Lattice Boltzmann Simulations*, (IGI Global, USA).

Van Albada, G. D., Van Leer, B. and Roberts, W. W. (1997). *A Comparative Study of Computational Methods in Cosmic Gas Dynamics*, eds. Hussaini, M. Y., Van Leer, B. and Van Rosendale, J., "Upwind and High-Resolution Schemes," (Springer, Germany) pp. 95–103.

Van Leer, B. (1977). Towards the ultimate conservative difference scheme. IV. A new approach to numerical convection, *J. Comput. Phys.*, 23, pp. 276–299.

Van Leer, B. (1979). Towards the ultimate conservative difference scheme. V. A second-order sequel to Godunov's method, *J. Comput. Phys.*, 32, pp. 101–136.

Van Leer, B., Thomas, J. L., Roe, P. L. and Newsome, R. W. (1987). A comparison of numerical flux formulas for the Euler and Navier–Stokes equations, *8th AIAA Comput. Fluid Dyn. Conf.*, pp. 1104.

Van Leer, B., Lee, W. T., Roe, P. L., Powell, K. G., Tai, C. H. (1992). Design of optimally smoothing multistage schemes for the Euler equations, *Commun. Appl. Numer. Methods*, 8, pp. 761–769.

Van Leer, B. (1997). *Flux-Vector Splitting for the Euler Equation*, eds. Hussaini, M. Y., Van Leer, B. and Van Rosendale, J., "Upwind and High-Resolution Schemes," (Springer, Germany) pp. 80–89.

Vassberg, J. C., Tinoco, E. N., Mani, M., Brodersen, O. P., Eisfeld, B., Wahls, R. A., Morrison, J. H., Zickuhr, T., Laflin, K. R. and Mavriplis, D. J. (2007). Summary of the third AIAA CFD drag prediction workshop, *45th AIAA Aerosp. Sci. Mtg. Exh.*, pp. 260.

Venkatakrishnan, V. (1995). Convergence to steady state solutions of the Euler equations on unstructured grids with limiters, *J. Comput. Phys.*, 118, pp. 120–130.

Verhaeghe, F., Luo, L. S. and Blanpain, B. (2009). Lattice Boltzmann modeling of microchannel flow in slip flow regime, *J. Comput. Phys.*, 228, pp. 147–157.

Versteeg, H. K. and Malalasekera, W. (2007). *An Introduction to Computational Fluid Dynamics: The Finite Volume Method*, 2nd Ed. (Pearson Education, UK).

Wada, Y. and Liou, M. S. (1997). An accurate and robust flux splitting scheme for shock and contact discontinuities, *SIAM J. Sci. Comput.*, 18, pp. 633–657.

Wang, Y., Shu, C. and Teo, C. J. (2014a). A fractional step axisymmetric lattice Boltzmann flux solver for incompressible swirling and rotating flows, *Comput. Fluids*, 96, pp. 204–214.

Wang, Y., Shu, C. and Teo, C. J. (2014b). Development of LBGK and incompressible LBGK-based lattice Boltzmann flux solvers for simulation of incompressible flows, *Int. J. Numer. Methods Fluids*, 75, pp. 344–364.

Wang, Y., Shu, C. and Teo, C. J. (2014c). Thermal lattice Boltzmann flux solver and its application for simulation of incompressible thermal flows, *Comput. Fluids*, 94, pp. 98–111.

Wang, Y., Shu, C., Huang, H. B. and Teo, C. J. (2015a). Multiphase lattice Boltzmann flux solver for incompressible multiphase flows with large density ratio, *J. Comput. Phys.*, 280, pp. 404–423.

Wang, Y., Shu, C. and Yang, L. M. (2015b). An improved multiphase lattice Boltzmann flux solver for three-dimensional flows with large density ratio and high Reynolds number, *J. Comput. Phys.*, 302, pp. 41–58.

Wang, Y., Shu, C., Shao, J. Y., Wu, J. and Niu, X. D. (2015c). A mass-conserved diffuse interface method and its application for incompressible multiphase flows with large density ratio, *J. Comput. Phys.*, 290, pp. 336–351.

Wang, Y., Shu, C., Teo, C. J. and Wu, J. (2015d). An immersed boundary-lattice Boltzmann flux solver and its applications to fluid-structure interaction problems, *J. Fluids Struct.*, 54, pp. 440–465.

Wang, Y., Shu, C., Teo, C. J., Wu, J. and Yang, L. M. (2015e). Three-dimensional lattice Boltzmann flux solver and its applications to incompressible isothermal and thermal flows, *Commu. Comput. Phys.*, 18, pp. 593–620.

Wang, Y., Yang, L. M. and Shu, C. (2015f). From lattice Boltzmann method to lattice Boltzmann flux solver, *Entropy*, 17, pp. 7713–7735.

Wang, Y., Shu, C. and Yang, L. M. (2016a). Boundary condition-enforced immersed boundary-lattice Boltzmann flux solver for thermal flows with Neumann boundary conditions, *J. Comput. Phys.*, 306, pp. 237–252.

Wang, Y., Shu, C., Teo, C. J. and Yang, L. M. (2016b). Numerical study on the freely falling plate: Effects of density ratio and thickness-to-length ratio, *Phys. Fluids*, 28, pp. 103603.

Wang, Y., Shu, C., Yang, L. M. and Sun, Y. (2017a). On the immersed boundary-lattice Boltzmann simulations of incompressible flows with freely moving objects, *Int. J. Numer. Methods Fluids*, 83, pp. 331–350.

Wang, Y., Shu, C., Yang, L. M. and Teo, C. J. (2017b). An immersed boundary-lattice Boltzmann flux solver in a moving frame to study three-dimensional freely falling rigid bodies, *J. Fluids Struct.*, 68, pp. 444–465.

Wang, Y., Shu, C., Yang, L. M. and Yuan, H. Z. (2018a). On the re-initialization of fluid interfaces in diffuse interface method, *Comput. Fluids*, 166, pp. 209–217.

Wang, Y., Shu, C., Yang, L. M. and Yuan, H. Z. (2018b). Development of axisymmetric lattice Boltzmann flux solver for complex multiphase flows, *Mod. Phys. Lett. B*, 32, pp. 1840005.

Wang, Y., Shu, C., Wang, T. G. and Valdivia y Alvarado, P. (2019). A generalized minimal residual method-based immersed boundary-lattice Boltzmann flux solver coupled with finite element method for non-linear fluid-structure interaction problems, *Phys. Fluids*, 31, pp. 103603.

Watari, M. and Tsutahara, M. (2003). Two-dimensional thermal model of the finite-difference lattice Boltzmann method with high spatial isotropy, *Phys. Rev. E*, 67, pp. 036306.

Wendt, J. F. and Anderson, J. D. (2008). *Computational Fluid Dynamics: An Introduction*, 3rd Ed. (Springer Science & Business Media, Germany).

White, F. M. (1991). Viscous Fluid Flow, 2nd Ed. (McGraw-Hill, USA).

Wieting, A. R. (1987). Experimental study of shock wave interface heating on a cylindrical leading edge, *NASA TM-100484*.

Williams, P. T. and Baker, A. J. (1997). Numerical simulations of laminar flow over a 3D backward-facing step, *Int. J. Numer. Methods Fluids*, 24, pp. 1159–1183.

Xi, H., Peng, G. and Chou, S. H. (1999). Finite-volume lattice Boltzmann method, *Phys. Rev. E*, 59, pp. 6202–6205.

Xiong, S., Zhong, C., Zhuo, C., Li, K., Chen, X. and Cao, J. (2011). Numerical simulation of compressible turbulent flow via improved gas-kinetic BGK scheme, *Int. J. Numer. Methods Fluids*, 67, pp. 1833–1847.

Xu, K., Martinelli, L. and Jameson, A. (1995). Gas-kinetic finite volume methods, flux-vector splitting, and artificial diffusion, *J. Comput. Phys.*, 120, pp. 48–65.

Xu, K. (1998). Gas-kinetic schemes for unsteady compressible flow simulations, *Lecture Series-Van Kareman Institute for Fluid Dynamics*, 3, pp. C1–C202.

Xu, K. and Lui, S. H. (1999). Rayleigh–Bénard simulation using the gas-kinetic Bhatnagar–Gross–Krook scheme in the incompressible limit, *Phys. Rev. E*, 60, pp. 464–470.

Xu, K. (2001). A gas-kinetic BGK scheme for the Navier–Stokes equations and its connection with artificial dissipation and Godunov method, *J. Comput. Phys.*, 171, pp. 289–335.

Xu, K. (2002). A well-balanced gas-kinetic scheme for the shallow-water equations with source terms, *J. Comput. Phys.*, 178, pp. 533–562.

Xu, K. and He, X. (2003). Lattice Boltzmann method and gas-kinetic BGK scheme in the low-Mach number viscous flow simulations, *J. Comput. Phys.*, 190, pp. 100–117.

Xu, K., Mao, M. and Tang, L. (2005). A multidimensional gas-kinetic BGK scheme for hypersonic viscous flow, *J. Comput. Phys.*, 203, pp. 405–421.

Xu, K. and Huang, J. C. (2010). A unified gas-kinetic scheme for continuum and rarefied flows, *J. Comput. Phys.*, 229, pp. 7747–7764.

Xu, K., Sun, Q. and Yu, P. (2010). Valid physical processes from numerical discontinuities in computational fluid dynamics, *Int. J. Hyperson.*, 1, pp. 157–172.

Yan, Y. Y. and Zu, Y. Q. (2007). A lattice Boltzmann method for incompressible two-phase flows on partial wetting surface with large density ratio, *J. Comput. Phys.*, 227, pp. 763–775.

Yang, J. Y. and Huang, J. C. (1995). Rarefied flow computations using nonlinear model Boltzmann equations, *J. Comput. Phys.*, 120, pp. 323–339.

Yang, L. M., Shu, C. and Wu, J. (2012). Development and comparative studies of three non-free parameter lattice Boltzmann models for simulation of compressible flows, *Adv. Appl. Math. Mech.*, 4, pp. 454–472.

Yang, L. M., Shu, C. and Wu, J. (2013a). A moment conservation-based non-free parameter compressible lattice Boltzmann model and its application for flux evaluation at cell interface, *Comput. Fluids*, 79, pp. 190–199.

Yang, L. M., Shu, C., Wu, J., Zhao, N. and Lu, Z. L. (2013b). Circular function-based gas-kinetic scheme for simulation of inviscid compressible flows, *J. Comput. Phys.*, 255, pp. 540–557.

Yang, L. M., Shu, C. and Wu, J. (2014a). A hybrid lattice Boltzmann flux solver for simulation of 3D compressible viscous flows, *8th Int. Conf. Comput. Fluid Dyn.*, pp. 14–18.

Yang, L. M., Shu, C. and Wu, J. (2014b). A simple distribution function-based gas-kinetic scheme for simulation of viscous incompressible and compressible flows, *J. Comput. Phys.*, 274, pp. 611–632.

Yang, L. M., Shu, C. and Wu, J. (2015). A three-dimensional explicit sphere function-based gas-kinetic flux solver for simulation of inviscid compressible flows, *J. Comput. Phys.*, 295, pp. 322–339.

Yang, L. M., Shu, C. and Wang, Y. (2016a). Development of a discrete gas-kinetic scheme for simulation of two-dimensional viscous incompressible and compressible flows, *Phys. Rev. E*, 93, pp. 033311.

Yang, L. M., Shu, C., Wang, Y. and Sun, Y. (2016b). Development of discrete gas kinetic scheme for simulation of 3D viscous incompressible and compressible flows, *J. Comput. Phys.*, 319, pp. 129–144.

Yang, L. M., Shu, C., Wu, J. and Wang, Y. (2016c). Numerical simulation of flows from free molecular regime to continuum regime by a DVM with streaming and collision processes, *J. Comput. Phys.*, 306, pp. 291–310.

Yang, L. M., Shu, C. and Wu, J. (2016d). A hybrid lattice Boltzmann flux solver for simulation of viscous compressible flows, *Adv. Appl. Math. Mech.*, 8, pp. 887–910.

Yang, L. M., Shu, C. and Wu, J. (2016e). Extension of lattice Boltzmann flux solver for simulation of 3D viscous compressible flows, *Comput. Math. Appl.*, 71, pp. 2069–2081.

Yang, L. M., Shu, C., Wu, J., Wang, Y. and Sun, Y. (2017a). Comparative study of 1D, 2D and 3D simplified gas kinetic schemes for simulation of inviscid compressible flows, *Appl. Math. Model.*, 43, pp. 85–109.

Yang, L. M., Shu, C., Yang, W. M. and Wang, Y. (2017b). A simplified circular function-based gas kinetic scheme for simulation of incompressible flows, *Int. J. Numer. Methods Fluids*, 85, pp. 583–598.

Yang, L. M., Shu, C., Yang, W. M. and Wu, J. (2017c). A simple gas kinetic scheme for simulation of 3D incompressible thermal flows, *Numer. Heat Tr. B-Fund.*, 72, pp. 450–468.

Yang, L. M., Shu, C., Yang, W. M., Wang, Y. and Wu, J. (2017d). An immersed boundary-simplified sphere function-based gas kinetic scheme for simulation of 3D incompressible flows, *Phys. Fluids*, 29, pp. 083605.

Yang, L. M., Shu, C., Yang, W. M. and Wu, J. (2018a). Development of an efficient gas kinetic scheme for simulation of two-dimensional incompressible thermal flows, *Phys. Rev. E*, 97, pp. 013305.

Yang, L. M., Shu, C., Yang, W. M., Chen, Z. and Dong, H. (2018b). An improved discrete velocity method (DVM) for efficient simulation of flows in all flow regimes, *Phys. Fluids*, 30, pp. 062005.

Yang, L. M., Shu, C., Yang, W. M., Wang, Y. and Lee, C. B. (2018c). An implicit simplified sphere function-based gas kinetic scheme for simulation of 3D incompressible isothermal flows, *Comput. Fluids*, 160, pp. 204–218.

Yang, L. M., Shu, C., Yang, W. M. and Wu, J. (2019). Simulation of conjugate heat transfer problems by lattice Boltzmann flux solver, *Int. J. Heat Mass Tran.*, 137, pp. 895–907.

Yarin, A. L. (2006). Drop impact dynamics: Splashing, spreading, receding, bouncing…., *Annu. Rev. Fluid Mech.*, 38, pp. 159–192.

Yoon, S. and Jameson, A. (1988). Lower-upper symmetric-Gauss-Seidel method for the Euler and Navier–Stokes equations, *AIAA J.*, 26, pp. 1025–1026.

Yoon, S. and Kwak, D. (1991). Three-dimensional incompressible Navier–Stokes solver using lower-upper symmetric-Gauss-Seidel algorithm, *AIAA J.*, 29, pp. 874–875.

Yu, Z. and Fan, L. S. (2009). An interaction potential based lattice Boltzmann method with adaptive mesh refinement (AMR) for two-phase flow simulation, *J. Comput. Phys.*, 228, pp. 6456–6478.

Yuan, P. and Schaefer, L. (2006). Equations of state in a lattice Boltzmann model, *Phys. Fluids*, 18, pp. 042101.

Yuan, R., Zhong, C. and Zhang, H. (2015). An immersed-boundary method based on the gas kinetic BGK scheme for incompressible viscous flow, *J. Comput. Phys.*, 296, pp. 184–208.

Yuan, H. Z., Wang, Y. and Shu, C. (2017). An adaptive mesh refinement-multiphase lattice Boltzmann flux solver for simulation of complex binary fluid flows, *Phys. Fluids*, 29, pp. 123604.

Zhang, T., Shi, B., Chai, Z. and Rong, F. (2012). Lattice BGK model for incompressible axisymmetric flows, *Commu. Comput. Phys.*, 11, pp. 1569–1590.

Zhang, P., Xia, Z. and Cai, Q. (2018). An h-adaptive implicit immersed boundary-lattice Boltzmann flux solver based on JASMIN AMR package, *Comput. Fluids*, 161, pp. 14–22.

Zhang, L. Q., Chen, Z., Shu, C. and Zhang, M. Q. (2019). A kinetic theory-based axisymmetric lattice Boltzmann flux solver for isothermal and thermal swirling flows, *J. Comput. Phys.*, 392, pp. 141–160.

Zheng, H. W., Shu, C. and Chew, Y. T. (2006). A lattice Boltzmann model for multiphase flows with large density ratio, *J. Comput. Phys.*, 218, pp. 353–371.

Zhou, D., Lu, Z. L., Guo, T. Q. and Yang, L. M. (2017). Development of circular function-based gas-kinetic scheme (CGKS) on moving grids for unsteady flows through oscillating cascades, *Int. J. Numer. Methods Fluids*, 84, pp. 715–736.

Zhou, D., Lu, Z. L. and Guo, T. Q. (2018). Improvement of computational efficiency of circular function-based gas kinetic scheme by using Jacobian-free Newton–Krylov method, *Comput. Fluids*, 161, pp. 121–135.

Zhu, Y., Zhong, C. and Xu, K. (2016). Implicit unified gas-kinetic scheme for steady state solutions in all flow regimes, *J. Comput. Phys.*, 315, pp. 16–38.

Zou, Q. and He, X. (1997). On pressure and velocity boundary conditions for the lattice Boltzmann BGK model, *Phys. Fluids*, 9, pp. 1591–1598.

Index

Lightning Source UK Ltd.
Milton Keynes UK
UKHW022042090720
366290UK00004B/250